UNITEXT for Physics

UNITEXT for Physics series, formerly UNITEXT Collana di Fisica e Astronomia, publishes textbooks and monographs in Physics and Astronomy, mainly in English language, characterized of a didactic style and comprehensiveness. The books published in UNITEXT for Physics series are addressed to graduate and advanced graduate students, but also to scientists and researchers as important resources for their education, knowledge and teaching.

More information about this series at http://www.springer.com/series/13351

Lorenzo Bianchini

Selected Exercises in Particle and Nuclear Physics

 Springer

Lorenzo Bianchini
Italian Institute for Nuclear Physics
Pisa
Italy

ISSN 2198-7882   ISSN 2198-7890   (electronic)
UNITEXT for Physics
ISBN 978-3-319-88948-1   ISBN 978-3-319-70494-4   (eBook)
https://doi.org/10.1007/978-3-319-70494-4

Printed on acid-free paper

This Springer imprint is published by Springer Nature
The registered company is Springer International Publishing AG
The registered company address is: Gewerbestrasse 11, 6330 Cham, Switzerland

To my wife Elisabetta,
1 Gv 4, 16

Preface

The realm of particle physics is vast: multidisciplinary knowledge across several domains of physics and mathematics is required to understand the reactions that occur when particles collide and to master the functioning of the experiments built to study these reactions: classical and quantum mechanics, special relativity, electrodynamics, thermodynamics, chemistry, atomic and nuclear physics, quantum field theory, electronics, analysis, geometry, group theory, probability, informatics, among the others. Large-scale particle experiments, like those hosted in the main laboratories around the world, are perhaps the best example of how multidisciplinary this field can become: the successful operation of these complex structures relies on the synergetic work of hundreds of scientists and engineers; it is only the combination of their individual expertise that makes it possible to cover all the needs.

Thanks to the maturity of this field (more than one hundred years old!), a huge collection of textbooks, topical schools, academic classes, and scientific literature is available, where both the theoretical and experimental foundations of particle physics can be elucidated to the desired level of detail. Yet, as for all the other domains of physics, particle physics should be more about solving problems rather than knowing concepts! The path towards a solid understanding of this discipline passes through the capability of solving exercises. This book collects a sample of about 240 solved problems about particle physics in general. About half of the exercises are drawn from the public exams that have been proposed by the Italian National Institute for Nuclear research (INFN) to select its scientific staff over the last decade. Additional material inspired by my personal experience as an undergraduate student at Scuola Normale Superiore di Pisa, researcher in the CMS experiment, and teaching assistant at ETH Zürich complements the selection. Throughout this book, the main emphasis is put on experimental problems, although some more theoretical ones are also included. Thus, this book is mostly addressed to experimentalists.

The proposed exercises span several subjects in particle physics, although I must acknowledge that it has not been possible to be truly exhaustive. Several topics have been unfortunately, yet necessarily, discarded or only marginally mentioned.

In particular, cosmology, dark matter, beyond-standard model theories will not be much discussed here. Also, a personal cultural bias towards an LHC-centric vision of the field may have driven the focus towards the high-energy frontier at the expenses of other equally lively sectors of research, like neutrino physics, rare decays, hadron spectroscopy, B physics. Much attention is devoted to the operational principles of particle detectors, from the more classical ones to the more recent technologies. Particle detectors cannot be understood without first mastering the basics of the interaction between particles and matter, which therefore represents another topic of foremost interest. Given that particle detectors typically provide electric outputs, which need to be processed, stored, and cleaned from the noise, electronics, informatics, and data analysis enter naturally into the game, and a basic knowledge of both subjects is therefore required. Furthermore, an experiment in particle physics usually starts by scattering particles: acceleration of particles in stable and repeatable beams is therefore another important topic. Finally, a proper scientific maturity demands also an overall picture of the field: what is known, what is still unknown but important to study, what are the technologies at hand, and what the future lines of research. Several exercises go along this direction by discussing the state of the art on the field, including ongoing or planned measurements and new experimental techniques.

The exercises are grouped by subject into five chapters, where the main topic of discussion is first introduced in an academic fashion. Within each chapter, the exercises are organised as much as possible according to a logical order, so that each exercise can be propaedeutic to those that follow. Some of the exercises are used as prototypes for a class of problems. In this case, the relevant concepts and the general-purpose formulas are derived once and recalled afterwards by pointing to the master exercise. Other exercises are instead chosen to introduce a particular topic, which is then explained in some more detail by dedicated mini-lectures. References to the scientific literature and topical textbooks are then provided to help the reader go into the various subjects in greater detail. Consistency of notation throughout the text is pursued to reduce at a minimum the confusion introduced by the abundance of acronyms and conventional symbols peculiar to this field. Some exercises require a few lines of calculations, others one or more pages. Whenever possible, one should always try to derive the symbolic solution analytically and carry out the numerical computation without the help of pocket calculators, as to train one's capability to handle simple calculations using approximations or order-of-magnitude estimates. Indeed, experience teaches that there exist a few constants and formulas that are really worth keeping in mind! In other situations, one should better rely on computers rather than try the analytical approach. In the latter case, examples of simple computer routines written in open-source programming languages are also proposed.

Per aspera ad astra: solving problems is the most difficult and painful task for students, but also one that unveils the true degree of comprehension of the subject. We hope that this book can serve as supporting material to back up existing and more complete textbooks on experimental and theoretical particle physics. At the same time, it should provide a test bench for undergraduate students and young researchers to validate their level of preparation and hopefully stimulate their curiosity on the field.

I am much indebted to INFN for allowing me to profit from a large number of the exercises contained in this book. The richness and variety of topics covered in this immense reservoir of knowledge have been fundamental to shape this work. I also want to heartily thank the Institute for Particle Physics of ETH Zürich for granting me the time to work on this book and for the fantastic teaching experience I enjoyed amid its brilliant and lively students.

Pisa, Italy   Lorenzo Bianchini
September 2017

Contents

Acronyms

ATLAS	A toroidal apparatus
BNL	Brookhaven National Laboratories
BR	Branching ratio
CC	Charged current
CDF	Collider Detector at Fermilab
CERN	Conseil européen pour la recherche nucléaire
CI	Confidence interval
CKM	Cabibbo–Kobayashi–Maskawa matrix
CL	Confidence level
CMB	Cosmic microwave background
CMS	Compact Muon Solenoid
DIS	Deep inelastic scattering
DT	Drift tube
EM	Electromagnetic
ES	Electron scattering
EW	Electroweak
FCNC	Flavour-changing neutral current
FWHM	Full width at half maximum
GEM	Gas electron multiplier
LEP	Large Electron–Positron collider
LHC	Large Hadron Collider
LINAC	Linear accelerator
LNF	Laboratory Nazionali di Frascati
LO	Leading order
LST	Limited streamer tubes
MC	Monte Carlo
MIP	Minimum ionising particle
MWPC	Multi-wire proportional chambers
NC	Neutral current
N(N...)LO	Next(-to-next-to-...) leading order

PDF	Parton density function
PDG	Particle Data Group
PEP	Positron Electron Project
PMNS	Pontecorvo–Maki–Nakagawa–Sakata matrix
QCD	Quantum chromodynamic
QED	Quantum electrodynamic
R&D	Research and development
RF	Radio frequency
RMS	Rootmeansquare
RPC	Resistive plate chambers
SM	Standard model
STP	Standard temperature and pressure
TDC	Time-to-digital converter
TOF	Time-of-flight
TPC	Time projection chamber
UL	Upper limit

Chapter 1
Kinematics

Abstract The first chapter is dedicated to the kinematics of relativistic particles. The starting point is the introduction of the Lorentz group through its representations. Large emphasis then is given to the transformation properties of velocities and angles. The centre-of-mass dynamics is studied in detail for two-to-two scattering and for two- and three-body decays. The last part of the chapter is devoted to the concept of cross section, which plays a central role in particle physics.

1.1 Lorentz Transformations

The set of space-time transformations under which the laws of physics are postulated to be invariant form the so-called *Poincaré group*: they comprise four space-time translations, three spatial rotations, and three velocity transformations. Space rotations and velocity transformations (the latter are often referred to as *boosts*) form the sub-group of *Lorentz* transformations. Rotations are determined by the three usual Euler angles, while boosts are determined by the three components of the velocity **v** of the new reference frame $\mathscr{R}'$ as measured by an observed at rest in the initial reference frame $\mathscr{R}$, or, equivalently, by the dimensionless *boost vector* $\boldsymbol{\beta} = \mathbf{v}/c$. Rotations and boosts change both the four-momentum and the spin vector of a particle. Since these transformations form a group, it is possible to find a representation of each element in terms of square matrices acting on vector spaces: the four-dimensional space of four-vectors and the $(2S + 1)$-dimensional space of spin vectors for a particle of spin S.

The space of four-vectors $p = (p_0, \ldots, p_3)$ in endowed with the *Minkowski norm* defined by:

$$p^2 \equiv p_\mu p_\nu g^{\mu\nu} = p_0^2 - \sum_{i=1}^{3} p_i^2, \tag{1.1}$$

with $g^{\mu\nu} = \mathrm{diag}(1, -1, -1, -1)$. By construction, the Lorentz transformations preserve the Minkowski norm of Eq. (1.1). Any function of four-vectors that has the same form in all reference frames related by a transformation of the Poincaré group is called an *invariant*: the squared norm of Eq. (1.1) provides an example.

© Springer International Publishing AG 2018
L. Bianchini, *Selected Exercises in Particle and Nuclear Physics*,
UNITEXT for Physics, https://doi.org/10.1007/978-3-319-70494-4_1

The three-momentum of a particle can be embedded into a four-vector $(E, \mathbf{p})$, called *four-momentum*. For a particle of mass m, the four-momentum is subject to the mass-shell constraint:

$$p^2 - m^2 = 0 \quad \Rightarrow \quad E = E_{\mathbf{p}} \equiv \sqrt{|\mathbf{p}|^2 + m^2}. \tag{1.2}$$

The velocity and gamma-factor of a particle are then defined as

$$\beta_{\mathbf{p}} \equiv \frac{|\mathbf{p}|}{E_{\mathbf{p}}}, \qquad \gamma_{\mathbf{p}} \equiv \left(1 - \beta_{\mathbf{p}}^2\right)^{-\frac{1}{2}} = \frac{E_{\mathbf{p}}}{m} \tag{1.3}$$

In the four-momentum space, each Lorentz transformation is represented by a 4×4 matrix which transforms p into a new four-momentum p'. The boost vector defines a privileged direction in space, and indeed the transformation distinguishes between the component parallel ($\parallel$) and orthogonal ($\perp$) to $\boldsymbol{\beta}$. In terms of these two components, the generic boost transformation is given by:

$$\begin{pmatrix} E' \\ p'_{\parallel} \end{pmatrix} = \begin{pmatrix} \gamma & -\beta\gamma \\ -\beta\gamma & \gamma \end{pmatrix} \begin{pmatrix} E \\ p_{\parallel} \end{pmatrix}, \qquad \mathbf{p}'_{\perp} = \mathbf{p}_{\perp} \tag{1.4}$$

with $\beta = |\boldsymbol{\beta}|$ and $\gamma = (1 - \beta)^{-\frac{1}{2}}$. Notice that γ is a function of β, although the explicit dependence is often omitted in the calculations. The variables at the left-hand side of Eq. (1.4) are the four-momenta components measured in the reference frame $\mathscr{R}'$ that moves with velocity $c\boldsymbol{\beta}$ with respect to $\mathscr{R}$. This way of writing the transformation corresponds to the so-called *passive transformation*, as opposed to the *active transformation* that changes the vector components in the same reference frame. The generic boost can be also written in a compact vectorial form, see Problem 1.17. For most of the applications it is however enough to remember the matrix version of Eq. (1.4).

Given that $\gamma^2 - (-\beta\gamma)^2 = 1$ and that $\gamma \geq 1$, the first of Eq. (1.4) can be equivalently written as:

$$\begin{pmatrix} E' \\ p'_{\parallel} \end{pmatrix} = \begin{pmatrix} \cosh\alpha & -\sinh\alpha \\ -\sinh\alpha & \cosh\alpha \end{pmatrix} \begin{pmatrix} E \\ p_{\parallel} \end{pmatrix} \tag{1.5}$$

with $\alpha = \operatorname{atanh}\beta$. Were not for the imaginary "angle" $\alpha = i\theta$ and the same-sign off-diagonal elements, Eq. (1.5) would be the transformation of a normal vector under a spatial rotation. The differences accounts for the fact that the transformation has to preserve the Minkowski norm $E^2 - p_{\parallel}^2$, and not the Euclidean norm $E^2 + p_{\parallel}^2$, as done by ordinary rotations.

A final word of caution: β and γ in Eq. (1.4) are the parameters of a transformation, and should not be confused with the velocity and gamma-factor of a particle as measured in a given reference frame: the suffix "$\mathbf{p}$" in the latter thus reminds that these quantities are different from the boost parameters. However, when no such ambiguity can arise, the suffix can be safely dropped to simplify the notation.

Problems

Problem 1.1 Express the MKS units [kg, m, s] in natural units. Using the results thus obtained, write down the following quantities in natural units:

1. a cross section $\sigma = 1$ pb;
2. the decay rate of the ϕ meson $1/\tau = 6.47 \times 10^{21}$ s^{-1};
3. the electric charge e.

Discussion

Natural units are of the greatest help in simplifying the calculations. In natural units, any dimensionful quantity is expressed in powers of the energy unit, taken to be the eV or one of its multiple. The underlying idea is to trade off the usual MKS units [kg, m, s] by the three dimensionful units [eV, $\hbar$, c], and then to "silence" the result with respect to $\hbar$ and c by setting them equal to unity. In order to convert the result back into the MKS system, one needs to multiply the number obtained by a factor $\hbar^\alpha c^\beta$ where α and β are chosen such that the overall dimension comes out right.

Solution

Remembering that $m c^2$ is an energy, we can write:

$$\text{kg}\, c^2 = \text{kg}\, (3 \times 10^8 \text{ m s}^{-1})^2 = 9 \times 10^{16} \text{ J} = \frac{9 \times 10^{16}}{1.6 \times 10^{-19}} \text{ eV} = 0.56 \times 10^{36} \text{ eV},$$

$$\text{kg} = 0.56 \times 10^{36} \text{ eV}\, c^{-2} \quad \Rightarrow \quad \text{kg} = 0.56 \times 10^{27} \text{ GeV}. \qquad (1.6)$$

To convert the metre, it proves useful to remember the MKS value of the constant $\hbar c$:

$$\hbar c = 197 \text{ MeV fm} = 1.97 \times 10^{-7} \text{ m eV},$$

$$\text{m} = 0.507 \times 10^7 \ \hbar c \, \text{eV}^{-1} \quad \Rightarrow \quad \text{m} = 0.507 \times 10^{16} \text{ GeV}^{-1}. \qquad (1.7)$$

Finally, we can convert the second by simply using that:

$$c = 3 \times 10^8 \text{ m s}^{-1}, \qquad \text{s} = \frac{3 \times 10^8}{c} (0.507 \times 10^7 \ \hbar c \, \text{eV}^{-1})$$

$$\text{s} = 1.52 \times 10^{15} \ \hbar \, \text{eV}^{-1} \quad \Rightarrow \quad \text{s} = 1.52 \times 10^{24} \text{ GeV}^{-1}. \qquad (1.8)$$

The inverse transformations into the MKS system laws are also very useful, since a theoretical calculation performed in natural unitswill yield the result as a power of GeV. In particular, one often needs to convert lengths, cross sections, and time intervals into MKS:

$$[\text{length}] \quad \text{GeV}^{-1} = 1.97 \text{ fm}$$
$$[\text{cross section}] \quad \text{GeV}^{-2} = 0.389 \text{ mbarn} \qquad (1.9)$$
$$[\text{time}] \quad \text{GeV}^{-1} = 0.66 \times 10^{-24} \text{ s}$$

Let's now apply the results above to the three cases of interest.

1. From Eq. (1.7) it follows that:

$$1 \text{ pb} = 10^{-40} \text{ m}^2 = 10^{-40} (0.507 \times 10^7 \text{ eV}^{-1})^2 = 0.26 \times 10^{-8} \text{ GeV}^{-2}.$$
$$(1.10)$$

2. From Eq. (1.8) we have:

$$\Gamma_\phi = 6.47 \times 10^{21} \text{ s}^{-1} = 6.47 \times 10^{21} (1.52 \times 10^{24})^{-1} \text{ GeV} = 4.26 \text{ MeV}.$$
$$(1.11)$$

3. From the definition of the fine structure in Heaviside–Lorentz units, such that the first of Maxwell's equation becomes $\nabla \cdot \mathbf{E} = \rho$ and $[e] = \text{kg}^{1/2} \text{ m s}$, we have:

$$\frac{e^2}{4\pi \hbar c} \approx \frac{1}{137} \quad \Rightarrow \quad e \approx \sqrt{\frac{4\pi}{137}} = 0.303. \qquad (1.12)$$

Problem 1.2 Prove that the two measures $d^4p \equiv dE\, d^3\mathbf{p}$ and $d^3\mathbf{p}/E_\mathbf{p}$, where $(E, \mathbf{p})$ form a four-vector and $E_\mathbf{p} = \sqrt{|\mathbf{p}|^2 + m^2}$, are invariant under a generic transformation of the Lorentz group.

Solution

Let's consider first the d^4p measure. As discussed in the introduction, the Lorentz transformations preserves the Minkowski norm of Eq. (1.1). Then, for every four-vector p and boost Λ it must hold:

$$\left(p^\mathrm{T} \Lambda^\mathrm{T}\right) g\, (\Lambda p) = p'_\mu p'_\nu g^{\mu\nu} = p_\mu p_\nu g^{\mu\nu} = p^\mathrm{T} g\, p \quad \Rightarrow \quad \Lambda^\mathrm{T} g\, \Lambda = g. \qquad (1.13)$$

The last equation implies $(\det \Lambda)^2 = 1$. In particular, the *proper* Lorentz transformation are defined by the condition $\det \Lambda = +1$, as it is the case for Eq. (1.4). Hence, under a generic boost we have:

$$d^4p' = |\det \Lambda|\, d^4p = d^4p, \qquad (1.14)$$

while under a space rotation, $d^3\mathbf{p}' = d^3\mathbf{p}$ and $dE' = dE$, which proves that the measure is indeed invariant under the Lorentz group.

Let's now consider the second measure. Without loss of generality, we can assume the x-axis to be aligned with the boost direction, so that:

$$\frac{d^3\mathbf{p}'}{E'_\mathbf{p}} = \frac{d\left[\gamma\,(p_x - \beta dE_\mathbf{p})\right]\,dp_y\,dp_z}{\gamma\,(E_\mathbf{p} - \beta p_x)} = \frac{\gamma\,(dp_x - \beta p_x/E_\mathbf{p}\,dp_x)\,dp_y\,dp_z}{\gamma\,(E_\mathbf{p} - \beta p_x)} = \frac{d^3\mathbf{p}}{E_\mathbf{p}}$$

(1.15)

Invariance under space rotation follows again from the fact that $d^3\mathbf{p}' = d^3\mathbf{p}$ and $E'_\mathbf{p} = E_\mathbf{p}$. Alternatively, we can write this measure in a way that if manifestly Lorentz-invariant. Indeed:

$$\Theta\,(E_\mathbf{p})\,\delta(p^2 - m^2)\,d^4p = \frac{d^3\mathbf{p}}{\left|\frac{\partial(p^2-m^2)}{\partial E}\right|_{+E_\mathbf{p}}} = \frac{d^3\mathbf{p}}{2E_\mathbf{p}},$$

(1.16)

where the Heaviside function Θ selects the positive root. Since the left-hand side of Eq. (1.16) is manifestly Lorentz-invariant,[1] such must be the right-hand side.

Problem 1.3 Prove that the scalar function:

$$I(p_1, p_2) = \sqrt{(E_2\mathbf{p}_1 - E_1\mathbf{p}_2)^2 - (\mathbf{p}_1 \times \mathbf{p}_2)^2}$$

(1.17)

where $p_i = (E_i, \mathbf{p}_i)$ are a pair of four-vectors, is invariant under rotations and boosts.

Solution

The invariance of the right-hand side of Eq. (1.17) under rotations follows from the fact that a rotation leaves unchanged both the 0th component of the four-vector and the relative angle between the three-vectors. The invariance under boosts is less trivial and has to be proved explicitly. The best way to do it is to find an equivalent expression for I that is manifestly Lorentz-invariant. To this purpose, it is convenient to take the square at both sides, obtaining:

$$\begin{aligned}
I(p_1, p_2)^2 &= (E_1\mathbf{p}_2 - E_2\mathbf{p}_1)^2 - (\mathbf{p}_1 \times \mathbf{p}_2)^2 = \\
&= E_1^2|\mathbf{p}_2|^2 + E_2^2|\mathbf{p}_1|^2 - 2E_1E_2\mathbf{p}_1 \cdot \mathbf{p}_2 - |\mathbf{p}_1|^2|\mathbf{p}_2|^2 + (\mathbf{p}_1 \cdot \mathbf{p}_2)^2 = \\
&= \underbrace{E_1^2E_2^2 - 2E_1E_2\mathbf{p}_1 \cdot \mathbf{p}_2 + (\mathbf{p}_1 \cdot \mathbf{p}_2)^2}_{(p_1p_2)^2} -E_1^2p_2^2 + E_2^2|\mathbf{p}_1|^2 - |\mathbf{p}_1|^2|\mathbf{p}_2|^2 = \\
&= (p_1p_2)^2 - (|\mathbf{p}_1|^2 + p_1^2)p_2^2 + (|\mathbf{p}_2|^2 + p_2^2)|\mathbf{p}_1|^2 - |\mathbf{p}_1|^2|\mathbf{p}_2|^2 = \\
&= (p_1p_2)^2 - p_1^2p_2^2.
\end{aligned}$$

(1.18)

---

[1] It can be proved that the sign of the 0th component is also a Lorentz invariant.

Since the left-hand side of Eq. (1.18) is invariant under boosts, so must be the $I(p_1, p_2)$. The special case $\mathbf{p}_1 \propto \mathbf{p}_2$ gives rise to a simpler expression, where the effect of a finite Lorentz transformation can be studied explicitly. By using the matrix form of Eq. (1.5), we get:

$$
\begin{aligned}
I(p_1', p_2') &= \left| E_1' p_{2x}' - E_2' p_{1x}' \right| = \\
&= |(\cosh \alpha\, E_1 + \sinh \alpha\, p_{1x})(\sinh \alpha\, E_2 + \cosh \alpha\, p_{2x}) - \\
&\quad - (\cosh \alpha\, E_2 + \sinh \alpha\, p_{2x})(\sinh \alpha\, E_1 + \cosh \alpha\, p_{1x})| \\
&= |(\cosh^2 \alpha - \sinh^2 \alpha) E_1 p_{2x} - (\cosh^2 \alpha - \sinh^2 \alpha) E_2 p_{2x}| = \\
&= |E_1 p_{2x} - E_2 p_{1x}| = I(p_1, p_2).
\end{aligned} \tag{1.19}
$$

Discussion

When the four-vectors are specialised to be the current densities[2] of two beams, i.e.: $j_i = (\rho_i, \rho_i \mathbf{v}_i)$, where ρ_i is a space density ($[\rho] = m^{-3}$) and $\rho_i \mathbf{v}_i$ is a density flux ($[\rho \mathbf{v}] = m^{-2} s^{-1}$), then Eq. (1.17) becomes:

$$
I(j_1, j_2) = \rho_1 \rho_2 \sqrt{\left(\boldsymbol{\beta}_1 - \boldsymbol{\beta}_2 \right)^2 - \left(\boldsymbol{\beta}_1 \times \boldsymbol{\beta}_2 \right)^2} \equiv \rho_1 \rho_2 v_{rel} \tag{1.20}
$$

where $\boldsymbol{\beta}_i$ are the particle velocities in units of c, and v_{rel} is called *relative velocity* between the two particles, although this is not a proper velocity (indeed, it can also exceed c in some reference frames). The interpretation of v_{rel} in terms of particle velocities will be elucidated in Problem 1.11. The invariant I finds application in the general formula of the cross section, see e.g. Eq. (1.293).

Problem 1.4 Prove the identity:

$$
\frac{dp_\mu}{d\tau} \frac{dp^\mu}{d\tau} = -m^2 \gamma^6 \left[\dot{\boldsymbol{\beta}}^2 - (\boldsymbol{\beta} \times \dot{\boldsymbol{\beta}})^2 \right], \tag{1.21}
$$

where p is the four-momentum of particle of mass m accelerated by external forces, τ is the proper time of the particle, while the time derivatives and the velocities are measured in a generic frame.

Solution

We first express the four-momentum components in terms of the particle velocity $\boldsymbol{\beta}$, namely:

---

[2] The fact that j is a Lorentz-vector can be proved by noticing that the continuity equation $\partial_t \rho + \mathrm{div}_x \rho \mathbf{v} = 0$ has to be invariant since it states the conservation of mass, which as to hold for any frame. The latter can be written in covariant notation as $\partial_\mu j^\mu = 0$, hence j_μ has to transform as a covariant vector since ∂_μ is contravariant.

$$p = (E_{\mathbf{p}}, \mathbf{p}) = (m\gamma, m\gamma\boldsymbol{\beta}). \tag{1.22}$$

Hence, we get:

$$\frac{dp_\mu}{d\tau}\frac{dp^\mu}{d\tau} = \left(\frac{dE_{\mathbf{p}}}{d\tau}\right)^2 - \left(\frac{d\mathbf{p}}{d\tau}\right)^2 = m^2\gamma^2\left[\left(\frac{d\gamma}{dt}\right)^2 - \left(\frac{d(\gamma\boldsymbol{\beta})}{dt}\right)^2\right], \tag{1.23}$$

where we have made use of the relation $dt = \gamma\,d\tau$. By applying the chain rule for the derivative of composite functions, we get:

$$\left(\frac{d\gamma}{dt}\right)^2 = \left(\gamma^3\boldsymbol{\beta}\cdot\dot{\boldsymbol{\beta}}\right)^2 = \gamma^6\left(\boldsymbol{\beta}\cdot\dot{\boldsymbol{\beta}}\right)^2 \tag{1.24}$$

$$\left(\frac{d(\gamma\boldsymbol{\beta})}{dt}\right)^2 = \left(\dot{\gamma}\boldsymbol{\beta} + \gamma\dot{\boldsymbol{\beta}}\right)^2 = \gamma^2\left(\gamma^2(\boldsymbol{\beta}\cdot\dot{\boldsymbol{\beta}})\boldsymbol{\beta} + \dot{\boldsymbol{\beta}}\right)^2 =$$
$$= \gamma^6\left(\boldsymbol{\beta}\cdot\dot{\boldsymbol{\beta}}\right)^2\boldsymbol{\beta}^2 + 2\gamma^4\left(\boldsymbol{\beta}\cdot\dot{\boldsymbol{\beta}}\right)^2 + \gamma^2\dot{\boldsymbol{\beta}}^2 =$$
$$= \gamma^4(\gamma^2 - 1)\left(\boldsymbol{\beta}\cdot\dot{\boldsymbol{\beta}}\right)^2\boldsymbol{\beta}^2 + 2\gamma^4\left(\boldsymbol{\beta}\cdot\dot{\boldsymbol{\beta}}\right)^2 + \gamma^2\dot{\boldsymbol{\beta}}^2 =$$
$$= \gamma^6\left(\boldsymbol{\beta}\cdot\dot{\boldsymbol{\beta}}\right)^2 + \gamma^4\left(\boldsymbol{\beta}\cdot\dot{\boldsymbol{\beta}}\right)^2 + \gamma^2\dot{\boldsymbol{\beta}}^2 \tag{1.25}$$

Inserting these identities in Eq. (1.23), we obtain:

$$\frac{dp_\mu}{d\tau}\frac{dp^\mu}{d\tau} = -m^2\gamma^4\left[\gamma^2\left(\boldsymbol{\beta}\cdot\dot{\boldsymbol{\beta}}\right)^2 + \dot{\boldsymbol{\beta}}^2\right] =$$
$$= -m^2\gamma^6\left[\boldsymbol{\beta}^2\dot{\boldsymbol{\beta}}^2 - (\boldsymbol{\beta}\times\dot{\boldsymbol{\beta}})^2 + \dot{\boldsymbol{\beta}}^2(1 - \boldsymbol{\beta}^2)\right] =$$
$$= -m^2\gamma^6\left[\dot{\boldsymbol{\beta}}^2 - (\boldsymbol{\beta}\times\dot{\boldsymbol{\beta}})^2\right], \tag{1.26}$$

which proves the identity of Eq. (1.21). Since the left-hand side is manifestly Lorentz-invariant, so has to be the right-hand side.

Discussion

Equation (1.21) gives the *Lieanard formula* for the power P emitted by a charged particle accelerated by an external force. Indeed, one can prove that $P \propto -e^2(dp/d\tau)^2$, where e is the electric charge of the particle [1].

Problem 1.5 Work out a heuristic representation of the boost generators $\mathbf{K}$ in dimension $d = 2$ starting from the finite boost transformation of Eq. (1.5).

Solution

Let's denote by (χ_1, χ_2) the four-vector components that transform non-trivially under boosts. We then rewrite Eq. (1.5) in an exponentiated form as to make the generators explicit:

$$\begin{pmatrix} \chi_1' \\ \chi_2' \end{pmatrix} = \begin{pmatrix} \cosh\frac{\alpha}{2} & -\sinh\frac{\alpha}{2} \\ -\sinh\frac{\alpha}{2} & \cosh\frac{\alpha}{2} \end{pmatrix} \begin{pmatrix} \chi_1 \\ \chi_2 \end{pmatrix} =$$

$$= \left[\cosh\frac{\alpha}{2} \cdot \begin{pmatrix} 1 & 0 \\ 0 & 1 \end{pmatrix} - \sinh\frac{\alpha}{2} \cdot \begin{pmatrix} 0 & 1 \\ 1 & 0 \end{pmatrix} + 0 \cdot \begin{pmatrix} 0 & -i \\ i & 0 \end{pmatrix} + 0 \cdot \begin{pmatrix} 1 & 0 \\ 0 & -1 \end{pmatrix} \right] \begin{pmatrix} \chi_1 \\ \chi_2 \end{pmatrix},$$

$$= \left[\cos\left(i\frac{\alpha}{2}\right) \mathbb{1} - i\sin\left(i\frac{\alpha}{2}\right) \boldsymbol{\sigma} \cdot \mathbf{n}_\alpha \right] \begin{pmatrix} \chi_1 \\ \chi_2 \end{pmatrix} = e^{i(-i\frac{\sigma}{2})\cdot\boldsymbol{\alpha}} \begin{pmatrix} \chi_1 \\ \chi_2 \end{pmatrix} \qquad (1.27)$$

where $|\mathbf{n}_\alpha| = 1$. By defining the transformation parameter as $\alpha/2$, we could get directly the correct normalisation for the generators. Since a generic element of a group can be always parametrised as $\exp\{i\mathbf{K} \cdot \boldsymbol{\alpha}\}$, where $\mathbf{K}$ are the generators, an immediate comparison with Eq. (1.27) yields the result:

$$\mathbf{K} = -i\frac{\boldsymbol{\sigma}}{2}. \qquad (1.28)$$

Discussion

The six generators of the Lorentz group satisfy the *Lie algebra*:

$$[J_i, J_j] = i\,\varepsilon_{ijk}J_k, \quad [J_i, K_j] = i\,\varepsilon_{ijk}K_k, \quad [K_i, K_j] = -i\,\varepsilon_{ijk}J_k \qquad (1.29)$$

where J_i are the generators of the space rotations and K_i of the boosts. The combinations $\mathbf{J}_\pm = (\mathbf{J} \pm i\mathbf{K})/2$ commute among themselves and satisfy individually the algebra of $SU(2)$. For $d = 2$, we know that the $\boldsymbol{\sigma}/2$ matrices provide a fundamental representation of the generators of $SU(2)$. Given that $\mathbf{J} = \mathbf{J}_+ + \mathbf{J}_-$ and $\mathbf{K} = -i(\mathbf{J}_+ - \mathbf{J}_-)$, and considering the commutation rules in Eq. (1.29), one can easily verify that $\mathbf{J} = \boldsymbol{\sigma}/2$ and $\mathbf{K} = -i\boldsymbol{\sigma}/2$ provide a representation of the generators. Notice that $\mathbf{K} = +i\boldsymbol{\sigma}/2$ provides an equally valid representation, since the commutation rules are all invariant under $\mathbf{K} \to -\mathbf{K}$.

The existence of two inequivalent representations of the boost vector in dimension $d = 2$, the so-called $(\frac{1}{2}, 0)$ and $(0, \frac{1}{2})$, which are related one-to-another by a parity operation,[3] has important implication for the quantisation of spin-1/2 fields (Weil spinors). The latter can indeed exist in two *chiralities*, depending under which representation of the Lorentz group they transform: right-handed (RH) and left-handed (LH) spinors.

Problem 1.6 Prove that all RH (LH) spin-1/2 particles of mass m have helicity $h = +1/2$ ($h = -1/2$) in the limit $|\mathbf{p}| \gg m$.

---

[3] $\mathbf{K}$ is a vector under rotations, see e.g. the second of Eq. (1.29), and it also transforms as a vector under parity transformations, since a parity operation must change the direction of the boost.

Discussion

The *helicity* operator is defined as:

$$h = \frac{\mathbf{S} \cdot \mathbf{p}}{|\mathbf{p}|}, \tag{1.30}$$

and it acts on the spin vector of the wave function. Notice that h is not a priori Lorentz-invariant, since, for a massive particle, it is always possible to find a reference frame where the momentum of the particle flips direction. This is not the case for massless particles, for which the helicity is instead a conserved quantum number.

Solution

As discussed in Problem 1.5 there are two inequivalent representation of the Lorentz group in dimension $d = 2$, which is the space suitable to construct the spin states for spin-1/2 particles. The generic passive transformation $\Lambda(\boldsymbol{\theta}, \boldsymbol{\alpha})$ can be written as:

$$\Lambda(\boldsymbol{\theta}, \boldsymbol{\alpha}) = e^{i(\mathbf{J} \cdot \boldsymbol{\theta} - \mathbf{K} \cdot \boldsymbol{\alpha})} = \begin{cases} \exp\left[i\frac{\boldsymbol{\sigma}}{2} \cdot \boldsymbol{\theta} - \frac{\boldsymbol{\sigma}}{2} \cdot \boldsymbol{\alpha}\right] & \text{RH} \\ \exp\left[i\frac{\boldsymbol{\sigma}}{2} \cdot \boldsymbol{\theta} + \frac{\boldsymbol{\sigma}}{2} \cdot \boldsymbol{\alpha}\right] & \text{LH} \end{cases} \tag{1.31}$$

The spin operator is represented by the matrix $\mathbf{S} = \boldsymbol{\sigma}/2$. Let's denote the spinor in the centre-of-mass frame by ξ^* for the RH fermion and by η^* for the LH fermion. Let's also chose the basis of eigenvectors of σ_z, where the z-axis is assumed to be aligned with the boost direction. The spinors ξ and η in the laboratory frame, where the particle three-momentum is $\mathbf{p}$, can be obtained by applying a boost with parameter $\boldsymbol{\alpha} = -\alpha\, \mathbf{e}_z$, giving:

$$\begin{cases} \xi = \Lambda(\boldsymbol{\alpha})\xi^* = \left(\cosh\frac{\alpha}{2}\mathbb{1} - \sinh\frac{\alpha}{2}\boldsymbol{\sigma} \cdot (-\mathbf{e}_z)\right)\xi^* = \left(\cosh\frac{\alpha}{2}\mathbb{1} + \sinh\frac{\alpha}{2}\sigma_z\right)\xi^* \\ \eta = \Lambda(\boldsymbol{\alpha})\eta^* = \left(\cosh\frac{\alpha}{2}\mathbb{1} + \sinh\frac{\alpha}{2}\boldsymbol{\sigma} \cdot (-\mathbf{e}_z)\right)\eta^* = \left(\cosh\frac{\alpha}{2}\mathbb{1} - \sinh\frac{\alpha}{2}\sigma_z\right)\eta^* \end{cases} \tag{1.32}$$

If we now take the spinor in the rest frame ξ^* and η^* to a generic admixture of $\pm\frac{1}{2}$ eigenstates, i.e. $(\cos\frac{\delta}{2}, \sin\frac{\delta}{2})$, the polarisation in the laboratory frame will be:

$$\frac{P_{+1/2}^{R,L} - P_{-1/2}^{R,L}}{P_{+1/2}^{R,L} + P_{-1/2}^{R,L}} = \frac{\cos^2\frac{\delta}{2}(\cosh\frac{\alpha}{2} \pm \cosh\frac{\alpha}{2})^2 - \sin^2\frac{\delta}{2}(\cosh\frac{\alpha}{2} \mp \cosh\frac{\alpha}{2})^2}{\cos^2\frac{\delta}{2}(\cosh\frac{\alpha}{2} \pm \cosh\frac{\alpha}{2})^2 + \sin^2\frac{\delta}{2}(\cosh\frac{\alpha}{2} \mp \cosh\frac{\alpha}{2})^2} =$$

$$= \frac{(\cosh^2\frac{\alpha}{2} + \sinh^2\frac{\alpha}{2})\cos\delta \pm 2\sinh\frac{\alpha}{2}\cosh\frac{\alpha}{2}}{\cosh^2\frac{\alpha}{2} + \sinh^2\frac{\alpha}{2} \pm 2\sinh\frac{\alpha}{2}\cosh\frac{\alpha}{2}\,\cos\delta} = \frac{\cos\delta \pm \tanh\alpha}{1 \pm \tanh\alpha\cos\delta} =$$

$$= \frac{\cos\delta \pm \beta}{1 \pm \beta\cos\delta}. \tag{1.33}$$

The last equality makes use of Eq. (1.5) to relate the boost parameter to the velocity in the laboratory frame. It can be noticed that this expression is identical to the transformation law of the cosine of the polar angle of a massless four-vector under a boost β, if one interprets δ as the polar angle in the centre-of-mass frame, see Eq. (1.40). In particular, if $\beta \to 1$, as it is the case for $|\mathbf{p}| \gg m$, the polarisation tends to ± 1 for $0 < \delta < \pi$ and to 0 for $\delta = 0, \pi$. For an unpolarised state in the centre-of-mass ($\delta = \pi/2$), the polarisation in the laboratory frame is given by $+\beta$ for RH fermions and by $-\beta$ for LH. Therefore, ultra-relativistic spin-1/2 particles of a given chirality will also have a net helicity, independently from their spin state in the centre-of-mass frame: the helicity is positive for right-handed fermions and negative for left-handed.

Suggested Readings

The reader is addressed to a more complete textbook on quantum field theory. See e.g. Chap. 3 of Ref. [2] or Chap. 9 of Ref. [3].

Problem 1.7 The $\Delta(1232)$ resonance can be produced by scattering pions of appropriate energy against a proton target. Assume the protons to be unpolarised. Determine the angular distribution of the scattered pions in the centre-of-mass frame.

Discussion

As discussed in the introduction, the Lorentz group comprises the set of spatial rotations. Invariance of the dynamics under rotations becomes an important selection rule when the particles involved carry spin, since the latter transforms non trivially under rotations. A useful tool to investigate the behaviour of spin states under rotations is provided by the so-called *rotation matrix*, $d^j_{m', m}(\theta)$, defined by:

$$\exp\left[-i\theta \mathbf{J} \cdot \mathbf{e}_y\right] |j, m\rangle = \sum_{m'} d^j_{m', m}(\theta) |j, m'\rangle \tag{1.34}$$

Here, the vectors $|j, m\rangle$ are the eigenstates of the angular momentum operators $\mathbf{J}^2$ and J_z. As made evident by Eq. (1.34), the elements of the rotation matrix are the linear coefficients of the rotated of the generic eigenstate by an angle θ around an axis orthogonal to the quantisation axis z. It can be also shown that:

$$d^j_{m', m} = (-1)^{m-m'} d^j_{m, m'} = d^j_{-m, -m'}, \tag{1.35}$$

see e.g. Ref. [4]. The rotation matrix proves very useful when one wants to analyse transition probabilities between states related by a spatial rotation.

Solution

When the centre-of-mass energy approaches 1.2 GeV, the πp scattering becomes resonant due to a baryonic bound-state, called Δ. The Δ baryons are members of the $j = 3/2$ decuplet. At the resonance, the dominant state participating in the scattering can be therefore assumed to have $j = 3/2$. The eigenvalue of $\hat{J}_z$, where z is the axis defined by the initial centre-of-mass momentum of the proton, $\mathbf{p}$, can take the values $m = \pm 1/2$, since only the proton can contribute to J_z through its spin. The assumption of unpolarised protons implies that the two configurations must be equally likely. When the Δ resonance is produced, it decays into a new πp state with centre-of-mass momentum $\mathbf{p}'$. Let the angle between $\mathbf{p}$ and $\mathbf{p}'$ be denoted by θ. The decay amplitude is fully determined by angular momentum conservation. Indeed, the final state corresponds to the rotated by θ of either $|\frac{3}{2}, +\frac{1}{2}\rangle$ or $|\frac{3}{2}, -\frac{1}{2}\rangle$. The two corresponds to orthogonal states, so the probability of decaying to an angle θ is given by the sum of the probabilities:

$$
\frac{1}{\Gamma}\frac{d\Gamma}{d\cos\theta} = \frac{1}{2}\cdot\underbrace{\frac{1}{\Gamma}\frac{d\Gamma_{m=-1/2}}{d\cos\theta}}_{\sum_{m'}\left|d^{3/2}_{m',-1/2}\right|^2} + \frac{1}{2}\cdot\underbrace{\frac{1}{\Gamma}\frac{d\Gamma_{m=+1/2}}{d\cos\theta}}_{\sum_{m'}\left|d^{3/2}_{m',+1/2}\right|^2} =
$$

$$
= \left(\frac{3\cos\theta - 1}{2}\cos\frac{\theta}{2}\right)^2 + \left(-\frac{3\cos\theta + 1}{2}\sin\frac{\theta}{2}\right)^2 =
$$

$$
= \frac{1}{4}\left[9\cos^2\theta + 1 - 6\cos\theta\left(\cos^2\frac{\theta}{2} - \sin^2\frac{\theta}{2}\right)\right] = \frac{1 + 3\cos^2\theta}{4}.
$$

(1.36)

Notice that we have made use of Eq. (1.35) to simplify the calculations. The angular distribution in the centre-of-mass frame will therefore feature a dependence on the polar angle θ of the form $\sim(1 + 3\cos^2\theta)$.

Suggested Readings

The reader is addressed to the original paper by E. Fermi et al. [5] about the evidence of the Δ resonance and to the determination of its spin based on the distribution of the scattering angles. A compendium of formulas for the rotation matrices can be found in Table 43 of Ref. [4].

Problem 1.8 A massless spin-1/2 particle scatters elastically against a much heavier particle that can be assumed to be always at rest. The two particles exchange force through an helicity-conserving interaction which does not change the spin of the target particle, if any. Prove that the light particle cannot be scattered exactly backwards.

Discussion

In the SM, the interaction between spin-$1/2$ particles and gauge bosons are helicity-conserving, since they only involve combinations of vector V and axial A currents. When the theory is not symmetric for RH and LH particles, it is said to be *chiral*, since it distinguishes between the two chiralities. The $SU(2)_L$ interaction that underlies the EWK force is chiral, since it only involves LH fermions. On the contrary, QCD and the EM interaction treat RH and LH fermions on the same footing, see Sect. 5.2.

Solution

A massless spin-$1/2$ particle can be in only one helicity state, depending on its chirality. If the interaction conserves the chirality, the particle will have the same helicity before and after the scattering. If the target particle does not participate to the interaction through its spin, it will also conserve its projection along the scattering axis. If the projectile were to scatter exactly backwards, it would imply a change of the overall angular momentum projection along the scattering axis by $|\Delta m| = 1$, which would violate angular momentum conservation. In terms of the rotation matrices of Eq. (1.34), we can argument that the scattering amplitude at an angle θ should be proportional to $d^{1/2}_{\pm 1/2, \pm 1/2} = \cos\frac{\theta}{2}$, which indeed vanishes at $\theta = \pi$.

Suggested Readings

The reader is addressed to Chap. 1 of Ref. [2] for a deeper discussion on this subject.

Problem 1.9 Determine the relativistic Doppler effect and the law of aberration of light for an observer moving with velocity $\mathbf{v} = \boldsymbol{\beta}c$ with respect to the light source.

Solution

The invariance under Lorentz transformations requires the phase of a light wave to be the same for two reference frames $\mathscr{R}$ and $\mathscr{R}'$, moving one with respect to the other with constant velocity $\boldsymbol{\beta}c$. In the frame $\mathscr{R}$, the frequency of the wave is ω and its direction of propagation $\mathbf{n}$, while in $\mathscr{R}'$, the same quantities are ω' and $\mathbf{n}'$. Synchronising the two clocks so that at the time $t = t' = 0$ the wave has a phase $\phi = 0$ at the origin $\mathbf{r} = \mathbf{r}' = 0$, we have:

$$\phi(t, \mathbf{r}) = \phi'(t', \mathbf{r}') \quad \Leftrightarrow \quad \omega(t - \mathbf{n} \cdot \mathbf{r}) = \omega'(t' - \mathbf{n}' \cdot \mathbf{r}'). \tag{1.37}$$

The coordinates $(t, \mathbf{r})$ can be expressed in terms of $(t', \mathbf{r}')$ by using the transformation of Eq. (1.4), with the only modification $\beta \rightarrow -\beta$, since $\mathscr{R}$, the frame where the light source is at rest, moves with velocity $-\boldsymbol{\beta}$ as seen from $\mathscr{R}'$. It is convenient to choose the coordinate system so that $\boldsymbol{\beta}$ is aligned along the x-axis. We also write

$\mathbf{n} = \cos\theta \mathbf{e}_x + \sin\theta \mathbf{e}_y$. By equating the coefficients of the space-time coordinates in Eq. (1.37), we obtain the following system of equations:

$$\begin{cases} \omega' = \omega\gamma\,(1 - \beta\cos\theta) \equiv \omega\gamma\,(1 - \boldsymbol{\beta} \cdot \mathbf{n}) \\ \omega'\cos\theta' = \omega\gamma\,(\cos\theta - \beta) \\ \omega'\sin\theta' = \omega\sin\theta \end{cases} \tag{1.38}$$

The first equation gives the relativistic *Doppler effect*. Taking the ratio of the last two equations:

$$\frac{\omega'\sin\theta'}{\omega'\cos\theta'} = \tan\theta' = \frac{\sin\theta}{\gamma(\cos\theta - \beta)}, \tag{1.39}$$

which describes the *aberration of light*. It is useful to write explicitly the two other trigonometric relations:

$$\cos\theta' = \frac{\cos\theta - \beta}{1 - \beta\cos\theta}, \qquad \sin\theta' = \frac{\sin\theta}{\gamma\,(1 - \beta\cos\theta)}, \tag{1.40}$$

which can be derived from Eq. (1.38), or even directly from Eq. (1.39) by using the trigonometric identity $\cos\theta = \pm(1 + \tan^2\theta)^{-\frac{1}{2}}$.

Problem 1.10 Derive the transformation law for the velocity $\mathbf{v}$ of a massive particle under a generic Lorentz transformation.

Solution

Let the particle velocity in the reference frame $\mathscr{R}$ be denoted by $\mathbf{v}$. The reference frame $\mathscr{R}'$ in which we want to calculate the particle velocity moves with velocity $\boldsymbol{\beta}$ as seen from $\mathscr{R}$. It's convenient to choose the coordinate system so that $\boldsymbol{\beta}$ is aligned along the x-axis. From the Lorentz transformations Eq. (1.4) applied to the four-vector $(t, \mathbf{r})$, one has:

$$\begin{cases} dt' = \gamma\,(dt - \beta dx) \\ dx' = \gamma\,(-\beta\,dt + dx) \\ dy' = dy \\ dz' = dz \end{cases} \tag{1.41}$$

Taking the ratios:

$$v'_x = \frac{dx'}{dt'} = \frac{-\beta\, dt + dx}{dt - \beta\, dx} = \frac{v_x - \beta}{1 - \beta v_x}$$

$$v'_y = \frac{dy'}{dt'} = \frac{dy}{\gamma\,(dt - \beta\, dx)} = \frac{v_y}{\gamma(1 - \beta v_x)} \qquad (1.42)$$

$$v'_z = \frac{dz'}{dt'} = \frac{dz}{\gamma\,(dt - \beta\, dx)} = \frac{v_z}{\gamma(1 - \beta v_x)}$$

from which we can derive the transformation of polar angle:

$$\tan \theta' = \frac{\sin \theta}{\gamma\,(\cos \theta - \beta/|\mathbf{v}|)}, \qquad (1.43)$$

which agrees with Eq. (1.39) for the case $|\mathbf{v}| = 1$.

Let's now consider the two distinct cases for $|\mathbf{v}| < 1$, namely: $|\mathbf{v}| < \beta$ and $|\mathbf{v}| \geq \beta$. For later consistency, we redefine the variables using the same notation that will be adopted to study the kinematics in the centre-of-mass frame, namely we replace $|\mathbf{v}| \to \beta^*$ and $\beta \to -\beta$. With this choice, Eq. (1.43) becomes:

$$\tan \theta = \frac{\sin \theta^*}{\gamma\,(\cos \theta^* + \beta/\beta^*)}. \qquad (1.44)$$

First, we notice that, for $\beta/\beta^* < 1$, we have:

$$\lim_{\theta^* \to \pi} \tan \theta = 0^-, \qquad (1.45)$$

which means that a particle moving backwards in $\mathscr{R}^*$ will also appear moving backwards in $\mathscr{R}$. This is not the case for $\beta/\beta^* > 1$, because, in the same limit, $\tan \theta \to 0^+$. Since $\tan \theta = 0$ for $\theta^* = 0$, by Rolle's theorem there must be an angle $\theta^*_{\max}$ giving the largest opening angle in $\mathscr{R}$. To find such an angle, we first use the relation $\cos x = \pm(1 + \tan^2 x)^{-\frac{1}{2}}$ to express $\cos \theta$ as a function of θ^*, and then set its first derivative to zero to find the maximum:

$$\cos \theta = \pm \frac{1}{\sqrt{1 + \tan^2 \theta}} = + \frac{\gamma\,(\cos \theta^* + \beta/\beta^*)}{\sqrt{\gamma^2(\cos \theta^* + \beta/\beta^*) + (1 - \cos^2 \theta^*)}} \qquad (1.46)$$

The choice of the "+" sign is motivated by the fact that for $\beta > \beta^*$, $\tan \theta$ is always positive, so that $\theta < \frac{\pi}{2}$ and $\cos \theta > 0$. For sake of notation, we define $\cos \theta^*_{\max} = x$ and $\beta/\beta^* = \xi$. Then, we get:

Fig. 1.1 Relation between the cosine of the polar angle in two reference frames related by a boost $\beta = 0.8$ for three different values of the particle velocity β^*: 0.6 (top), 0.9 (middle), and 1 (bottom)

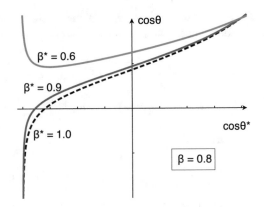

$$0 = \left.\frac{d\cos\theta}{dx}\right|_{\theta^*_{max}} = \frac{\sqrt{\gamma^2(x+\xi)^2 + 1 - x^2} - (x+\xi)\frac{2\gamma^2(x+\xi)-2x}{2\sqrt{\gamma^2(x+\xi)^2+1-x^2}}}{\gamma^2(x+\xi)^2 + 1 - x^2},$$
$$0 = \gamma^2(x+\xi)^2 + 1 - x^2 - (x+\xi)[\gamma^2(x+\xi) - x] = 1 + x\xi, \tag{1.47}$$

from which:

$$\cos\theta^*_{max} = -\frac{\beta^*}{\beta}. \tag{1.48}$$

At this centre-of-mass angle, the opening angle in the laboratory frame is given by:

$$\tan\theta_{max} = \sqrt{1 - \left(\frac{\beta^*}{\beta}\right)^2}\left[\gamma\left(-\frac{\beta^*}{\beta} + \frac{\beta}{\beta^*}\right)\right]^{-1} = \frac{\beta^*}{\gamma\sqrt{\beta^2 - \beta^{*2}}}. \tag{1.49}$$

Figure 1.1 shows a graph of $\cos\theta$ as a function of $\cos\theta^*$ for a boost $\beta = 0.8$ and three representative values of β^*.

Problem 1.11 Consider two particles with velocity $\mathbf{v}_A$ and $\mathbf{v}_B$. Determine the velocity of particle B in the rest frame of A.

Solution

It is convenient to first write Eq. (1.42) in a vectorial form. Taking into account both the longitudinal and transverse components, we have:

$$\mathbf{v}' = \frac{1}{\gamma(1 - \mathbf{v}\cdot\boldsymbol{\beta})}\left[\mathbf{v} + \left(\frac{\gamma-1}{\beta^2}\mathbf{v}\cdot\boldsymbol{\beta} - \gamma\right)\boldsymbol{\beta}\right]. \tag{1.50}$$

To express the velocity $\mathbf{v}_{B|A}$ of the particle B in the rest frame of A, we need to perform a boost with parameter $\boldsymbol{\beta} = \mathbf{v}_A$, so that Eq. (1.50) becomes:

$$\mathbf{v}_{B|A} = \frac{1}{\gamma_A(1 - \mathbf{v}_A \cdot \mathbf{v}_B)}\left[\mathbf{v}_B + \left(\frac{\gamma_A - 1}{|\mathbf{v}_A|^2}\mathbf{v}_A \cdot \mathbf{v}_B - \gamma_A\right)\mathbf{v}_A\right],$$

$$\gamma_A^2(1 - \mathbf{v}_A \cdot \mathbf{v}_B)^2|\mathbf{v}_{B|A}|^2 = |\mathbf{v}_B|^2 + |\mathbf{v}_A|^2\left[\frac{(\gamma_A - 1)^2}{|\mathbf{v}_A|^4}(\mathbf{v}_A \cdot \mathbf{v}_B)^2 - \frac{2\gamma_A(\gamma_A - 1)}{|\mathbf{v}_A|^2}\mathbf{v}_A \cdot \mathbf{v}_B + \gamma_A^2\right]$$

$$+ 2(\mathbf{v}_A \cdot \mathbf{v}_B)^2\frac{\gamma_A - 1}{|\mathbf{v}_A|^2} - 2\gamma_A\mathbf{v}_A \cdot \mathbf{v}_B =$$

$$= (\mathbf{v}_A \cdot \mathbf{v}_B)^2\left[\frac{(\gamma_A - 1)^2}{|\mathbf{v}_A|^2} + \frac{2(\gamma_A - 1)}{|\mathbf{v}_A|^2}\right] - 2\mathbf{v}_A \cdot \mathbf{v}_B\left[\gamma_A(\gamma_A - 1) + \gamma_A\right] + |\mathbf{v}_B|^2 + |\mathbf{v}_A|^2\gamma_A^2 =$$

$$= \gamma_A^2\left[(1 - \mathbf{v}_A \cdot \mathbf{v}_B)^2 - (1 - |\mathbf{v}_A|^2)(1 - |\mathbf{v}_B|^2)\right],$$

$$|\mathbf{v}_{B|A}| = \sqrt{1 - \frac{(1 - |\mathbf{v}_A|^2)(1 - |\mathbf{v}_B|^2)}{(1 - \mathbf{v}_A \cdot \mathbf{v}_B)^2}} = \frac{\sqrt{(\mathbf{v}_A - \mathbf{v}_B)^2 - (\mathbf{v}_A \times \mathbf{v}_B)^2}}{(1 - \mathbf{v}_A \cdot \mathbf{v}_B)}. \tag{1.51}$$

The last equality is easy to prove since $|\mathbf{v}_A \times \mathbf{v}_B|^2 = |\mathbf{v}_A|^2|\mathbf{v}_B|^2 - (\mathbf{v}_A \cdot \mathbf{v}_B)^2$. This expression is symmetric with respect to $A \leftrightarrow B$, therefore we also have $|\mathbf{v}_{B|A}| = |\mathbf{v}_{A|B}|$.

Discussion

From an immediate comparison with Eq. (1.20), we notice that:

$$|\mathbf{v}_{B|A}| = \frac{v_{\text{rel}}}{1 - \mathbf{v}_A \cdot \mathbf{v}_B}. \tag{1.52}$$

If we now multiply this expression by the density flux of B times the density of A in the rest frame of A, we obtain:

$$\rho_{A|A}\,\rho_{B|A}\,|\mathbf{v}_{B|A}| = \frac{\gamma_A\,\rho_{A|A}\,\rho_{B|A}}{\gamma_A(1 - \mathbf{v}_A \cdot \mathbf{v}_B)}\sqrt{(\mathbf{v}_A - \mathbf{v}_B)^2 - (\mathbf{v}_A \times \mathbf{v}_B)^2} =$$

$$= \rho_A\,\rho_B\sqrt{(\mathbf{v}_A - \mathbf{v}_B)^2 - (\mathbf{v}_A \times \mathbf{v}_B)^2}, \tag{1.53}$$

where we have used the fact that $\rho_{B|A}$ is the transformed of the density ρ_B in the rest frame of A, whereas $\gamma_A\,\rho_{A|A}$ is the density of A in the laboratory frame. This can be seen as another proof that the scalar variable I of Eq. (1.20) is Lorentz-invariant, since all the quantities at the left-hand side of Eq. (1.53) are defined in a *particular* reference frame, i.e. the rest frame of either A or B.

Problem 1.12 Determine the minimum and the maximum opening angle between the two massless particles produced in the decay of a particle of mass m and momentum $\mathbf{p}$.

Fig. 1.2 Representation of a boost $\boldsymbol{\beta}$ from the laboratory frame $\mathcal{R}$ to the centre-of-mass frame $\mathcal{R}^*$

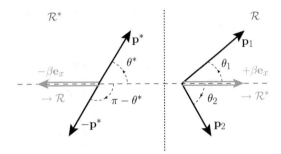

Solution

In the centre-of-mass frame, the two massless particles have momenta $\pm \mathbf{p}^*$ and energy $E^* = |\mathbf{p}^*| = m/2$. We choose the reference frame so that the x-axis is aligned along the direction of $\mathbf{p}$ and the decay takes place in the x–y plane. Let θ^* be the polar angle of the photon moving along the $y > 0$ direction, so that the other photon makes an angle of $\pi - \theta^*$. Seen from the laboratory frame, the centre-of-mass frame moves with velocity $\beta = |\mathbf{p}|/E_\mathbf{p}$, see Fig. 1.2. The momentum components in the laboratory frame are therefore given by Eq. (1.4), with the modification $\beta \to -\beta$:

$$
\begin{cases}
E_{1,2} = \gamma \left(E^* \pm \beta |\mathbf{p}^*| \cos \theta^* \right) = \gamma E^* (1 \pm \beta \cos \theta^*) \\
p_{1,2}^x = \gamma \left(\beta E^* \pm |\mathbf{p}^*| \cos \theta^* \right) = \gamma E^* (\beta \pm \cos \theta^*) \\
p_{1,2}^y = |\mathbf{p}^*| \sin \theta^*
\end{cases}
\tag{1.54}
$$

$$
\Rightarrow \quad \tan \theta_{1,2} = \frac{p_{1,2}^x}{p_{1,2}^y} = \frac{\sin \theta^*}{\gamma (\beta \pm \cos \theta^*)},
\tag{1.55}
$$

which agrees with Eq. (1.39). It is convenient to express $\cos \theta_{1,2}$ in terms of $\cos \theta^*$. To this purpose, we could either use Eq. (1.40) directly, or notice that $\cos \theta_{1,2} = p_{1,2}^x / E_{1,2}$, so that:

$$
\cos \theta_1 = \frac{\beta + \cos \theta^*}{1 + \beta \cos \theta^*}, \quad \sin \theta_1 = \frac{\sin \theta^*}{\gamma (1 + \beta \cos \theta^*)}
\tag{1.56}
$$

$$
\cos \theta_2 = \frac{\beta - \cos \theta^*}{1 - \beta \cos \theta^*}, \quad \sin \theta_2 = \frac{\sin \theta^*}{\gamma (1 - \beta \cos \theta^*)}
\tag{1.57}
$$

Let's define the opening angle between the two particles in the laboratory frame by ϕ. Then:

$$
\cos \phi = \cos (\theta_1 + \theta_2) = \cos \theta_1 \cos \theta_2 - \sin \theta_1 \sin \theta_2 =
$$

$$
= \frac{\beta^2 - \cos^2 \theta^*}{1 - \beta^2 \cos^2 \theta^*} - (1 - \beta^2) \frac{\sin^2 \theta^*}{1 - \beta^2 \cos^2 \theta^*} = \frac{2\beta^2 - 1 - \beta^2 \cos^2 \theta^*}{1 - \beta^2 \cos^2 \theta^*} =
$$

$$
= 1 - 2 \left(\frac{1 - \beta^2}{1 - \beta^2 \cos^2 \theta^*} \right).
\tag{1.58}
$$

The right-hand side of Eq. (1.58) in a monotonously decreasing function of $\cos^2 \theta^* \in [0, 1]$, so that:

$$\cos(\phi_{max}) = \min_{\theta^*}(\cos \phi) = 1 - 2 \frac{1 - \beta^2}{1 - \beta^2 \cos^2 \theta^*} \Big|_{\cos^2 \theta^* = 1} = -1$$

$$\Rightarrow \quad \phi_{max} = \phi(\theta^* = 0, \pi) = \pi \tag{1.59}$$

$$\cos(\phi_{min}) = \max_{\theta^*}(\cos \phi) = 1 - 2 \frac{1 - \beta^2}{1 - \beta^2 \cos^2 \theta^*} \Big|_{\cos^2 \theta^* = 0} = 2\beta^2 - 1$$

$$\Rightarrow \quad \phi_{min} = \phi(\theta^* = \frac{\pi}{2}) = \arccos(2\beta^2 - 1) = 2 \arccos(\beta) \tag{1.60}$$

An interesting case is when $|\mathbf{p}| \gg m$, so that $\beta \approx 1$ and ϕ_{min} is small. By Taylor-expanding the cosine around zero, we obtain:

$$2\beta^2 - 1 = \cos \phi_{min} \approx 1 - \frac{\phi_{min}^2}{2} \quad \Rightarrow \quad \phi_{min} = 2\sqrt{1 - \beta^2} = \frac{2}{\gamma}. \tag{1.61}$$

For large boosts, the opening angle between the two photons is therefore contained in the range $[2/\gamma, \pi]$. Notice that, for massless particles, it is always possible for one of them to move backwards with respect to the direction of the mother particle. For large boosts, however, the backward-emitted photon gets increasingly red-shifted, see the first of Eq. (1.54), or equivalently the first of Eq. (1.38), so that it eventually becomes of vanishing energy for $\gamma \gg 1$.

Problem 1.13 Determine the minimum and the maximum opening angle between two particles with mass m_1 and m_2 produced in the decay of a particle of mass m and momentum $\mathbf{p}$.

Solution

This exercise is analogous to Problem 1.12. We can therefore start from Eq. (1.54) for the more general case m_1, $m_2 > 0$, giving an equation for $\tan \theta_{1,2}$ as in Eq. (1.44). From this expression, we can compute the tangent of the opening angle ϕ:

$$\tan(\phi) = \frac{\tan \theta_1 + \tan \theta_2}{1 - \tan \theta_1 \tan \theta_2} =$$

$$= \frac{\frac{\beta}{\gamma} \frac{\beta_1^* + \beta_2^*}{\beta_1^* \beta_1^*} \sqrt{1 - \cos^2 \theta^*}}{-\beta^2 \cos^2 \theta^* + \beta \frac{\beta_1^* - \beta_2^*}{\beta_1^* \beta_2^*} \cos \theta^* + \left(\frac{\beta^2}{\beta_1^* \beta_2^*} - 1 + \beta^2 \right)}. \tag{1.62}$$

where $\beta_{1,2}^* = |\mathbf{p}^*|/E_{1,2}^*$ are the velocities of two particles in the centre-of-mass frame. The denominator $D(\cos \theta^*)$ at the right-hand side of Eq. (1.67) is a second-degree polynomial with negative concavity. Let's study its value for $\cos \theta^* = \pm 1$:

$$D(\pm 1) = \frac{\beta^2}{\beta_1^* \beta_2^*} \pm \frac{\beta}{\beta_2^*} \mp \frac{\beta}{\beta_1^*} - 1. \qquad (1.63)$$

These expressions are two parabola in β with roots $[-\beta_1^*, \beta_2^*]$ and $[-\beta_2^*, \beta_1^*]$, respectively. Since the concavity is positive, we have:

$$\begin{cases} D(+1) > 0, \ D(-1) > 0 & \text{if } \beta^* > \beta_{1,2}^* \\ D(+1) > 0, \ D(-1) < 0 & \text{if } \beta^* > \beta_1^* \text{ and } \beta^* < \beta_1^* \\ D(+1) < 0, \ D(-1) > 0 & \text{if } \beta^* < \beta_2^* \text{ and } \beta^* > \beta_1^* \\ D(+1) < 0, \ D(-1) < 0 & \text{if } \beta^* < \beta_{1,2}^* \end{cases} \qquad (1.64)$$

Hence, if $\beta > \max\{\beta_1^*, \beta_2^*\}$, then $D(\cos\theta^*) > 0$ for every angle θ^*. Consequently, $\tan\phi$ in Eq. (1.62) is limited and has a global maximum for some value $\cos\theta_{max}^*$. Conversely, if at least one among β_1^* and β_2^* is larger than β, the denominator has to vanish at some point, as for Rolle's theorem, and then it flips sign, so that $\tan\phi$ eventually approaches 0^- as $\cos\theta^* \to \pm 1$. If instead both β_1^* and β_2^* exceed β, then D can be either always negative, or vanish twice. We can summarise the various cases as follows:

- $\beta > \max\{\beta_1^*, \beta_2^*\}$: the maximum opening angle ϕ_{max} corresponds to atan (x_{max}), where $\cos\theta_{max}^*$ is the value that maximises the right-hand side of Eq. (1.62). We can easily see that such value does not correspond, in general, to $\theta^* = \pi/2$. One can find a numerical solution for ϕ_{max}, for example using Newton's method to iteratively maximise $\tan\phi$. Appendix 1.3 provides an example of how to determine numerically argmax $\{\tan\phi\}$ by using a computer program in Python.[4] The case $\beta_1^* = \beta_2^* = \beta^*$ allows to simplify further Eq. (1.67). First one notices that $\tan(\phi)$ becomes an even function of $\cos\theta^*$, hence we the maximum has to occur at $\cos\theta^* = 0$. At this angle, we have:

$$\tan(\phi_{max})_{\beta_1^* = \beta_2^*} = \frac{2\gamma\beta\beta^*}{\gamma^2\beta^2 - \beta^{*2}}. \qquad (1.65)$$

The minimum opening angle is $\phi_{min} = 0$, corresponding to a pair of collinear particles in the laboratory frame.

- $\beta_{2(1)}^* > \beta$, $\beta_{1(2)}^* < \beta$: the backward emission in the centre-of-mass frame of a particles with velocity larger than β, corresponds to a backward-propagating particle in the laboratory frame, hence $\phi_{max} = \pi$. Instead, since $\beta^* < \beta$ for the other particle, the solution with $\cos\theta^* = -1$ corresponds instead to a pair of collinear particles in the laboratory frame, hence $\phi_{min} = 0$.

- $\beta < \min\{\beta_1^*, \beta_2^*\}$. In this case, $\phi_{max} = \pi$ like the previous scenario, while ϕ_{min} is strictly larger than zero, and it can be computed, again numerically, starting from Eq. (1.62).

---

[4]For example, using the input values $\beta = 0.8$, $\beta_1^* = 0.3$ and $\beta_2^* = 0.5$, one gets $\cos\theta_{max}^* = 0.392$, which is in agreement with the numerical evaluation in Fig. 1.3.

Fig. 1.3 The trigonometric tangent of the opening angle ϕ in the laboratory frame between the decay products of a massive particle with velocity $\beta = 0.8$, shown as a function of $\cos\theta^*$, where θ^* is the polar angle in the centre-of-mass frame. Three different cases are considered: $\beta_{1,2}^* = 1 > \beta$, $\beta_2^* > \beta > \beta_1^*$, and $\beta_{1,2}^* > \beta$, where $\beta_{1,2}^*$ are the centre-of-mass velocities of the two particles

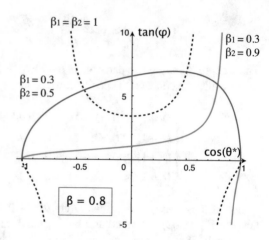

Figure 1.3 shows $\tan\phi$ as a function of $\cos\theta^*$, assuming $\beta = 0.8$ and for three representative cases: $\beta_{1,2}^* = 1 > \beta$, $\beta_2^* > \beta > \beta_1^*$ and $\beta_{1,2}^* > \beta$.

Bando n. 18211/2016

Problem 1.14 An electromagnetic calorimeter is able to separate the showers induced by high-energy photons when the separation angle between the two photons is larger than 5°. The calorimeter is then used to detect π^0's. What is the largest π^0 energy such that any π^0 decay can be reconstructed as a pair of distinct photons?

Solution

Given that a detector resolution of 5° is small, we can use the approximate formula of Eq. (1.61) to express the minimum opening angle between the two photons as a function of the π^0 energy. The condition that all the π^0's get reconstructed as two separate photons amounts to require that the minimum opening angle exceeds the resolution of the detector, that is to say:

$$\phi_{\min} = \frac{2}{\gamma} = \frac{2\,m_{\pi^0}}{E} > 5° \cdot \frac{\pi}{180°} \quad \Rightarrow \quad E < \frac{2\,m_{\pi^0}}{0.87 \times 10^{-1}} \approx 3.1\ \text{GeV}, \quad (1.66)$$

where we have used the PDG value $m_{\pi^0} = 135$ MeV [4].

Problem 1.15 A particle of mass m and momentum $\mathbf{p}$ decays to a pair of massless particles. In the rest frame of the mother particle, the angular distribution of the decay products is described by the probability density $(\Gamma^*)^{-1}d\Gamma^*/d\cos\theta^*$, where θ^* is the polar angle with respect to $\mathbf{p}$. What is the corresponding density $\Gamma^{-1}d\Gamma/d\cos\theta$ in the laboratory frame? Find out an approximate formula valid for an isotropic decay in the limit $|\mathbf{p}| \gg m$.

Solution

If the decay distribution in the centre-of-mass frame is described by the density $(\Gamma^*)^{-1} d\Gamma^*/d\cos\theta^*$, in the laboratory frame one has:

$$\frac{1}{\Gamma} \frac{d\Gamma}{d\cos\theta} = \frac{1}{\Gamma^*} \frac{d\Gamma^*}{d\cos\theta^*} \left| \frac{d\cos\theta^*}{d\cos\theta} \right|. \tag{1.67}$$

The relation between $\cos\theta^*$ and $\cos\theta$ can be obtained from Eq. (1.56), giving:

$$\left| \frac{d\cos\theta^*}{d\cos\theta} \right| = \frac{|-(\beta\cos\theta - 1) - (\beta - \cos\theta)\beta|}{(1 - \beta\cos\theta)^2} =$$

$$= \frac{1 - \beta^2}{(1 - \beta\cos\theta)^2} = \frac{1}{\gamma^2} \frac{1}{(1 - \beta\cos\theta)^2}, \tag{1.68}$$

where $\beta = |\mathbf{p}|/\sqrt{|\mathbf{p}|^2 + m^2}$. Inserting this expression into the right-hand side of Eq. (1.67), we obtain:

$$\frac{1}{\Gamma} \frac{d\Gamma}{d\cos\theta} = \left[\frac{1}{\gamma^2} \frac{1}{(1 - \beta\cos\theta)^2} \right] \frac{1}{\Gamma^*} \frac{d\Gamma^*}{d\cos\theta^*}. \tag{1.69}$$

The Jacobian factor at the right-hand side of Eq. (1.68) implies that the angular distribution in the laboratory frame will be skewed towards the boost direction. The ratio between the Jacobian factor for backward- and forward-emitted photons is $(1 - \beta)^2/(1 + \beta)^2 = [(1 + \beta)\gamma]^{-4}$: already for $\gamma = 5$, this ratio is 10^{-4}.

Let's now consider in more detail the case $|\mathbf{p}| \gg m$. In this limit, we can approximate:

$$1 - \beta = 1 - \sqrt{1 - \frac{1}{\gamma^2}} \approx \frac{1}{2\gamma^2}. \tag{1.70}$$

As already discussed, Eq. (1.68) implies a forward-peaked angular distribution, so that, for all practical purposes, we can assume $\theta \ll 1$ and consider only the first-order Taylor expansion of $\cos\theta$. With this approximation, Eq. (1.67) can be simplified to:

$$\frac{1}{\Gamma} \frac{d\Gamma}{d\cos\theta} \approx \frac{1}{\gamma^2} \frac{1}{\left(1 - \beta + \beta\frac{\theta^2}{2}\right)^2} \frac{1}{\Gamma^*} \frac{d\Gamma^*}{d\cos\theta^*} \approx \frac{4\gamma^2}{(1 + \gamma^2\theta^2)^2} \frac{1}{\Gamma} \frac{d\Gamma}{d\cos\theta^*}. \tag{1.71}$$

A case of special interest is for an isotropic angular distribution, which is the appropriate case for spin-0 particles or unpolarised beams. The angular distribution and the energy of the particle in the laboratory frame are then given respectively by:

$$\frac{1}{\Gamma}\frac{d\Gamma}{d\theta} \approx \frac{2\gamma^2\theta}{\left(1+\gamma^2\theta^2\right)^2}, \qquad E(\theta) = \frac{E^*}{\gamma(1-\beta\cos\theta)} \approx \frac{2\gamma}{1+\gamma^2\theta^2}E^*, \qquad (1.72)$$

where E^* is the energy of the massless particle in the centre-of-mass frame. The mean angle $\langle\theta\rangle$ and RMS σ_θ can be computed from the first of Eq. (1.72) by assuming that the approximation is valid up to the maximum laboratory angle $\theta = \pi$:

$$\langle\theta\rangle = \int_0^\pi d\theta\,\theta\left(\frac{1}{\Gamma}\frac{d\Gamma}{d\theta}\right) = \frac{\pi}{2\gamma} + \mathcal{O}\left(\frac{1}{\gamma}\right) \qquad (1.73)$$

$$\langle\theta^2\rangle = \int_0^\pi d\theta\,\theta^2\left(\frac{1}{\Gamma}\frac{d\Gamma}{d\theta}\right) = \frac{1}{\gamma^2}\left[\log(\gamma^2\pi+1)-1+\mathcal{O}\left(\frac{1}{\gamma^2}\right)\right] \qquad (1.74)$$

$$\sigma_\theta = \sqrt{\langle\theta^2\rangle - \langle\theta\rangle^2} = \frac{1}{\gamma}\left[\log(\gamma^2\pi+1)-1+\frac{\pi^2}{4}\right] + \mathcal{O}\left(\frac{1}{\gamma^2}\right) \qquad (1.75)$$

The cumulative distribution $F(\theta)$ can be obtained by integrating the differential distribution in Eq. (1.72). It can be used to determine the laboratory angle containing a given fraction α of the decay particles:

$$F(\theta) = 1 - \frac{1}{\gamma^2\theta^2+1} \qquad \Rightarrow \qquad \theta_\alpha = \frac{1}{\gamma}\sqrt{\frac{\alpha}{1-\alpha}}. \qquad (1.76)$$

Problem 1.16 A particle of mass M and three-momentum along the z-axis decays into a pair of massless particles. Let the angular distribution of the decay products in the rest frame be described by the density $(\Gamma^*)^{-1}d\Gamma^*/d\cos\theta^*$. Show that the transverse momentum distribution $\Gamma^{-1}d\Gamma/d|\mathbf{p_T}|$ develops an integrable singularity at $|\mathbf{p_T}| = |\mathbf{p}^*|$, where the trasverse momentum $\mathbf{p_T}$ is defined as the projection of $\mathbf{p}$ onto the plane perpendicular to the z-axis, and $\mathbf{p}^*$ is the centre-of-mass momentum of the decay particle. Consider now the transverse mass m_T of the two daughter particles, defined as:

$$m_T^2 = (E_{T1}+E_{T2})^2 - (\mathbf{p_{T1}}+\mathbf{p_{T2}})^2 = 2|\mathbf{p_{T1}}||\mathbf{p_{T2}}|(1-\cos(\Delta\phi)) \qquad (1.77)$$

Show that the distribution $\Gamma^{-1}d\Gamma/dm_T$ features a singularity at $m_T = M$.

Solution

The assumption that the momentum of the decaying particle purely longitudinal allows to relate the centre-of-mass kinematics to the transverse variables in laboratory frame through simple formulas, since:

$$|\mathbf{p}_{T1}| = |\mathbf{p}_{T2}| \equiv |\mathbf{p}_T| = |\mathbf{p}^*|\sqrt{1 - \cos^2\theta^*}$$

$$\frac{1}{\Gamma}\frac{d\Gamma}{d|\mathbf{p}_T|} = \frac{1}{\Gamma^*}\frac{d\Gamma^*}{d\cos\theta^*}\left|\frac{d\cos\theta^*}{d|\mathbf{p}_T|}\right| = \frac{1}{\Gamma^*}\frac{d\Gamma^*}{d\cos\theta^*}\frac{|\mathbf{p}_T|}{|\mathbf{p}^*|^2\sqrt{1 - |\mathbf{p}_T|^2/|\mathbf{p}^*|^2}} =$$

$$= \frac{1}{\Gamma^*}\frac{d\Gamma^*}{d\cos\theta^*}\left(\frac{4|\mathbf{p}_T|}{M^2\sqrt{1 - 4|\mathbf{p}_T|^2/M^2}}\right) \tag{1.78}$$

where we have used the relation $|\mathbf{p}^*| = M/2$, see Eq. (1.90). Therefore, the transverse momentum distribution features a singularity at $|\mathbf{p}_T| = |\mathbf{p}^*| = M/2$, which is entirely due to the change of variables, hence the name of *Jacobian peak*. Notice that the singularity is integrable, since $d\Gamma/d|\mathbf{p}_T| \sim (1 - 4|\mathbf{p}_T|^2/M^2)^{-\frac{1}{2}}$, which in the neighbourhood of the singularity goes like $\sim\varepsilon^{-\frac{1}{2}}$, where we have put $\varepsilon = |\mathbf{p}^*| - |\mathbf{p}_T|$. In terms of the transverse mass of Eq. (1.77), we have:

$$m_T^2 = 2|\mathbf{p}_{T1}||\mathbf{p}_{T2}|(1 - \cos(\Delta\phi)) = 4|\mathbf{p}_T|^2$$

$$\frac{1}{\Gamma}\frac{d\Gamma}{dm_T} = \frac{1}{\Gamma}\frac{d\Gamma}{d|\mathbf{p}_T|}\left|\frac{d|\mathbf{p}_T|}{dm_T}\right| = \frac{1}{\Gamma^*}\frac{d\Gamma^*}{d\cos\theta^*}\left(\frac{m_T}{M^2\sqrt{1 - m_T^2/M^2}}\right). \tag{1.79}$$

Thus, the transverse mass features a Jacobian peak located at the mass of the decaying particle.

Discussion

The appearance of a Jacobian peak in both the transverse momentum and the transverse mass distribution of the decay products of a heavy resonance provides a powerful handle to distinguish such events from a non-resonant background. For example, the transverse mass has been extensively used at hadron colliders as the single most-efficient signature to identify the decay of W bosons into a charged lepton and a neutrino. While the charged lepton momentum can be fully reconstructed (if the lepton is either e or μ), the neutrino does not interact with the detector. An indirect evidence of its production is however provided by the momentum imbalance in the transverse plane, which, in the absence of other invisible particles, is just given by the neutrino transverse momentum. Thus, even in the presence of invisible particle, the transverse mass of Eq. (1.77) can be computed.

Two remarks are due here. Firstly, in real experiments, the Jacobian peak is smeared by the finite detector resolution and by the natural width of the decaying particle, see e.g. Ref. [6] for the effect of the W boson width. Secondly, the expressions in Eqs. (1.78) and (1.79) have been derived under the assumption that the momentum of the decaying particle is purely longitudinal: if that is not true, i.e. if the particle has a momentum component orthogonal to the z-axis, the formula change. Differently from the transverse momentum, the transverse mass variable is less affected by a finite transverse momentum of the decaying particle. This is further elaborated in Problem 1.17.

Problem 1.17 Show that the transverse momentum distribution $\Gamma^{-1}d\Gamma/d|\mathbf{p_T}|$ of Eq. (1.78) for a particle of mass M decaying to massless particles, receives $\mathcal{O}(\beta)$ corrections when the decaying particle velocity β has a transverse component, while for the transverse mass distribution $\Gamma^{-1}d\Gamma/dm_T$ such corrections start at $\mathcal{O}(\beta^2)$.

Solution

All transverse variables are invariant under longitudinal boosts, as one can easily verify. We then study their properties under transverse boosts. We first write Eq. (1.4) in a vectorial form as:

$$\mathbf{p}' = (\gamma\, \mathbf{p} \cdot \hat{\boldsymbol{\beta}})\hat{\boldsymbol{\beta}} - \beta\,\gamma\,E\,\hat{\boldsymbol{\beta}} + \mathbf{p} - (\mathbf{p} \cdot \hat{\boldsymbol{\beta}})\hat{\boldsymbol{\beta}} = \mathbf{p} + \left[\frac{(\gamma - 1)}{\beta^2}\mathbf{p} \cdot \boldsymbol{\beta} - \gamma\,E\right]\boldsymbol{\beta}$$

$$(1.80)$$

$$E' = \gamma\,E - \gamma\,\mathbf{p} \cdot \boldsymbol{\beta} \tag{1.81}$$

If $\boldsymbol{\beta}$ is a small transverse vector, and the momentum $\mathbf{p}$ is almost longitudinal, we can work out the transformation properties of $\mathbf{p_T}$:

$$\mathbf{p'_T} \approx \mathbf{p_T} - E\boldsymbol{\beta} \quad \Rightarrow \quad \delta|\mathbf{p_T}| = -E\boldsymbol{\beta} \cdot \frac{\mathbf{p_T}}{|\mathbf{p_T}|}. \tag{1.82}$$

hence the transverse momentum changes already at the first order in β. The transverse mass for two massless particles is defined as:

$$m_T^2 = 2|\mathbf{p_{T1}}||\mathbf{p_{T2}}| - 2\mathbf{p_{T1}} \cdot \mathbf{p_{T2}}. \tag{1.83}$$

This expression resembles closely the invariant mass squared $m^2 = 2p_1p_2$ for two transverse vectors, but it has not the same properties under transverse Lorentz boosts. This can be proved by noticing that

$$\sqrt{|\mathbf{p_{T1}}|^2 + p_{z1}^2}\sqrt{|\mathbf{p_{T2}}|^2 + p_{z2}^2} - \mathbf{p_{T1}} \cdot \mathbf{p_{T2}} - p_{z1}p_{z2} \tag{1.84}$$

is invariant under both longitudinal and transverse boosts, since it coincides with the invariant mass squared $(p_1 + p_2)^2$. Since p_z is invariant under transverse boosts, then it must hold:

$$\delta\,(\mathbf{p_{T1}} \cdot \mathbf{p_{T2}}) = \delta\left(\sqrt{|\mathbf{p_{T1}}|^2 + p_{z1}^2}\sqrt{|\mathbf{p_{T2}}|^2 + p_{z2}^2}\right)$$

$$= \frac{|\mathbf{p_{T1}}|\,\delta|\mathbf{p_{T1}}|}{E_1}E_2 + \frac{|\mathbf{p_{T2}}|\,\delta|\mathbf{p_{T2}}|}{E_2}E_1. \tag{1.85}$$

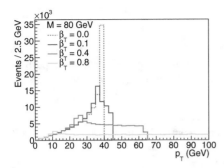 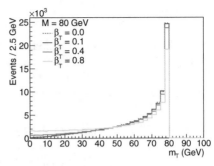

Fig. 1.4 Distribution of the transverse momentum $|\mathbf{p}_T|$ (left) and transverse mass m_T (right) obtained from a set of toy MC events where an unpolarised resonance of mass M decays into a pair of massless particles. Four different velocities $\boldsymbol{\beta}$ of the mother particle are assumed, differing by the component on the transverse plane. The dashed histogram corresponds to a purely longitudinal velocity, while the other three distributions correspond to increasingly larger transverse boosts β_T

Using Eq. (1.82) and (1.85), we get:

$$\delta m_T^2 = (|\mathbf{p}_{T2}|E_1 - |\mathbf{p}_{T1}|E_2)\,\boldsymbol{\beta} \cdot \left(\frac{\mathbf{p}_{T2}}{|\mathbf{p}_{T2}|} - \frac{\mathbf{p}_{T1}}{|\mathbf{p}_{T1}|}\right), \qquad (1.86)$$

which vanishes because, in the centre-of-mass frame, $E_1 = E_2$ and $|\mathbf{p}_{T1}| = |\mathbf{p}_{T2}|$.

Hence, at first order in β, the transverse mass does not change. This is however not true at second order, as one can readily verify by considering the $\mathcal{O}(\beta^2)$ term.

The response of the transverse momentum and transverse mass under boosts can be studied by using toy events generated with MC techniques, see Problem 4.5, in which the decay of an unpolarised resonance of mass M is simulated. A simple ROOT macro that performs the toy generation is illustrated in Appendix 1.3. The results are shown in Fig. 1.4. As expected, the m_T variable is found to be significantly more stable against transverse boosts compared to $|\mathbf{p}_T|$. An other interesting feature of the m_T variable is that the location of the Jacobian peak location is not affected by boosts; this is clearly not the case for the transverse momentum, whose peak value is significantly smeared by transverse boosts.

Problem 1.18 A particle of mass m and momentum $\mathbf{p}$ scatters against an identical particle of mass m, initially at rest. Compute the minimum opening angle between the two particle directions after the scattering.

Discussion

This problem is a prototype for studying the kinematics of a fixed-target experiment. To be more general, we can assume the two masses to be different, with particle (1) having momentum $\mathbf{p}$ and particle (2) at rest. The energy and momenta in the centre-of-mass frame are given by:

$$\sqrt{s} = E_1^* + E_2^* = \sqrt{m_1^2 + m_2^2 + 2E_1 m_2} \tag{1.87}$$

$$|\mathbf{p}^*| = \sqrt{E_1^{*2} - m_1^2} = \sqrt{E_2^{*2} - m_2^2}. \tag{1.88}$$

The last equation, in particular, implies:

$$E_1^{*2} - m_1^2 = E_2^{*2} - m_2^2, \qquad (E_1^* + E_2^*)(E_1^* - E_2^*) = m_1^2 - m_2^2,$$

$$\sqrt{s}\,(E_1^* - \sqrt{s} + E_1^*) = m_1^2 - m_2^2 \quad \Rightarrow \quad E_1^* = \frac{s + m_1^2 - m_2^2}{2\sqrt{s}}. \tag{1.89}$$

An analogous result for E_2^* can be obtained by changing $1 \leftrightarrow 2$. Inserting Eq. (1.89) into the second of (1.87), and symmetrising the expression for $1 \leftrightarrow 2$ exchange, we get:

$$
\begin{aligned}
|\mathbf{p}^*|^2 &= \frac{1}{2}\left[\frac{(s + m_1^2 - m_2^2)^2}{4s} - m_1^2 + \frac{(s + m_2^2 - m_1^2)^2}{4s} - m_2^2\right] = \\
&= \frac{1}{8s}\left[2s^2 + 2(m_1^2 - m_2^2)^2 - 4s\,m_1^2 - 4s\,m_2^2\right] = \\
&= \frac{1}{4s}\left[s^2 + (m_1 - m_2)^2(m_1 + m_2)^2 - 2s\,m_1^2 - 2s\,m_2^2\right] = \\
&= \frac{1}{4s}\left[\left(s + (m_1 + m_2)^2\right)\left(s + (m_1 - m_2)^2\right)\right],
\end{aligned}
\tag{1.90}
$$

from which:

$$|\mathbf{p}^*| = \frac{\sqrt{\left(s - (m_1 + m_2)^2\right)\left(s - (m_1 - m_2)^2\right)}}{2\sqrt{s}} = \frac{\sqrt{(s - m_1^2 - m_2^2)^2 - 4m_1^2 m_2^2}}{2\sqrt{s}}. \tag{1.91}$$

A special case is provided by $m_1 = m_2 = m$, for which:

$$|\mathbf{p}^*| = \frac{\sqrt{s}}{2}\sqrt{1 - 4\frac{m^2}{s}}, \qquad E^* = \frac{\sqrt{s}}{2}, \qquad \beta^* = \sqrt{1 - 4\frac{m^2}{s}}. \tag{1.92}$$

Notice that, after doing the algebra, the numerator of Eq. (1.91) assumes a completely symmetric form under exchange of $s \leftrightarrow m_1^2 \leftrightarrow m_2^2$:

$$|\mathbf{p}^*| = \frac{\sqrt{s^2 + m_1^4 + m_2^4 - 2s^2 m_1^2 - 2s^2 m_1^2 - 2m_2^2 m_1^2}}{2\sqrt{s}} = \frac{\sqrt{\lambda(s, m_1^2, m_2^2)}}{2\sqrt{s}} \quad (1.93)$$

where $\lambda(a, b, c)$ is known as the *triangular function*. The boost parameter to the centre-of-mass frame, β, can be easily obtained by noticing that, in this frame, particle (2), which has momentum component along the boost direction $-|\mathbf{p}^*|$ and energy E_2^*, has to come to a rest under an inverse boost $-\beta$, that is to say:

$$0 = \gamma(-|\mathbf{p}^*| + \beta E_2^*) \quad \Rightarrow \quad \beta = \frac{|\mathbf{p}^*|}{E_2^*} = \beta_2^*, \quad (1.94)$$

which implies that β is also the velocity of particle (2) in the centre-of-mass frame, as one would have expected. Besides, by definition of centre-of-mass frame, the two particles must have equal and opposite momentum under this boost, i.e.:

$$|\mathbf{p}^*| = \gamma(|\mathbf{p}| - \beta E_1) = \gamma \beta m_2, \quad (1.95)$$

from which we can derive the two relations:

$$\beta = \frac{|\mathbf{p}|}{E_1 + m_2}, \quad \gamma = \frac{E_1 + m_2}{\sqrt{s}} \quad (1.96)$$

The latter follows from:

$$\gamma = \frac{1}{\sqrt{1 - \beta^2}} = \frac{E_1 + m_2}{\sqrt{m_1^2 + m_2^2 + 2m_2 E_1}} = \frac{E_1 + m_2}{\sqrt{s}}. \quad (1.97)$$

Equation (1.96) thus implies that the boost to the centre-of-mass is the velocity of a "particle" of mass $\sqrt{s}$ and total momentum $\mathbf{p}$. Furthermore, combining Eqs. (1.95) and (1.96), one can relate the centre-of-mass momentum to the momentum in the laboratory frame as:

$$|\mathbf{p}^*| = \gamma \beta m_2 = m_2 \frac{|\mathbf{p}|}{\sqrt{s}}. \quad (1.98)$$

Solution

We can use directly Eq. (1.43) to express the velocities in terms of the momenta and energies. Also, since $m_1 = m_2 = m$, we have $E_1^* = \sqrt{|\mathbf{p}^*| + m^2} = E_2^* \equiv E^*$. Let's define the opening angle by ϕ. With reference to Fig. 1.5, we then get:

$$\tan \theta_1 = \frac{|\mathbf{p}^*| \sin \theta_1^*}{\gamma(\beta E^* + |\mathbf{p}^*| \cos \theta^*)}, \quad \tan \theta_2 = \frac{|\mathbf{p}^*| \sin \theta_1^*}{\gamma(\beta E^* - |\mathbf{p}^*| \cos \theta^*)}, \quad (1.99)$$

Fig. 1.5 Elastic scattering between two identical particles of mass m

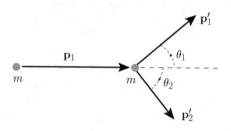

$$\tan \phi = \tan (\theta_1 + \theta_2) = \frac{\tan \theta_1 + \tan \theta_2}{1 - \tan \theta_1 \tan \theta_2} =$$
$$= \frac{2\beta\gamma E^* |\mathbf{p}^*| \sin \theta^*}{\gamma^2 (\beta^2 E^{*2} - |\mathbf{p}^*|^2) + |\mathbf{p}^*|^2 (\gamma^2 - 1) \sin^2 \theta^*} \tag{1.100}$$

By using Eq. (1.94), we see that the first term at the denominator vanishes. The right-hand side is a monotonously decreasing function of θ^*, hence the minimum of ϕ occurs for $\theta^* = \frac{\pi}{2}$. Still using Eq. (1.94), and the first of Eq. (1.96), the expression can be simplified to:

$$\tan (\phi_{\min}) = \frac{2\gamma}{\gamma^2 - 1} = \frac{2\sqrt{1 - \beta^2}}{\beta^2} = \frac{2\sqrt{1 - |\mathbf{p}^*|^2/(m + E^*)^2}}{|\mathbf{p}^*|^2/(m + E^*)^2} = \frac{2\sqrt{2m^2 + 2mE^*}}{E^* - m}. \tag{1.101}$$

Finally, using a trigonometric identity, we can further simplify the expression as:

$$\cos \phi_{\min} = \frac{1}{\sqrt{1 + \tan^2 \phi_{\min}}} = \sqrt{\frac{(E^* - m)^2}{E^{*2} - 2mE^* + m^2 + 8mE^* + 8m^2}} = \frac{E^* - m}{E^* + 3m}. \tag{1.102}$$

Problem 1.19 Consider the decay of the K_S^0 meson into a pair of opposite-charge pions: calculate the energy and momenta of the charged pions in the K_S^0 rest frame. Consider now a monochromatic beam of K_S^0 decaying as before. Determine the energy distribution of the charged pions in the laboratory frame as a function of the beam momentum.

Solution

Since the K_S^0 particle decays to a pair of particles of identical mass, we can use Eq. (1.92) with $\sqrt{s} = m_K$ to derive the energy E^* and momentum $|\mathbf{p}^*|$ of the pions in the centre-of-mass frame:

$$E^* = \frac{m_K}{2} = 248 \text{ MeV}, \qquad |\mathbf{p}^*| = \frac{m_K}{2}\sqrt{1 - 4\left(\frac{m_\pi}{m_K}\right)^2} = 206 \text{ MeV}, \quad (1.103)$$

where we have used the PDG values $m_K = 497$ MeV and $m_\pi = 139$ MeV [4].

The laboratory energy E_π of either of the two pions can be expressed in terms of the centre-of-mass polar angle θ^* via:

$$E_\pi(\cos\theta^*) = E^* \gamma (1 + \beta\beta^* \cos\theta^*) = \frac{1}{2}(E + |\mathbf{p}|\beta^* \cos\theta^*), \quad (1.104)$$

where E and $|\mathbf{p}|$ are the laboratory energy and momentum of the kaon, and $\beta^* = |\mathbf{p}^*|/E^* \approx 0.829$ is the centre-of-mass velocity of the pions. Since the K^0 is a spin-0 particle, it's decay into pions is isotropic in the rest frame, see Problem 1.15. Furthermore, Eq. (1.104) shows that the energy in the laboratory frame is a linear function of $\cos\theta^*$, so that it's distribution will be also uniformly distributed in the range $[E_\pi(-1), E_\pi(1)]$:

$$\frac{1}{\Gamma}\frac{d\Gamma}{dE_\pi} = \begin{cases} (|\mathbf{p}|\beta^*)^{-1} & \text{if } \frac{1}{2}(E - |\mathbf{p}|\beta^*) \leq E_\pi \leq \frac{1}{2}(E + |\mathbf{p}|\beta^*) \\ 0 & \text{otherwise} \end{cases} \quad (1.105)$$

Discussion

The fact that an isotropic distribution in the rest frame for a decay $1 \to 2$ gives rise to a rectangular distribution for the energy in the laboratory frame holds irrespectively of the mother energy and of the daughter mass, since it only depends on E_π being a linear function of $\cos\theta^*$. Notice that the momentum $|\mathbf{p}|$ is not uniformly distributed, since E is not a linear function of $|\mathbf{p}|$, whereas the kinetic energy $T = E - m$ is. If the mother particle is relativistic, i.e. $E \gg m$, and the mass of the daughter particles is small compared to m, then Eq. (1.105) becomes:

$$\frac{1}{\Gamma}\frac{d\Gamma}{dE_\pi} \approx \begin{cases} \frac{1}{E_\pi} & \text{if } 0 \lesssim E_\pi \leq E \\ 0 & \text{otherwise} \end{cases} \quad (1.106)$$

For example, the τ lepton has a mass of about 1.7 GeV, and can decay via $\tau \to \pi\, \nu_\tau$, with $m_\pi, m_\nu \ll m_\tau$. At colliders, τ leptons are abundantly produced from the decay of Z^0 or W bosons, so that $E_\tau \gtrsim 40$ GeV $\gg m_\tau$. Although it is a spin-1/2 particle and the EWK dynamics is chiral, see Problem 1.8, one can usually consider unpolarised ensembles of τ leptons, for example by averaging over the pion charge, so that the decay distribution in the τ rest frame can be still considered as isotropic. Then, the energy spectrum of the charged pion features an approximate rectangular distribution as in Eq. (1.106).

Another intriguing property of the energy distribution of Eq. (1.105) will be discussed in Problem 1.20.

Problem 1.20 Show that for a two-body decay $B \rightarrow A\,a$ of an unpolarised particle B of mass m_B, into a pair of particles A and a, where the latter is assumed massless, the energy spectrum of a in the laboratory frame has a global maximum at E_a^* irrespectively of the momentum of B.

Solution

Let's consider the decay $B \rightarrow A\,a$ in the rest frame of B, and let's denote by θ^* the polar angle of a with respect to the direction of flight of B in the laboratory. It follows that:

$$E_a = \gamma \, E_a^* (1 + \beta \cos \theta^*) \tag{1.107}$$

where $\gamma \equiv \gamma_B$ and $\beta \equiv \beta_B$ are the gamma-factor and velocity of B in the laboratory frame. The centre-of-mass energy of a is given by Eq. (1.89), namely $E_a^* = (m_B^2 - m_A^2)/2m_A$. Since E_a is a linear function of $\cos \theta^*$, it follows that

$$E_a \in [\gamma \, E_a^*(1 - \beta), \, \gamma \, E_a^*(1 + \beta)],$$
$$x \equiv \frac{E_a}{E_a^*} \in I_\gamma \equiv \left[\gamma - \sqrt{\gamma^2 - 1}, \, \gamma + \sqrt{\gamma^2 - 1} \right] \tag{1.108}$$

Furthermore, since $\gamma - \sqrt{\gamma^2 - 1} < 1$ and $\gamma + \sqrt{\gamma^2 - 1} > 1$ for any γ, it follows that $x = 1$ is contained in all intervals I_γ. It is also the only value featuring this property, since for $x \neq 1$, one can always find a γ such that $x \notin I_\gamma$. Indeed, for a fixed x, the condition $x \in I_\gamma$ can be obtained by solving the system:

$$\begin{cases} x > \gamma - \sqrt{\gamma^2 - 1} \\ x < \gamma + \sqrt{\gamma^2 - 1} \end{cases} \quad \Rightarrow \quad \gamma > \frac{1}{2}\left(x + \frac{1}{x} \right). \tag{1.109}$$

The assumption that B is unpolarised implies that E_a has a rectangular distribution, see Problem 1.19, hence x is uniformly distributed in I_γ. If we assume that the boost factors of B are described by a distribution $g(\gamma)$, the distribution of x will be given by:

$$\frac{1}{\Gamma}\frac{d\Gamma}{dx} = \int d\gamma \, f(x|\gamma) \cdot g(\gamma) = \int_{\frac{1}{2}(x+\frac{1}{x})}^{\infty} d\gamma \, \frac{g(\lambda)}{2\sqrt{\gamma^2 - 1}}. \tag{1.110}$$

The derivative of Eq. (1.110) is given by:

$$\frac{\partial}{\partial x}\left(\frac{1}{\Gamma}\frac{d\Gamma}{dx}\right) = -\frac{g\left(\frac{1}{2}\left(x+\frac{1}{x}\right)\right)}{2\sqrt{\left(\frac{1}{2}\left(x+\frac{1}{x}\right)\right)^2 - 1}}\left[\frac{1}{2}\left(1-\frac{1}{x^2}\right)\right] =$$

$$= \frac{\operatorname{sign}(1-x)}{2x}g\left(\frac{1}{2}\left(x+\frac{1}{x}\right)\right). \tag{1.111}$$

Therefore: if $g(1) = 0$, then $x = 1$ is the unique maximum of Eq. (1.110); if $g(1) \neq 0$, the derivative flips sign at $x = 1$, hence this point represents a cusp in the distribution of x. In both cases, $x = 1$ is a global maximum. An alternative way to convince oneself that $x = 1$ is indeed a maximum is to notice that this value is the only one that is contained by all intervals I_y, hence it must have the highest probability density. The same conclusion would hold, under some more restrictive conditions, also for $m_a > 0$. In the latter case, Eq. (1.109) gets modified to:

$$x \equiv \frac{E_a}{E_a^*} \in I_y \equiv \left[\gamma - \sqrt{\gamma^2 - 1}\frac{\sqrt{\gamma^{*2} - 1}}{\gamma^*}, \gamma + \sqrt{\gamma^2 - 1}\frac{\sqrt{\gamma^{*2} - 1}}{\gamma^*}\right], \tag{1.112}$$

with $\gamma^* = E_a^*/m_a$. The condition $1 \in I_y$ is then satisfied provided that:

$$\gamma - \sqrt{\gamma^2 - 1}\frac{\sqrt{\gamma^{*2} - 1}}{\gamma^*} < 1 \quad \Leftrightarrow \quad \gamma < 2\gamma^{*2} - 1. \tag{1.113}$$

As expected, this condition is satisfied for any γ in the limit $m_a \to 0$, since $\gamma^* \to +\infty$ and the inequality becomes $\gamma < +\infty$.

Discussion

This subtle property offers the possibility of measuring the mass of the parent particle m_B regardless of both the kinematic of B and A, which plays here no role other than determining the centre-of-mass energy E_a^*. The latter can be directly measured from the mode of the distribution of E_a from the relation:

$$m_B = E_a^* + \sqrt{m_A^2 - m_a^2 + (E_a^*)^2}. \tag{1.114}$$

Figure 1.6 shows the simulated spectrum of $\log E_a$, for the case where particle a is the b-jet produced in the decay of a top quark.

Suggested Readings

This problem is inspired by Ref. [8], from which the notation and the mathematical proof have been also taken. The idea of using the peak position of the energy spectrum

Fig. 1.6 Fitted log E_a distribution in a simulated sample of $t\bar{t}$ events with a mass hypothesis of 172.5 GeV. The Gaussian fit yields a log E_a peak position of 4.199 ± 0.002, corresponding to an uncalibrated value of $m_t = 171.01 \pm 0.25$ GeV using Eq. (1.114) (taken from Ref. [7])

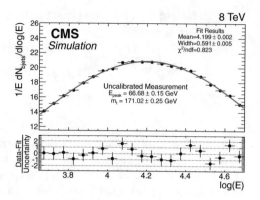

of the b-jets to measure the mass of the top quark, which decays via $t \rightarrow b\,W^+$, has been first suggested by the same authors and first pursued at the CMS experiment [7].

Problem 1.21 A K^{*-} meson with momentum $|\mathbf{p}| = 5.5$ GeV decays via $K^{*-} \rightarrow K^-\pi^0$. In the rest frame of the mother particle, the K^- momentum forms an angle $\theta^* = 55°$ with respect to $\mathbf{p}$. What is the opening angle between the K^- and π^0 momenta in the laboratory frame? Which are the centre-of-mass angles $\theta^*_{\max}$ and $\theta^*_{\min}$ for which the opening angle is, respectively, the largest and smallest?

Solution

We can use the results derived in Problem 1.12. In particular, the opening angle ϕ depends on the centre-of-mass angle θ^* and on the velocity of the mother particle in the laboratory frame, β_{K^*}, and the velocities of the two daughter particles in the rest frame of the mother particle, β_K^* and β_π^*. The velocity and gamma factor of the K^{*-} meson in the laboratory frame are given respectively by:

$$\beta = \left[1 - \left(\frac{m_{K^*}}{|\mathbf{p}|}\right)^2\right]^{-\frac{1}{2}} = 0.987, \qquad \gamma = \left[1 + \left(\frac{|\mathbf{p}|}{m_{K^*}}\right)^2\right]^{\frac{1}{2}} = 6.25, \quad (1.115)$$

where we have used the PDG value $m_{K^*} = 892$ MeV [4]. The velocities of the daughter particles in the rest frame of the mother particle can be computed from Eqs. (1.89) and (1.91). In order to make the numerical computation less error prone when using a calculator, it is convenient to write the formula in terms of the dimensionless ratios $r_\pi \equiv m_K/m_{K^*} = 0.554$ and $r_\pi \equiv m_\pi/m_{K^*} = 0.151$, that is:

$$\beta^*_{K,\pi} = \frac{|\mathbf{p}^*|}{E^*_{K,\pi}} = \frac{\sqrt{\left(m^2_{K^*} - (m_\pi + m_K)^2\right)\left(m^2_{K^*} - (m_\pi - m_K)^2\right)}}{m^2_{K^*} \pm m^2_K \mp m^2_\pi} =$$

$$= \frac{\sqrt{\left(1 - (r_\pi + r_K)^2\right)\left(1 - (r_\pi - r_K)^2\right)}}{1 \pm r^2_K \mp r^2_\pi} = \begin{cases} 0.502 & K \\ 0.906 & \pi \end{cases} \quad (1.116)$$

We can then use Eq. (1.67) to derive ϕ as a function of θ^*, or equivalently use Eq. (1.44) to calculate θ_K and θ_π separately. Following the latter approach, we get:

$$\tan \theta_{K,\pi} = \frac{\sin 55°}{6.25 \cdot \left(\pm \cos 55° + \frac{0.987}{\beta^*_{K,\pi}}\right)} = \begin{cases} 0.0516 & K \\ 0.254 & \pi \end{cases}$$

$$\phi = \operatorname{atan}(0.0516) + \operatorname{atan}(0.254) = 17.2°. \quad (1.117)$$

Since $\beta_{K^*} > \min\{\beta^*_K, \beta^*_\pi\}$, there must be a maximum opening angle $\phi_{\max} < \pi$, whose tangent can be found by numerically maximising Eq. (1.67) with respect to $\cos \theta^*$. By using the numerical routine of Appendix 1.3, we get a value:

$$\operatorname{argmax}\{\tan \phi\} = 0.906 \quad \Rightarrow \quad \phi_{\max} = 86°. \quad (1.118)$$

Bando n. 13705/2010

Problem 1.22 A beam of K^+ mesons with energy E propagates along the z-axis. Consider the decay $K^+ \to \mu^+ \nu_\mu$ with massless neutrinos. Determine:

1. the angular distribution of the muons in the centre-of-mass frame;
2. the polar angle that contains 50% of the neutrinos;
3. the fraction of neutrinos emitted with negative velocity;
4. the beam energy threshold for which all muons move forwards;
5. which are the implications of the weak interaction for the muon helicity when the latter is emitted forwards or backwards along the z axis.

Solution

1. Since the K^+ meson has spin-0, the decay products are isotropically distributed in the centre-of-mass frame, i.e.:

$$\frac{1}{\Gamma^*} \frac{d\Gamma^*}{d\cos\theta^* \, d\phi^*} = \frac{1}{4\pi}. \quad (1.119)$$

2. As seen in Problem 1.10, and in particular Eq. (1.56), for massless particles the polar angle in the laboratory frame, θ, is a monotonously increasing function of θ^*. Indeed:

$$\frac{d \cos \theta}{d \cos \theta^*} = \frac{1 - \beta^2}{(1 + \beta \cos \theta^*)^2} > 0, \qquad (1.120)$$

where β is the K^+ velocity in the laboratory frame. Therefore, the transformed to the laboratory frame of the polar angle θ^* giving the 50% quantile in the centre-of-mass frame, which is trivially $\pi/2$ for an isotropic decay, will also give the same quantile in the laboratory frame, since a monotonous mapping preserves the quantiles. Hence:

$$\theta_{50\%} = \mathrm{acos} \left(\frac{\cos \frac{\pi}{2} + \beta}{\beta \cos \frac{\pi}{2} + 1} \right) = \mathrm{acos}\, \beta = \mathrm{acos} \sqrt{1 - \left(\frac{m_K}{E} \right)^2}. \qquad (1.121)$$

3. For the neutrino to be emitted backwards, one has the condition:

$$p_z^\nu = \gamma \left(p_z^* + \beta E_\nu^* \right) = \gamma E_\nu^* (\cos \theta^* + \beta) < 0 \quad \Rightarrow \quad \cos \theta^* < -\beta. \quad (1.122)$$

Hence, the fraction α_B of backward-emitted neutrinos as a function of E is given by:

$$\alpha_B = \int_{-1}^{-\beta} d \cos \theta^* \frac{1}{\Gamma} \frac{d\Gamma}{d \cos \theta^*} = \frac{1}{2} \int_{-1}^{-\beta} d \cos \theta^* = \frac{1 - \sqrt{1 - \left(\frac{m_K}{E} \right)^2}}{2}$$

$$(1.123)$$

4. For the muon to always move forward, one needs:

$$p_z^\mu = \gamma \left(p_z^* + \beta E_\mu^* \right) = \gamma E_\mu^* (\beta_\mu^* \cos \theta^* + \beta) > 0 \quad \forall \theta^* \quad \Leftrightarrow \quad \beta > \beta_\mu^*,$$

$$\sqrt{1 - \left(\frac{m_K}{E} \right)^2} > \frac{m_K^2 - m_\mu^2}{m_K^2 + m_\mu^2}, \qquad E > E_{\mathrm{th}} \equiv \frac{m_\mu}{2} \left(1 + \left(\frac{m_K}{m_\mu} \right)^2 \right) = 1.20 \text{ GeV}.$$

$$(1.124)$$

5. The $V - A$ structure of the charged-weak interaction implies that the neutrino is a pure left-handed particle, and since it is massless, it is also in a helicity eigenstate with eigenvalue $h_\nu = -1/2$, see Problem 1.6. Since the K^+ is a spin-0 particle, the muon and neutrino needs to be in opposite helicity eigenstate in the centre-of-mass frame, as to conserve the angular momentum in the z direction. If the muon moves backwards along z in the laboratory frame, then it must have $h_\mu = -1/2$ as to compensate for the forward-moving neutrino with $h_\nu = -1/2$. If instead the muon moves forward, we should consider separately the case where the K^+ energy is below the threshold E_{th} of Eq. (1.124) (in which case, the neutrino moves backwards and hence $h_\mu = -1/2$), from the case $E > E_{\mathrm{th}}$. In the latter case, the neutrino can move either backwards or forwards, depending on the observed muon momentum, and then h_μ can take both values of $\pm 1/2$.

Problem 1.23 Consider the decay chain $C \to B\, b$, followed by $B \to A\, a$, where a and b are massless particles, whereas A, B, and C have non-zero masses m_A, m_B, and m_C, respectively. Determine the lower and upper bounds on the invariant mass m_{ab} as a function of the mass of the three other particles. Assume now that the decay of B is isotropic in its rest frame: what is the expected distribution of m_{ab} in the laboratory frame?

Solution

We study the problem in the rest frame of C, where b has a fixed energy: since $m_{ab}^{\max}$ is an invariant, the results obtained in this particular frame will hold true for any other frame. The minimum invariant mass $m_{ab}^{\min}$ corresponds to a and b moving collinear, which can always happen if $m_a = m_b = 0$, see Problem 1.10. Hence: $m_{ab}^{\min} = 0$.

The maximum invariant mass corresponds instead to a and b moving back-to-back, since this configuration will also maximise the energy of a. We can use Eq. (1.89) to express the energy of b, which is identical to the momentum of B since the former is assumed massless:

$$E_b = |\mathbf{p}_B| = \frac{m_C^2 - m_B^2}{2m_C}, \qquad E_B = \frac{m_C^2 + m_B^2}{2m_C}. \qquad (1.125)$$

The same equation with the replacement $C \to B$ and $B \to A$ will also give the energy E_a^* of particle a in the rest frame of B. We can transform it back to the C rest frame by applying a boost of magnitude β_B, which is the velocity of B in the rest frame of C:

$$E_a^{\max} = \gamma_B\, E_a^* \,(1 + \beta_B) = \left(\frac{m_C^2 + m_B^2}{2m_C m_B}\right)\left(\frac{m_B^2 - m_A^2}{2m_B}\right)\left(1 + \frac{m_C^2 - m_B^2}{m_C^2 + m_B^2}\right) =$$

$$= \left(\frac{m_C}{m_B}\right)\frac{m_B^2 - m_A^2}{2m_B}, \qquad (1.126)$$

from which we get the result:

$$m_{ab}^{\max} = 2\sqrt{E_a^{\max} E_b} = \frac{\sqrt{(m_C^2 - m_B^2)(m_B^2 - m_A^2)}}{m_B}. \qquad (1.127)$$

To study the distribution of m_{ab}, we first write it explicitly as a function of the kinematics of a and b:

$$m_{ab}^2 = 2\, E_a\, E_b\, (1 + \cos\theta_a), \qquad (1.128)$$

where θ_a is the angle of a with respect to the direction of B in the rest frame of C, see Fig. 1.7. Then, we write $\cos\theta_a$ as a function of E_a, as to obtain an expression which depends only on the latter. We do so because we know that E_a is uniformly

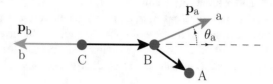

Fig. 1.7 Representation of the decay chain $C \rightarrow B\,b$, followed by $B \rightarrow A\,a$ in the rest frame of C

distributed, if such is the angular distribution in the centre-of-mass frame, see e.g. Eq. (1.105). Given that E_a^* is constant, we get:

$$E_a^* = \gamma_B\, E_a\,(1 - \beta_B \cos \theta_a)\,, \qquad \cos \theta_a = \frac{\gamma_B\, E_a - E_a^*}{\beta_B\, \gamma_B\, E_a} = \frac{E_B\, E_a - m_B\, E_a^*}{|\mathbf{p}_B|\, E_a}.$$
$$(1.129)$$

Inserting this expression into Eq. (1.128) we get:

$$m_{ab}^2 = 2\, E_a\, E_b \left(\frac{|\mathbf{p}_B|\, E_a + E_B\, E_a - m_B\, E_a^*}{|\mathbf{p}_B|\, E_a} \right) = \frac{2 E_b}{|\mathbf{p}_B|} \left[E_a\,(E_B + |\mathbf{p}_B|) - m_B\, E_a^* \right] =$$
$$= 2\, m_C\, E_a - (m_B^2 - m_A^2),$$
$$(1.130)$$

where the last equality has been obtained by means of Eqs. (1.89) and (1.125). It is easy to verify that Eq. (1.128) can be recovered by using the expression at the right-hand side of Eq. (1.126). Since m_{ab}^2 is a linear function of E_a, which is uniformly distributed, it follows that m_{ab} has a triangular distribution:

$$\frac{1}{\Gamma}\frac{d\Gamma}{dm_{ab}} = \begin{cases} \left(\dfrac{\sqrt{2}}{m_{ab}^{\max}}\right)^2 m_{ab} & \text{if } 0 \le m_{ab} \le m_{ab}^{\max} \\ 0 & \text{otherwise} \end{cases}$$
$$(1.131)$$

Discussion

Searching for end-points in the invariant mass spectrum of light particles provides an experimental technique to measure new heavy particles that decay to intermediate states, for example the supersymmetric partners of the SM particles. Figure 1.8 shows the spectrum of opposite-sign dilepton masses measured by the CMS experiment in proton-proton collisions at $\sqrt{s} = 8$ TeV [9]. A putative signal like the one discussed in this exercise would manifest itself as an *edge* in the mass distribution (dashed green line). Furthermore, the edge location provides a constraint on the mass-scale of the new particles as shown by Eq. (1.127).

Fig. 1.8 Invariant mass distribution of same-flavour opposite-sign lepton pairs $(e^+e^-, \mu^+\mu^-)$ measured by the CMS experiment in pp collisions at $\sqrt{s} = 8$ TeV. The contribution from a putative signal is shown as a green dashed-line histogram featuring a peculiar triangular shape: the end-point of the distribution is related to the particle masses. Notice that no other known SM process produces a similar invariant mass shape (taken from Ref. [9])

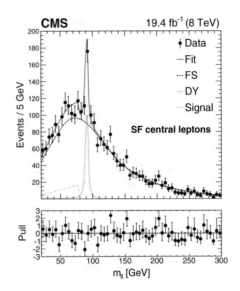

Problem 1.24 A non-relativistic particle of mass m and initial velocity $\mathbf{v}$ scatters elastically against a particle of mass $A\,m$, where A is a positive coefficient, assumed to be initially at rest. Determine, as a function of A, what is the range of kinetic energy T' in which the projectile particle can be found after the scattering.

Solution

We first consider the generic scattering in the centre-of-mass frame. Here, the kinematics is fully specified by the polar angle θ^* with respect to $\mathbf{v}$. The velocity of the centre-of-mass in the laboratory frame, $\mathbf{v}_{CM}$, is given by:

$$\mathbf{r}_{CM} = \frac{m \cdot \mathbf{r}_1 + A\,m \cdot \mathbf{r}_A}{(A+1)\,m} \quad \Rightarrow \quad \mathbf{v}_{CM} = \frac{\mathbf{v}}{A+1}. \tag{1.132}$$

In the c.o.m frame, the velocity of the incoming particle before and after the collision are $\mathbf{v}^*$ and $\mathbf{v}^{*\prime}$, and since the collision is elastic, we have:

$$|\mathbf{v}^*| = |\mathbf{v}^{*\prime}| = |\mathbf{v} - \mathbf{v}_{CM}| = \frac{A}{A+1}|\mathbf{v}|. \tag{1.133}$$

Expressing $\mathbf{v}'$ back into the laboratory frame, we obtain:

$$\mathbf{v}' = \mathbf{v}^{*\prime} + \mathbf{v}_{CM}, \qquad |\mathbf{v}'|^2 = |\mathbf{v}^{*\prime}|^2 + |\mathbf{v}_{CM}|^2 + 2|\mathbf{v}^{*\prime}||\mathbf{v}_{CM}|\cos\theta^*,$$

$$|\mathbf{v}'|^2 = \left(\frac{A}{A+1}\right)^2 |\mathbf{v}|^2 + \left(\frac{1}{A+1}\right)^2 |\mathbf{v}|^2 + \frac{2A}{A+1}|\mathbf{v}|^2 \cos\theta^*,$$

$$T' = \frac{A^2 + 1 + 2A\cos\theta^*}{(A+1)^2} T \quad \Rightarrow \quad T' \in \left[T\left(\frac{A-1}{A+1}\right)^2, T\right]. \tag{1.134}$$

Equation (1.134) implies that the maximum energy transfer, i.e. the largest recoil energy transferred to the target particle, is:

$$\Delta T_{max} = T - T_{min} = \left[1 - \left(\frac{A-1}{A+1}\right)^2\right] T = \frac{4A}{(A+1)^2} T. \tag{1.135}$$

Let's now specify the relative energy exchange for the cases $A \ll 1$, $A = 1$, and $A \gg 1$:

$$\frac{\Delta T_{max}}{T} = \begin{cases} 4/A & \text{if } A \gg 1 \\ 1 & \text{if } A = 1 \\ 4A & \text{if } A \ll 1 \end{cases} \tag{1.136}$$

In particular, $\Delta T_{max}/T \to 0$ for $A \gg 1$, is maximal for $A = 1$, and is proportional to A for low values of the target mass. In particular, the last limit implies that the maximum velocity of the target after the scattering is twice the velocity of the incoming particle. This is easy to prove, since we have:

$$\frac{1}{2}(A\,m)|\mathbf{v}'_A|^2 = 4A \cdot \frac{1}{2}m|\mathbf{v}|^2 \quad \Rightarrow \quad |\mathbf{v}'_A| = 2|\mathbf{v}|. \tag{1.137}$$

Discussion

The dependence of the relative energy transfer on the mass ratio between the projectile and the target has important implications on the possibility of slowing-down particles by elastic collisions, like e.g. neutrons produced in fission reactions. Indeed, if $A \gg 1$, an elastic scattering implies only a very small energy loss per binary collision: massive elements are not efficient velocity moderators. On the contrary, elements with comparable mass are more effective in reducing the energy of the incoming particles, since the energy transfer ΔT can reach larger values, as for from Eq. (1.134), so that it will take on average fewer binary collisions to achieve the desired moderation. For example, H_2O is a good neutron moderator thanks to the presence of free protons in the molecule.

Suggested Readings

For further details on this subject, the reader is addressed to Sect. 2.8.1 of Ref. [10].

Problem 1.25 Consider the elastic scattering of a massless particle against a particle of mass m, initially at rest. Determine the largest energy transfer from the massless particle to the target.

Solution

Differently from Problem 1.24, the initial particle is now. It is most convenient to use the covariant formalism. For this purpose, let the four-momenta of the massless particle before and after the scattering be denoted by $k = (E, \mathbf{k})$ and $k' = (E', \mathbf{k}')$, respectively, and let θ the angle between the two momenta. The target initially has four-momentum $P = (m, \mathbf{0})$, which becomes P' after the scattering. Conservation of energy-momentum implies:

$$k + P = k' + P', \qquad P' = k - k' + P, \qquad m^2 = m^2 + 2kk' + 2m(E - E'),$$
$$0 = -2EE'(1 - \cos\theta') + 2m(E - E') = E'[m + (1 - \cos\theta)E] - mE,$$
$$E' = \frac{E}{1 + \frac{E}{m}(1 - \cos\theta)}. \tag{1.138}$$

The energy E' is at a minimum for back-scattering, $\cos\theta = -1$, and the corresponding energy transfer is:

$$\Delta T_{\max} = E - E'(\theta = \pi) = E\left(1 - \frac{1}{1 + 2E/m}\right) = E\frac{2E/m}{1 + 2E/m} \equiv E\frac{2k}{1 + 2k}. \tag{1.139}$$

with $k \equiv E/m$.

Discussion

The *Compton scattering* of a photon against atomic electrons falls into this class of problems. The existence of a maximum energy transfer leads to a characteristic threshold in the energy distribution of the recoil electrons, called *Compton peak*, see Sect. 2.1 for more details.

Suggested Readings

For further details on this subject, the reader is addressed to Sect. 2.7 of Ref. [10].

Problem 1.26 A particle of mass M and momentum $\mathbf{p}$ interacts elastically with a particle of mass m, initially at rest. Determine the maximum possible energy transfer involved in the scattering. In particular, specialise the formula for the case of a non-relativistic particle, $M \gg |\mathbf{p}|$, and for the case of an initial massless particle, $M = 0$.

Discussion

The results derived in this exercise will be useful when discussing the energy lost by a charged particle in the collision with the atomic electrons, as described by the so-called *Bethe formula*, see Eq. (2.1). The maximum energy transfer to an atomic electron, usually denoted by W_{max}, provides a natural energy scale for this kind of problems, the other being the work needed to extract an electron from its orbital. It is also interesting to see how the maximum energy transfer in the non-relativistic approximation and in the Compton scattering, whose derivation Problems 1.24 and 1.25 was based on specific assumptions (classical kinematics for the former, massless initial particle for the latter), can be obtained as a special case of a more general formula, valid in all regimes, provided that $m > 0$ (otherwise there would be no frame where the target is at rest).

Solution

It's convenient to study the kinematics of the scattering in the centre-of-mass frame. Here the projectile has a momentum $\mathbf{p}^*$ and the target has opposite momentum $-\mathbf{p}^*$. The largest energy transfer to the target in the laboratory frame corresponds to its back-scattering in the centre-of-mass frame. We can then calculate the corresponding energy in the laboratory frame by making a Lorentz transformation with boost parameter β. We can use the results obtained in Problem 1.18, and in particular Eqs. (1.89), (1.96), and (1.98), to express all quantities in terms of the energy-momentum in the laboratory frame. The projectile energy in the laboratory frame is given by $E = \sqrt{|\mathbf{p}|^2 + M^2}$. In the laboratory frame, the target energy for the case of back-scattering is given by:

$$
\begin{aligned}
E_2 &= \gamma \left(E_2^* + \beta |\mathbf{p}^*| \right) = \frac{E + m}{\sqrt{s}} \left(\frac{s - M^2 + m^2}{2\sqrt{s}} + \frac{|\mathbf{p}|}{E + m} |\mathbf{p}| \frac{m}{\sqrt{s}} \right), \\
&= \frac{E + m}{\sqrt{s}} \left(\frac{2m^2 + 2Em}{\sqrt{s}} + \frac{m|\mathbf{p}|^2}{(E + m)\sqrt{s}} \right) = \frac{m}{s} \left((E + m)^2 + |\mathbf{p}|^2 \right), \\
&= \frac{m}{s} \left(E^2 + 2mE + m^2 + |\mathbf{p}|^2 + M^2 - M^2 \right) = \frac{m}{s} \left(s + 2|\mathbf{p}|^2 \right) \\
&= m \left(1 + 2\frac{|\mathbf{p}|^2}{s} \right).
\end{aligned}
\tag{1.140}
$$

Hence, the maximum energy transfer is given by:

$$\Delta T_{\max} = \frac{2\,m\,|\mathbf{p}|^2}{s} = \frac{2\,m\,(\beta_{\mathbf{p}}\gamma_{\mathbf{p}})^2}{1 + (m/M)^2 + 2\,(m/M)\,\sqrt{(\beta_{\mathbf{p}}\gamma_{\mathbf{p}})^2 + 1}}. \tag{1.141}$$

Let's now specialise Eq. (1.141) to the following cases:

1. $M \gg |\mathbf{p}|$. In this approximation, $E \approx M$ and $|\mathbf{p}| \approx M|\mathbf{v}|$, where $\mathbf{v}$ is the projectile velocity the laboratory frame. Hence:

$$\Delta T_{\max} \approx \frac{2\,m\,(M|\mathbf{v}|)^2}{(M+m)^2} = \left(\frac{1}{2}M|\mathbf{v}|^2\right) 4\,\frac{m/M}{(m/M + 1)^2}, \tag{1.142}$$

which agrees with Eq. (1.135) since $A = m/M$ in the current notation. It's worth noticing that no assumption was made on the target being relativistic or not after the scattering. Indeed, Eq. (1.135) was derived under the sole assumption that only the projectile is non-relativistic, while the target plays no role other than to specify the boost parameter.

2. $M = 0$. In this case, $|\mathbf{p}| = E$ and Eq. (1.141) becomes:

$$\Delta T_{\max} = \frac{2\,m\,E^2}{m^2 + 2\,m\,E} = E\,\frac{2\,E/m}{1 + 2\,E/m}, \tag{1.143}$$

which agrees with Eq. (1.139). Figure 1.9 shows the ratio $\Delta T_{\max}/T$ between the maximal energy transfer and the kinetic energy of the projectile as a function of E/m. Four different cases are considered: $\gamma_{\mathbf{p}} \approx 1$, $\gamma_{\mathbf{p}} = 2$, $\gamma_{\mathbf{p}} = 10$, and the case $M = 0$.

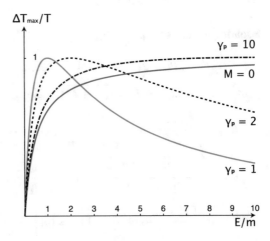

Fig. 1.9 Ratio $\Delta T_{\max}/T$ between the maximal energy transfer and the total energy of the projectile as a function of E/m, where m is the mass of the target. Four different cases are considered: $\gamma_{\mathbf{p}} \approx 1$ (non-relativistic limit), $\gamma_{\mathbf{p}} = 2$, $\gamma_{\mathbf{p}} = 10$, and the case of massless projectile, $M = 0$

Suggested Readings

The reader is addressed to Chap. 33 of Ref. [4], the PDG review dedicated to the passage of particles through matter, where the maximum energy transfer in a single collision is discussed. See also Sect. 2.2.2 of Ref. [10].

Problem 1.27 A charged particle of unknown mass M and momentum $|\mathbf{p}|$ much larger than the electron mass m_e, crosses a detector. At a certain depth inside the active material, an atomic electron is knocked out by the incoming particle, for which the polar angle θ_e with respect to $\mathbf{p}$ and the energy E_e are measured. Show that the mass M can be estimated from the following formula:

$$M = |\mathbf{p}| \left[\frac{E_e + m_e}{E_e - m_e} \cos^2 \theta_e - 1 \right]^{\frac{1}{2}}. \tag{1.144}$$

Solution

Let the four-momentum of the unknown particle be P, and the four-momenta of the electron before and after the scattering be k and k'. After the scattering, the incoming particle will have four-momentum P'. With a convenient choice of the reference frame, we can write:

$$
\begin{aligned}
P &= (\sqrt{|\mathbf{p}|^2 + M^2},\ |\mathbf{p}|,\ 0,\ 0) \\
k &= (m_e,\ 0,\ 0,\ 0) \\
k' &= (E_e,\ \sqrt{E_e^2 - m_e^2} \cos \theta_e,\ \sqrt{E_e^2 - m_e^2} \sin \theta_e,\ 0) \\
P' &= P + k - k'
\end{aligned}
\tag{1.145}
$$

Squaring the last of Eq. (1.145), we get rid of the unknown kinematics, obtaining:

$$P'^2 = P^2 + 2Pk - 2Pk' + k^2 + k'^2 - 2kk',$$

$$M^2 = M^2 + 2(m_e - E_e)\sqrt{|\mathbf{p}|^2 + M^2} + 2|\mathbf{p}|\sqrt{E_e^2 - m_e^2} \cos \theta_e + 2m_e(m_e - E_e),$$

$$0 = \sqrt{|\mathbf{p}|^2 + M^2} - |\mathbf{p}|\sqrt{\frac{E_e + m_e}{E_e - m_e}} \cos \theta_e + m_e,$$

$$0 = \sqrt{1 + \frac{M^2}{|\mathbf{p}|^2}} - \sqrt{\frac{E_e + m_e}{E_e - m_e}} \cos \theta_e + \frac{m_e}{|\mathbf{p}|}$$

$$M \approx |\mathbf{p}| \left[\frac{E_e + m_e}{E_e - m_e} \cos^2 \theta_e - 1 \right]^{\frac{1}{2}} = |\mathbf{p}| \left[\frac{T_e + 2m_e}{T_e} \cos^2 \theta_e - 1 \right]^{\frac{1}{2}}, \tag{1.146}$$

where the factor $m_e/|\mathbf{p}|$ has been neglected, and we have introduced the kinetic energy $T_e = E_e' - m_e$.

Discussion

This formula was used by Leprince-Ringuet and collaborators to estimate the mass of a new long-lived particles discovered in cosmic rays. A magnetised nuclear emulsion had been exposed to cosmic rays on the top of the french Alps. Among the others, it recorded an event that could be interpreted as a charged particle, identified as such by the bubble density of the track impressed on the emulsion, transferring a sizable fraction of its momentum to a single atomic electron, a so-called δ-ray. The latter was identified as a spiraling track originating from the kink of the primary track, and its momentum, like the momentum of the unknown particle, could be measured from the radius of curvature of the track. With these measurements at hand, the mass of the unknown could be measured with enough accuracy to establish the discovery of a new particle, later identified as the charged kaon.

Suggested Readings

The reader is addressed to the paper by Leprince-Ringuet and Crussard [11], reporting the first evidence for a particle with a mass of about one-thousand times the electron mass contained in cosmic rays. The rather precise estimation of the particle mass was based on the kinematics of a fully-reconstructed event. This experiment is also discussed in Ref. [12].

Problem 1.28 A high-energy positron beam on a fixed-target experiment can produce the reaction $e^+ e^- \to f \bar{f}$, where f and $\bar{f}$ have the same mass M. Show that M can be estimated using the formula:

$$M = \frac{1}{2}\left[2m_e|\mathbf{p}|\left(1 - \left(\frac{\theta_1 - \theta_2}{\theta_1 + \theta_2}\right)^2\right)\left(1 - \frac{|\mathbf{p}|}{2m_e}\theta_1\theta_2\right)\right]^{\frac{1}{2}}, \qquad (1.147)$$

where $|\mathbf{p}|$ is the beam momentum, assumed to be much larger than m_e and M, and θ_i are the polar angles of f and $\bar{f}$ with respect to the beam direction.

Solution

Let's denote the momenta of the outgoing particles by $\mathbf{p}_1$ and $\mathbf{p}_2$. Momentum conservation along the beam and its orthogonal axis implies:

$$\begin{cases} |\mathbf{p}_1| \sin\theta_1 = |\mathbf{p}_2| \sin\theta_2 \\ |\mathbf{p}_1| \cos\theta_1 + |\mathbf{p}_2| \cos\theta_2 = |\mathbf{p}| \end{cases} \qquad (1.148)$$

We now work in the assumption $\theta_i \ll 1$, which is indeed justified if $|\mathbf{p}| \gg m_e, M$ because of the large boost of the centre-of-mass frame, see Problem 1.15. With this approximation:

$$\begin{cases} |\mathbf{p}_1|\theta_1 = |\mathbf{p}_2|\theta_2 \\ |\mathbf{p}_1|(1 - \frac{\theta_1^2}{2}) + |\mathbf{p}_2|(1 - \frac{\theta_2^2}{2}) = |\mathbf{p}| \end{cases} \tag{1.149}$$

Inserting the first equation into the second:

$$|\mathbf{p}_2| \left[\frac{\theta_2}{\theta_1} - \frac{\theta_1 \theta_2}{2} + 1 - \frac{\theta_2^2}{2} \right] = |\mathbf{p}_2| \left[\frac{\theta_1 + \theta_2}{\theta_1} + \mathcal{O}(\theta_i^2) \right] = |\mathbf{p}|,$$

$$\Rightarrow \begin{cases} |\mathbf{p}_1| = \frac{\theta_2}{\theta_1 + \theta_2} |\mathbf{p}| \\ |\mathbf{p}_2| = \frac{\theta_1}{\theta_1 + \theta_2} |\mathbf{p}| \end{cases} \tag{1.150}$$

Let's now denote the four-momentum of the positron, electron, and of two outgoing particles by k, P, p_1, p_2, respectively. We can treat the electron/positron as massless. Energy-momentum conservation implies:

$$p_1 + p_2 = k + P, \qquad 2M^2 + 2p_1 p_2 = 2m_e |\mathbf{p}|,$$

$$M^2 + (\sqrt{|\mathbf{p}_1|^2 + M^2})(\sqrt{|\mathbf{p}_2|^2 + M^2}) - |\mathbf{p}_1||\mathbf{p}_2|\cos(\theta_1 + \theta_2) = m_e |\mathbf{p}|,$$

$$M^2 + \left[\left(|\mathbf{p}_1| + \frac{M^2}{2|\mathbf{p}_1|} \right) \left(|\mathbf{p}_2| + \frac{M^2}{2|\mathbf{p}_2|} \right) \right] - |\mathbf{p}_1||\mathbf{p}_2|(1 - \frac{(\theta_1 + \theta_2)^2}{2}) = m_e |\mathbf{p}|,$$

$$M^2 \left[1 + \frac{1}{2} \left(\frac{|\mathbf{p}_2|}{|\mathbf{p}_1|} + \frac{|\mathbf{p}_1|}{|\mathbf{p}_2|} \right) \right] = m_e |\mathbf{p}| - |\mathbf{p}_2||\mathbf{p}_2| \frac{(\theta_1 + \theta_2)^2}{2},$$

$$M^2 \left[1 + \frac{1}{2} \left(\frac{\theta_1}{\theta_2} + \frac{\theta_2}{\theta_1} \right) \right] = m_e |\mathbf{p}| \left(1 - \frac{|\mathbf{p}|}{2m_e} \theta_1 \theta_2 \right),$$

$$M^2 \frac{(\theta_1 + \theta_2)^2}{2\theta_1 \theta_2} = m_e |\mathbf{p}| \left(1 - \frac{|\mathbf{p}|}{2m_e} \theta_1 \theta_2 \right),$$

$$2M^2 \left[1 - \left(\frac{\theta_1 - \theta_2}{\theta_1 + \theta_2} \right)^2 \right]^{-1} = m_e |\mathbf{p}| \left(1 - \frac{|\mathbf{p}|}{2m_e} \theta_1 \theta_2 \right),$$

$$M = \frac{1}{2} \left[2m_e |\mathbf{p}| \left(1 - \left(\frac{\theta_1 - \theta_2}{\theta_1 + \theta_2} \right)^2 \right) \left(1 - \frac{|\mathbf{p}|}{2m_e} \theta_1 \theta_2 \right) \right]^{\frac{1}{2}}. \tag{1.151}$$

Discussion

This formula was used by the NA7 Collaboration in a fixed-target experiment aiming at measuring the pion form factor in the time-like region in the range $0.1 < q^2/\text{GeV}^2 < 0.18$, where $q^2 = 2m_e|\mathbf{p}|$ [13]. These values of q^2 correspond to beam energies ranging from 100 up to 175 GeV. The experimental apparatus consisted in a liquid hydrogen target followed by a planes of MWPC, see Problem 2.52, giving an angular resolution of 0.02 mrad for particles emerging from the target within 7 mrad from the beam direction. A magnetic spectrometer, complemented by electromagnetic calorimeters, allowed to separate electrons/positrons from muons and pions. The latter two were further separated by performing an angular analysis of the two polar angles θ_1 and θ_2, since the two angles are correlated by the mass of the produced particle. The comparison between the cross sections for $e^+ e^- \to \mu^+ \mu^-$ and $e^+ e^- \to \pi^+ \pi^-$ allowed to extract the pion form factor $F_\pi(q^2)$ in the q^2 region corresponding to the available beam energies.

Suggested Readings

The reader is addressed to the NA7 paper [13], where the angular analysis used to disentangle between pions and muons is discussed in detail.

Problem 1.29 The rapidity of a particle of momentum **p** and mass m is defined as:

$$y = \frac{1}{2} \ln \left(\frac{E + p_z}{E - p_z} \right). \tag{1.152}$$

1. How does y transform under a boost $\boldsymbol{\beta} = \beta \mathbf{e}_z$?
2. Write down the phase-space measure $d^3\mathbf{p}/E_\mathbf{p}$ in terms of new set of variables $(|\mathbf{p_T}|, \phi, y)$, where $|\mathbf{p_T}|^2 = p_x^2 + p_y^2$ and ϕ is the azimuthal angle around the z-axis.
3. Show that, in the limit $|\mathbf{p}| \gg m$, the rapidity reduces to the purely geometrical quantity, η, called *pseudorapidity*. Express η in terms of the polar angle θ.

Solution

We first derive a set of equations that will prove useful in the following. The rapidity y can be also expressed as:

$$y = \frac{1}{2} \ln \left(\frac{E + p_z}{E - p_z} \right), \qquad \frac{p_z}{E} = \frac{e^y - e^{-y}}{e^y + e^{-y}} = \tanh y, \qquad y = \operatorname{atanh} \left(\frac{p_z}{E} \right). \tag{1.153}$$

By introducing the *transverse mass* $m_T \equiv \sqrt{|\mathbf{p_T}|^2 + m^2}$, we also get:

$$y = \frac{1}{2} \ln \left(\frac{E + p_z}{E - p_z} \right) = \frac{1}{2} \ln \left(\frac{(E + p_z)^2}{E^2 - p_z^2} \right) = \ln \left(\frac{E + p_z}{m_T} \right),$$

$$m_T e^y - p_z = \sqrt{m_T^2 + p_z^2}, \qquad -2 p_z e^y = m_T (1 - e^{2y}), \qquad p_z = m_T \sinh y. \tag{1.154}$$

If the particle momentum **p** is aligned with the z-axis, Eq. (1.153) gives:

$$y = \frac{1}{2} \ln \left(\frac{1 + \beta_\mathbf{p}}{1 - \beta_\mathbf{p}} \right) = \text{atanh} \, \beta_\mathbf{p}, \tag{1.155}$$

from which:

$$\beta_\mathbf{p} = \tanh y, \qquad \gamma_\mathbf{p} = \cosh y \qquad \beta_\mathbf{p}\gamma_\mathbf{p} = \sinh y. \tag{1.156}$$

1. After a boost $\boldsymbol{\beta} = \beta \mathbf{e}_z$, the rapidity expressed in the new reference frame becomes:

$$y' = \frac{1}{2} \ln \left(\frac{E' + p'_z}{E' - p'_z} \right) = \frac{1}{2} \ln \left(\frac{\gamma \, (E - \beta p_z) + \gamma \, (-\beta E + p_z)}{\gamma \, (E - \beta p_z) - \gamma \, (-\beta E + p_z)} \right) =$$

$$= \frac{1}{2} \ln \left(\left(\frac{E + p_z}{E - p_z} \right) \left(\frac{1 - \beta}{1 + \beta} \right) \right) = y - \frac{1}{2} \ln \left(\frac{1 + \beta}{1 - \beta} \right) = y - \text{atanh} \, \beta, \tag{1.157}$$

so that the difference Δy between the rapidity of two particles is an invariant under longitudinal boosts, as are differences between azimuthal angles, $\Delta\phi$.

2. We want to find the Jacobian $J = |\partial(p_x, p_y, p_z)/\partial(|\mathbf{p_T}|, \phi, y)|$ of the transformation:

$$\begin{cases} p_x = |\mathbf{p_T}| \cos \phi \\ p_x = |\mathbf{p_T}| \sin \phi \\ p_z = \sqrt{m^2 + |\mathbf{p_T}|^2} \sinh y \end{cases} \tag{1.158}$$

To this purpose, we first notice that $dp_x \, dp_y = |\mathbf{p_T}| d|\mathbf{p_T}| \, d\phi$, so that all we are left to do is to compute $\partial p_z / \partial y$. By using the fact that $\partial E_\mathbf{p} / \partial p_i = p_i / E_\mathbf{p}$, we get:

$$\frac{\partial y}{\partial p_z} = \frac{1}{2} \left(\frac{E_\mathbf{p} - p_z}{E_\mathbf{p} + p_z} \right) \frac{(p_z/E_\mathbf{p} + 1)(E_\mathbf{p} - p_z) - (E_\mathbf{p} + p_z)(p_z/E_\mathbf{p} - 1)}{(E_\mathbf{p} - p_z)^2} = \frac{1}{E_\mathbf{p}}, \tag{1.159}$$

hence:

$$\frac{d\mathbf{p}}{E_\mathbf{p}} = |\mathbf{p_T}| d|\mathbf{p_T}| \, d\phi \, dy = \frac{1}{2} d|\mathbf{p_T}|^2 \, d\phi \, dy = \pi \, d|\mathbf{p_T}|^2 \, dy \tag{1.160}$$

The last equality in Eq. (1.160) holds under integration over the azimuthal angle.

3. Let's now consider the case $E \gg m$. The momentum $|\mathbf{p}|$ can be Taylor-expanded around E to give:

$$y = \frac{1}{2} \ln \frac{E + E \cos \theta - \frac{m^2}{2E^2} \cos \theta + \dots}{E - E \cos \theta + \frac{m^2}{2E^2} \cos \theta + \dots} \approx \frac{1}{2} \ln \left(\frac{1 + \cos \theta}{1 - \cos \theta} \right) = \frac{1}{2} \ln \left(\frac{\cos^2 \frac{\theta}{2}}{\sin^2 \frac{\theta}{2}} \right) =$$

$$= -\ln\left(\tan\frac{\theta}{2}\right) \equiv \eta \tag{1.161}$$

From Eq. (1.161), we also get two more useful trigonometric relations:

$$\eta = \frac{1}{2}\ln\left(\frac{1+\cos\theta}{1-\cos\theta}\right) = \mathrm{atanh}\,(\cos\theta), \qquad \cos\theta = \tanh y, \qquad \sin\theta = \frac{1}{\cosh y}. \tag{1.162}$$

Discussion

Rapidity is a useful variable in hadron colliders thanks to its transformation properties: a partonic differential cross section $d\sigma/dy$, computed in any reference frame (e.g. the centre-of-mass frame of the parton-parton scattering), has the same form in any other frame, provided one replaces y by the linearly transformed value as for Eq. (1.157).

Differential cross sections at hadron colliders are often forward-peaked due to the large longitudinal momentum of the initial-state partons, see Problem 1.15. When expressed in terms of the rapidity, differential cross sections become more smooth for large boost factors γ. Indeed, for an isotropic process in the centre-of-mass frame, and assuming $y \approx \eta$, we have:

$$\frac{d\sigma}{dy} = \frac{d\sigma}{d\cos\theta}\left|\frac{d\cos\theta}{dy}\right| = \frac{1}{\gamma^2}\frac{1}{(1-\beta\cos\theta)^2}\frac{d\sigma}{d\cos\theta^*}\cdot\frac{d\tanh y}{dy} =$$

$$= \frac{1}{2\gamma^2}\frac{1}{(1-\beta\tanh y)^2}\frac{1}{\cosh^2 y} =$$

$$= \frac{2}{\gamma^2}\frac{1}{e^{2y}(1-\beta)^2 + e^{-2y}(1+\beta)^2 - 2(1+\beta^2) + 4}, \tag{1.163}$$

where we have used Eq. (1.161). This expression simplifies in the limit $\beta \to 1$ to:

$$\frac{d\sigma}{dy} \approx \frac{2}{\gamma^2}\frac{1}{(1-\beta)^2 e^{2y} + 4e^{-2y}}, \tag{1.164}$$

which goes to zero for $y \gg 1$ and has a maximum at $y = \ln(2\gamma)$. Figure 1.10 shows $d\sigma/d\cos\theta$ and $d\sigma/dy$, superimposed on the same axis for illustration purposes. A comparison between the two shows that, for a scattering process characterised by a large boost factor γ, the differential cross section in y is a smooth and broad function, whereas the differential cross section in $\cos\theta$ is squeezed around $\cos\theta = 1$.

Another reason of interest for y (or η) at hadron colliders is due to the fact that a variety of soft processes, like the productions of particles in minimum bias hadron-hadron collisions, turns out to be almost uniformly distributed in η, see e.g. by Fig. 1.11. This variable then becomes the relevant metric when designing a detector

Fig. 1.10 Comparison between the differential cross section $d\sigma/d\cos\theta$ and $d\sigma/dy$ for a scattering process which is isotropic in the centre-of-mass frame, superimposed on the same axis for illustration. The boost factor of the centre-of-mass is taken to be $\gamma = 5$

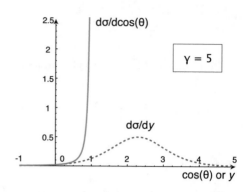

Fig. 1.11 Distributions of the pseudorapidity density of charged hadrons in the region $|\eta| < 2$ in inelastic pp collisions at 13 TeV measured in data (markers) and predicted by two of LHC event generators (curves). This plot has been taken from Ref. [14]

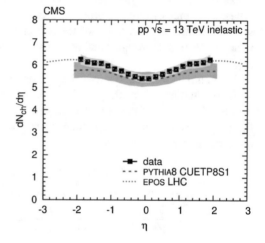

for occupancy and radiation hardness.

Problem 1.30 Write down the invariant mass m_{12}^2 of two particles with four-momentum p_1 and p_2 in terms of their transverse momenta p_T, transverse mass m_T, and rapidity y with respect to the same axis. Find out an approximate formula for m_{12} valid in the case that the mass of the two particles is small compared to their energy and the momenta are almost collinear.

Solution

We can replace the canonical variables by the transverse variables by means of Eq. (1.154), giving:

$$m_{12}^2 = m_1^2 + m_2^2 + 2(E_1 E_2 - \mathbf{p}_1 \cdot \mathbf{p}_2) = m_1^2 + m_2^2 +$$
$$+ 2\left[m_{T1}(e^{y_1} - \sinh y_1) m_{T2}(e^{y_2} - \sinh y_2) - m_{T1} m_{T2} \sinh y_1 \sinh y_2 - \mathbf{p}_{T1} \cdot \mathbf{p}_{T1} \right]$$
$$= m_1^2 + m_2^2 + 2\left[m_{T1} m_{T2}\left(e^{y_1} e^{y_2} - \frac{e^{y_2} - e^{-y_2}}{2} e^{y_1} - \frac{e^{y_1} - e^{-y_1}}{2} e^{y_2} \right) - \mathbf{p}_{T1} \cdot \mathbf{p}_{T2} \right]$$
$$= m_1^2 + m_2^2 + 2\left[m_{T1} m_{T2} \cosh(y_1 - y_2) - \mathbf{p}_{T1} \cdot \mathbf{p}_{T2} \right]. \tag{1.165}$$

In the limit $|\mathbf{p}_i| \gg m_i$ and $\theta_{12} \ll 1$, we have:

$$m_{12} \approx \left[2p_{T1} p_{T2} \left(\underbrace{\cosh(\Delta\eta)}_{1 + \frac{\Delta\eta^2}{2} + \dots} - \underbrace{\cos(\Delta\phi)}_{1 - \frac{\Delta\phi^2}{2} + \dots} \right) \right]^{\frac{1}{2}} \equiv \sqrt{p_{T1} p_{T2}}\, R_{12}, \tag{1.166}$$

where we have exploited the fact that $y_i \to \eta_i$ in this limit, see Problem 1.29, and we have introduced the Euclidean metric R_{12} in the (η, ϕ) space, defined by:

$$R_{12} = \sqrt{(\eta_1 - \eta_2)^2 + (\phi_1 - \phi_2)^2}. \tag{1.167}$$

Discussion

The Euclidean metric R_{ij} in the (η, ϕ) space is widely used at hadron colliders to define the geometric "distance" between two particles. Jet-clustering algorithm often rely on R_{ij} to quantify the distance between two particles such that they can be associated with the same jet. The level of isolation of a particle is usually defined by energy collected within a cone of radius R centered around the particle direction.

Suggested Readings

For an application of the R metric in jet-clustering algorithms, the reader is addressed to Ref. [15].

Problem 1.31 A particle of mass M and momentum $\mathbf{p}$ parallel to the z-axis is produced at the interaction point of an accelerator and then decays to a pair of identical particles of mass m. The interaction point is surrounded by a cylindrical detector around the z-axis, whose geometrical coverage is however limited to the pseudorapidity region $|\eta| \leq \eta_{\text{acc}}$. Determine the largest rapidity of the mother particle for which the experiment is sensitive to the mother particle, assuming two definitions of acceptance:

1. at least one of the daughter particles is within acceptance;
2. both daughter particles are within acceptance.

Solution

Since the problem has cylindrical symmetry, it's enough to consider the case $p_z > 0$ and get an upper bound on y: symmetry will then imply that the same bound, in absolute value, will hold for $|y|$. The velocity β^* of the daughter particles in the centre-of-mass frame in given by Eq. (1.92) with $\sqrt{s} = M$, while the velocity of the mother particle in the laboratory frame is β, and the rapidity is $y = \operatorname{atanh} \beta$, see Eq. (1.155).

If $\beta < \beta^*$, it is always possible to find a polar angle θ^* in the centre-of-mass frame such that one of the two particles emerges at an angle larger than $\theta_{\mathrm{acc}} \equiv 2 \tan^{-1}\left(e^{-\eta_{\mathrm{acc}}}\right)$ in the laboratory frame, see e.g. Problem 1.10. Since the other particle is emitted at an angle $\pi - \theta^*$ in the centre-of-mass frame, it will emerge at an even larger polar angle in the laboratory frame: the experiment therefore is sensitive to a non-zero fraction of decays up to $y = \operatorname{atanh} \beta^*$, regardless of the definition of acceptance.

When $\beta > \beta^*$, the polar angle θ is bounded by a maximum angle $\theta_{\max}$ satisfying Eq. (1.49). The condition that at least one of the daughter particles is within the detector acceptance amounts to require

$$\theta_{\mathrm{acc}} \leq \max\{\theta_1, \theta_2\} \leq \theta_{\max}. \tag{1.168}$$

By using a trigonometric identity and Eq. (1.47), the condition that none of the daughter particles fall within the acceptance translates to $\theta_{\max} < \theta_{\mathrm{acc}}$, or:

$$-\ln\left(\tan\frac{\theta_{\max}}{2}\right) = \ln\frac{\tan\theta_{\max}}{\sqrt{1 + \tan^2\theta_{\max}} - 1} = \ln\frac{\frac{1}{\gamma}\frac{\beta^*}{\sqrt{\beta^2 - \beta^{*2}}}}{\sqrt{1 + \frac{1}{\gamma^2}\frac{\beta^{*2}}{\beta^2 - \beta^{*2}}} - 1} =$$

$$\ln\frac{\gamma^*\beta^*}{\gamma\beta - \gamma\gamma^*\sqrt{\beta^2 - \beta^{*2}}} = \ln\frac{\gamma^*\beta^*}{\sinh y - \gamma^*\sqrt{\tanh^2 y - \beta^{*2}}\cosh y} > \eta_{\mathrm{acc}}, \tag{1.169}$$

where Eq. (1.156) has been used to express γ and $\beta\gamma$ in terms of the rapidity y.

To obtain the maximum rapidity y such that both particles have $\theta > \theta_{\mathrm{acc}}$, we notice that $\theta^* = \pi/2$ plays a special role, since:

- if $\theta^* \in [0, \pi/2]$, then $\theta_1 \leq \theta(\theta^* = \frac{\pi}{2}) \equiv \theta_\perp$;
- if $\theta^* \in (\pi/2, \theta_{\max}^*]$, then $\theta_1 > \theta_\perp$, but $\theta_2 < \theta_\perp$;
- if $\theta^* \in (\theta_{\max}^*, \pi)$, then $\theta_2 < \theta_\perp$.

Hence, in order to have both daughter particles within the acceptance, one needs:

$$\theta_{\mathrm{acc}} \leq \min\{\theta_1, \theta_2\} \leq \theta_\perp. \tag{1.170}$$

The condition that none of the daughter particles fall within the acceptance translates to $\theta_\perp < \theta_{\mathrm{acc}}$, or:

Fig. 1.12 Maximum
detector pseudorapidity η_{acc}
as a function of the particle
rapidity y_{max} such that the
detector is sensitive to
$y > y_{max}$, for two definitions
of acceptance: requiring that
both daughter particles have
$\eta < \eta_{acc}$ (solid curve) or
requiring that at least one has
$\eta < \eta_{acc}$ (dashed curve)

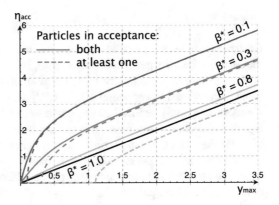

$$-\ln\left(\tan\frac{\theta_\perp}{2}\right) = \ln\frac{\tan\theta_\perp}{\sqrt{1+\tan^2\theta_\perp}-1} = \ln\frac{\beta^*/\gamma\beta}{\sqrt{1+\frac{\beta^{*2}}{\gamma^2\beta^2}}-1} =$$

$$= \ln\frac{\beta^*}{\sqrt{\sinh^2 y + \beta^{*2}}-\sinh y} > \eta_{acc}. \tag{1.171}$$

Figure 1.12 shows a graph of the functions at the left-hand side of Eqs. (1.169)
and (1.171) for a few values of β^*. The case $\beta^* = 1$ (massless particles) corresponds
to $y < \eta_{acc}$: the maximum particle rapidity for which both daughter particles are
within the acceptance coincides with the pseudorapidity range of the detector. For
smaller centre-of-mass velocities, the maximum rapidity becomes smaller than η_{acc}.

Discussion

Although we have arrived at an analytical result, this exercise offers an example of
how problems in particle physics are often better tackled by using Monte Carlo methods. Indeed, the problem could have been studied by a MC program that (1) generates
particle decays in the centre-of-mass for random values of θ^* and for discrete values
of β^*, (2) applies the Lorentz boost to the laboratory frame for some values of β,
and (3) saves the values (η_1, η_2) in a binned 2D-histogram, or in a n-tuple. For a
sufficiently large number of toy events, we can then get the acceptance map for the
tested values of (β, β^*) by studying the joint distribution of (η_1, η_2).

Problem 1.32 A beam made of pions and muons of momentum $|\mathbf{p}| = 170$ MeV are
produced from a short proton pulse against a fixed target. The beams are detected
by a detector located $d = 21$ m downstream of the primary target. Determine which
fraction of undecayed pions and muons arrives to the detector, and the difference
between the time of flight and kinetic energy of the two particles.

Discussion

Unstable particles are associated with a mean life-time τ, defined as the inverse of the decay probability per unit of *proper time*, i.e. the time measured in the rest frame of the particle. It is customary to define *particle* any quantum state with τ above some conventional lower bound (e.g. $\sim 10^{-18}$ s), while states with smaller lifetimes are more frequently referred to as *resonance*, and the natural width $\Gamma = \hbar/\tau$, is reported instead. In some cases, it is more convenient to quote the length $c\tau$ in place of τ. While τ is a constant, a time interval is however not a Lorentz invariant, and one often wants to express a survival probability, which, for the proper time obeys an exponential law with time constant τ, in a different reference frame, like the laboratory frame where particles have $|\mathbf{p}| > 0$.

Solution

In the rest frame of an unstable particle, the probability $P(t^* \mid \tau)$ that the particle survive up to time t^*, given that at time $t^* = 0$ it has not decayed, is provided by the cumulative of the exponential p.d.f.:

$$P(t^* \mid \tau) = \exp\left(-t^*/\tau\right). \tag{1.172}$$

In our case, we need to compute the probability that the beam particles have survived over the time t that takes them to travel the whole length d in the laboratory frame. This time interval corresponds to a proper time:

$$t^* = \frac{t}{\gamma} = \frac{1}{\gamma}\left(\frac{d}{\beta c}\right) = \frac{m\,d}{m\,\beta\gamma\,c} = \frac{m}{|\mathbf{p}|}d. \tag{1.173}$$

It is easy to see that a consistent use of $c\tau$ together with masses in units of GeV/c^2 and momenta in units of GeV/c gives the correct numerical result:

$$\exp\left(-\frac{m}{|\mathbf{p}|}\frac{d}{\tau}\right) = \begin{cases} \exp\left(-\frac{106 \text{ MeV}}{170 \text{ MeV}}\frac{21 \text{ m}}{6.6\times10^2 \text{ m}}\right) = 0.98 & \mu \\ \exp\left(-\frac{139 \text{ MeV}}{170 \text{ MeV}}\frac{21 \text{ m}}{7.8 \text{ m}}\right) = 0.11 & \pi \end{cases} \tag{1.174}$$

where we used the values $c\tau = 6.6 \times 10^2$ m for muons and 7.8 m for pions, and masses $m_\mu = 106$ MeV and $m_{\pi^-} = 139$ MeV [4].

Finally we can compute the difference between the TOF for the two beams, which is given by:

$$t = \frac{d}{\beta c} = \frac{d}{c}\sqrt{1 + \left(\frac{m}{|\mathbf{p}|}\right)^2} = \frac{21 \text{ m}}{3 \times 10^8 \text{ m/s}} \cdot \begin{cases} 1.18 = 82.5 \text{ ns} & \mu \\ 1.29 = 90.4 \text{ ns} & \pi \end{cases} \tag{1.175}$$

from which $|\Delta t| = 8.2$ ns. The kinetic energy is given by:

$$T = E - m = m \left[\sqrt{1 + \left(\frac{|\mathbf{p}|}{m} \right)^2} - 1 \right] = \begin{cases} 106 \text{ MeV} \cdot 0.89 = 94.3 \text{ MeV} & \mu \\ 139 \text{ MeV} \cdot 0.58 = 80.6 \text{ MeV} & \pi \end{cases}$$

(1.176)

from which $|\Delta T| = 14.1$ MeV.

Bando n. 18211/2016

Problem 1.33 A neutrino beam is produced by a 120 GeV proton accelerator. A magnetic selector filters positively charged particles of momentum $|\mathbf{p}| = 20 \pm 5$ GeV, which then decay inside a pipe filled with helium. We want the beam to be as pure as possible in ν_μ. Estimate the length of the pipe and at least one process that can reduce the beam purity.

Solution

The selected beam consists of π^+ and K^+. Both mesons decay with lifetimes of order 10^{-8} s. The main decay reactions giving rise to a ν_μ are

$$\pi^+ \to \mu^+ \nu_\mu, \qquad K^+ \to \mu^+ \nu_\mu, \qquad K^+ \to \pi^0 \mu^+ \nu_\mu, \qquad (1.177)$$

where the last decay ($K^+_{\mu 3}$) is suppressed with respect to the direct decay by a factor of 20. For a fixed beam momentum, the probability of decay per unit length is constant, thus giving a beam profile and probability of survival up to distance d:

$$\text{Prob}\,[x \geq d] = \exp \left[-\frac{m}{|\mathbf{p}|\, c\tau} d \right] \qquad (1.178)$$

If we want this probability to be small, thus allowing for a sizable fraction of the mesons to be decayed, the pipe length L needs to be larger than the decay length

$$L \gtrsim \frac{|\mathbf{p}|}{m} c\tau = \begin{cases} \frac{20 \pm 5 \text{ GeV}}{0.139 \text{ GeV}} \cdot 7.8 \text{ m} = 1100 \pm 300 \text{ m} & \pi^+ \\ \frac{20 \pm 5 \text{ GeV}}{0.497 \text{ GeV}} \cdot 3.7 \text{ m} = 150 \pm 40 \text{ m} & K^+ \end{cases} \qquad (1.179)$$

The beam purity can be affected by the presence of $\bar{\nu}_\mu$ and ν_e. These particles are produced in the decay reaction:

$$\mu^+ \to e^+ \, \nu_e \, \bar{\nu}_\mu. \qquad (1.180)$$

The mean muon energy in the two main decays can be estimated from Eq. (1.109) giving

$$\langle E_\mu \rangle \approx \gamma_{\pi,K} \frac{m_{\pi,K}}{2} \left[1 + \left(\frac{m_\mu}{m_{\pi,K}} \right)^2 \right] \approx \begin{cases} 16\,\text{GeV} & \pi^+ \\ 10\,\text{GeV} & K^+ \end{cases} \tag{1.181}$$

The probability of muon decay over a distance of order L as given by Eq. (1.179) is therefore:

$$\left(\frac{E_\mu}{m_\mu} \right) \frac{c\tau_\mu}{L} \approx \begin{cases} 1\% & \pi^+ \\ 0.2\% & K^+ \end{cases} \tag{1.182}$$

where we have used $c\tau_\mu = 660$ m. Electron neutrinos can be also produced directly by the helicity-suppressed decays $\pi^+ \to e^+ \, v_e$ and $K^+ \to e^+ \, v_e$, although with a probability about $10^{-4} \div 10^{-5}$ smaller than for the corresponding muonic decays. For kaons, the decay reaction $K^+ \to \pi^0 \, e^+ \, v_e$ (BR $= 5\%$) represents the largest source of background.

Bando n. 18211/2016

Problem 1.34 A proton and an electron, both of energy $E = 2$ GeV, pass through two scintillators separated by a distance $d = 30$ m. What is the time of flight between the two scintillators for the two particles?

Solution

Since the velocity is constant, the TOF is simply given by:

$$t = \frac{d}{\beta c} = \frac{d}{c} \left[1 - \left(\frac{m}{E} \right)^2 \right]^{-\frac{1}{2}} = \frac{30\,\text{m}}{3 \times 10^8\,\text{m/s}} \cdot \begin{cases} 1.00 & \text{electrons} \\ 1.13 & \text{protons} \end{cases} \tag{1.183}$$

giving approximately $0.100\ \mu$s and $0.113\ \mu$s for electrons and protons, respectively.

1.2   Center-of-Mass Dynamics and Particle Decays

The position and three-momenta of the particles $(\mathbf{r}_i, \mathbf{p}_i)$ define a set of canonical *phase-space* variables. In practice one is usually concerned with the measurement or the prediction of transition probabilities per unit volume, and between free-particle states, i.e. quantum states of definite three-momentum $\mathbf{p}_i$. Therefore, the measure in the phase-space $(\mathbf{r}_i, \mathbf{p}_i)$ comes out quite naturally in calculations. The relativistic-invariant phase-space measure for a system of n particles of prescribed energy and momentum is defined by

$$d\Phi_n(P;\, p_1, \ldots, p_n) = (2\pi)^4 \delta^4 \left(P - \sum p_i \right) \prod_{i=1}^{n} \frac{d\mathbf{p}_i}{(2\pi)^3 2E_{\mathbf{p}_i}}. \tag{1.184}$$

The factor of $2E_\mathbf{p}$ at the denominator comes from the conventional free-particle normalisation:

$$\langle \mathbf{p}|\mathbf{p}' \rangle = 2E_\mathbf{p} (2\pi)^3 \delta^3(\mathbf{p} - \mathbf{p}'), \tag{1.185}$$

which makes $d\Phi_n$ Lorentz-invariant, see Problem 1.2. The volume factor $d^3\mathbf{r}_i$ is here omitted because it usually cancels and is not much relevant in practice. The presence of a δ^4 function in Eq. (1.184) implies that not all of the $3n$ variables are independent, since the allowed states only live in a hyper-surface such that $\sum p_i - P = 0$: all other states have null measure, i.e. they cannot be "visited" by the system.

The differential decay width of a particle with total four-momentum P in the initial state $|i\rangle$ into a channel $|f\rangle$ is given by the formula:

$$d\Gamma_{i\to f}(P; p_1, \ldots, p_n) = \frac{1}{2E} |\mathcal{M}_{fi}(P, p_1, \ldots, p_n)|^2 \, d\Phi_n(P; p_1, \ldots, p_n) \tag{1.186}$$

where:

- $\mathcal{M}_{fi}$ is the relativistic matrix element of the interaction Hamiltonian between the initial and final state. When the theory is perturbative, Feynman's rules help organising the perturbative expansion of $i\mathcal{M}$ and calculate the various terms up to the desired perturbative order.
- $d\Phi_n$ is the phase-space measure of Eq. (1.184).

Due to the factor of $(2E)^{-1}$ at the denominator of Eq. (1.186), the differential width transforms like the inverse of a time. The *total width* Γ is obtained by integrating the differential width in the rest mass of the particle over the full phase-space and over all decay channels. Given a particle decay, the probability of falling into a given channel $|f\rangle$ is called *branching ratio* (BR):

$$\Gamma = \sum_f \int d\Gamma_{i\to f}, \qquad \mathrm{BR}(i \to f) = \frac{\Gamma_{i\to f}}{\Gamma} \tag{1.187}$$

Problems

Problem 1.35 Express the relativistic two-body phase-space measure $d\Phi_2$ as a function of:

1. the solid angle of particle (1) in the centre-of-mass frame, and in particular for the two cases $p_1^2 = p_2^2$ and $p_1^2 \neq 0, p_2^2 = 0$;
2. the solid angle of particle (1) in the laboratory frame;
3. the Mandelstam invariant $t = (p_c - p_a)^2$ for a scattering $a + b \to c + d$.

Solution

At the price of adding some burden to the notation, we will write $E^2_{\mathbf{p}_i, m_i} = |\mathbf{p}_i|^2 + m_i^2$, and use the notation E_i^* for the centre-of-mass energy of the i-th particle as in Eq. (1.89). In order to reduce the phase-space measure into a measure on a subset of phase-space variables, one has to integrate out the remaining ones and perform suitable changes of variables.

1. In the centre-of-mass frame, $|\mathbf{p}_1| = |\mathbf{p}_2| = |\mathbf{p}^*|$, so that Eq. (1.184) becomes:

$$d\Phi_2(P;\ p_1, p_2) = (2\pi)^{-2}\delta(\sqrt{s} - E_{\mathbf{p}_1, m_1} - E_{\mathbf{p}_2, m_2})\delta(\mathbf{p}_1 + \mathbf{p}_2)\frac{d\mathbf{p}_1}{2E_{\mathbf{p}_1, m_1}}\frac{d\mathbf{p}_2}{2E_{\mathbf{p}_2, m_2}}$$

$$\int \ldots d^3\mathbf{p}_2 = \frac{1}{(2\pi)^2 4E_{\mathbf{p}^*, m_1} E_{-\mathbf{p}^*, m_2}}\delta\left(\sqrt{s} - E_{\mathbf{p}^*, m_1} - E_{-\mathbf{p}^*, m_2}\right)d^3\mathbf{p}_1 =$$

$$= \frac{1}{16\pi^2 E_1^* E_2^*}\delta\left(\sqrt{s} - \sqrt{|\mathbf{p}^*|^2 + m_1^2} - \sqrt{|\mathbf{p}^*|^2 + m_2^2}\right)|\mathbf{p}^*|^2\,d|\mathbf{p}^*|\,d\Omega^*$$

$$\int \ldots d|\mathbf{p}^*| = \frac{1}{16\pi^2 E_1^* E_2^*}\frac{|\mathbf{p}^*|^2}{\left|\frac{|\mathbf{p}^*|}{E_1^*} + \frac{|\mathbf{p}^*|}{E_2^*}\right|}d\Omega^* = \frac{1}{16\pi^2}\frac{|\mathbf{p}^*|}{\sqrt{s}}d\Omega^*. \qquad (1.188)$$

Two special cases are worth being considered:

$$d\Phi_2 = \begin{cases} \sqrt{1 - 4\frac{m^2}{s}}\frac{d\Omega^*}{32\pi^2} = \beta^*\frac{d\Omega^*}{32\pi^2} & \text{for } m_1 = m_2 = m \\ \left(1 - \frac{m^2}{s}\right)\frac{d\Omega^*}{32\pi^2} & \text{for } m_1 = 0,\ m_2 = m \end{cases} \qquad (1.189)$$

where β^* is the velocity of both particles in the centre-of-mass frame, see Eq. (1.92).

2. This second parametrisation is useful for treating e.g. fixed-target scatterings, where one is interested in one particle only (the projectile), and wants to integrate-out the degrees of freedom of the struck target. Let the total four-momentum be $(E, \mathbf{P})$. Starting from Eq. (1.184) we get:

$$d\Phi_2(P;\ p_1, p_2) = (2\pi)^{-2}\delta(E - E_{\mathbf{p}_1, m_1} - E_{\mathbf{p}_2, m_2})\delta(\mathbf{P} - \mathbf{p}_1 - \mathbf{p}_2)\frac{d\mathbf{p}_1}{2E_{\mathbf{p}_1, m_1}}\frac{d\mathbf{p}_2}{2E_{\mathbf{p}_2, m_2}}$$

$$\int \ldots d^3\mathbf{p}_2 = \frac{1}{(2\pi)^2 4E_{\mathbf{p}_1, m_1} E_{\mathbf{P} - \mathbf{p}_1, m_2}}\delta\left(E - E_{\mathbf{p}, m_1} - E_{\mathbf{P} - \mathbf{p}_1, m_2}\right)d^3\mathbf{p}_1 =$$

$$= \frac{1}{16\pi^2 E_1 E_2}\delta\left(E - \sqrt{|\mathbf{p}_1|^2 + m_1^2} - \sqrt{|\mathbf{P} - \mathbf{p}_1|^2 + m_2^2}\right)|\mathbf{p}_1|^2\,d|\mathbf{p}_1|\,d\Omega_1$$

$$\int \ldots d|\mathbf{p}_1| = \frac{1}{16\pi^2 E_1 E_2}\frac{|\mathbf{p}_1|^2}{\left|\frac{|\mathbf{p}_1|}{E_1} + \frac{|\mathbf{p}_1| - |\mathbf{P}|\cos\theta_1}{E_2}\right|}d\Omega_1 = \frac{1}{16\pi^2}\frac{|\mathbf{p}_1|}{E - E_1\frac{|\mathbf{P}|}{|\mathbf{p}_1|}\cos\theta_1}d\Omega_1.$$

$$(1.190)$$

It remains to express $|\mathbf{p}_1|$ as a function of $\cos\theta_1$. This is best achieved by using the equation of conservation of four-momentum, and squaring it in order to get rid of the struck target kinematics:

$$p_2 = P - p_1, \qquad m_2^2 = s + m_1^2 - 2p_1 P,$$

$$E\sqrt{|\mathbf{p}_1|^2 + m_1^2} = |\mathbf{p}_1||\mathbf{P}|\cos\theta_1 + \frac{s + m_1^2 - m_2^2}{2},$$

$$0 = |\mathbf{p}_1|^2\left(1 - \frac{|\mathbf{P}|^2}{E^2}\cos^2\theta_1\right) - 2|\mathbf{p}_1|\left(\frac{|\mathbf{P}|}{E}\right)\cos\theta_1 E_1^* - |\mathbf{p}^*|^2,$$

$$|\mathbf{p}_1| = \frac{\frac{|\mathbf{P}|}{E}E_1^*\cos\theta_1 + \sqrt{|\mathbf{p}^*|^2 + \frac{|\mathbf{P}|^2}{E^2}m_1^2\cos^2\theta_1}}{1 - \frac{|\mathbf{P}|^2}{E^2}\cos^2\theta_1}, \tag{1.191}$$

where E_1^* and $|\mathbf{p}^*|$ are given by Eqs. (1.89) and (1.91), respectively.

It is interesting to study Eq. (1.190) in the non-relativistic limit $|\mathbf{p}_1| \ll E \approx m_2$. To fix the ideas, we can imagine that the projectile is a classical particle, so that $E_1 \approx m_1$, and $dE_1/d|\mathbf{p}_1| \approx 2E_1/|\mathbf{p}_1|$ while the target is a heavy nucleus at rest of mass m_2. In this case:

$$d\Phi_2(P; p_1, p_2) \approx \frac{1}{16\pi^2}\frac{|\mathbf{p}_1|}{m_2}d\Omega_1 = \left[\frac{(2\pi)^4}{2m_2(2\pi)^3}\right]\frac{1}{2m_1}\frac{m_1|\mathbf{p}_1|}{(2\pi)^3}d\Omega_1. \tag{1.192}$$

Modulo the numerical factor within parentheses, which accounts for the phase-space of the struck particle and the $(2\pi)^4$ factor in front of the delta function, and the factor of $(2m_1)^{-1}$ that comes from the relativistic normalisation of the wave function of Eq. (1.185), this expression coincides with the relativistic version of the density of states dN_{NR}/dE for a non-relativistic particle of momentum $\mathbf{p}$, since:

$$\frac{dN_{\mathrm{NR}}}{dE} = \frac{|\mathbf{p}_1|^2\,d\Omega_1}{(2\pi)^3}\frac{d|\mathbf{p}_1|}{dE_1} = \frac{|\mathbf{p}_1|^3}{2E_1(2\pi)^3}d\Omega_1 = \frac{m_1|\mathbf{p}_1|}{(2\pi)^3}d\Omega_1. \tag{1.193}$$

3. In the centre-of-mass frame, the momentum magnitude before and after the scattering is in general different if there is some inelasticity. Indicating the centre-of-mass momentum before (after) the scattering by $\mathbf{p}_{ab}^*$ ($\mathbf{p}_{cd}^*$), we have:

$$t = (p_c - p_a)^2 = m_a^2 + m_b^2 - 2|\mathbf{p}_{cd}^*||\mathbf{p}_{ab}^*|\left(\frac{1}{\beta_a^*\beta_c^*} - \cos\theta^*\right),$$

$$dt = 2|\mathbf{p}_{cd}^*||\mathbf{p}_{ab}^*|\,d\cos\theta^*. \tag{1.194}$$

We then use Eq. (1.188) and get rid of $\cos \theta^*$ in favour of t, obtaining:

$$d\Phi_2 = \frac{1}{16\pi^2} \frac{|\mathbf{p}_{cd}^*|}{\sqrt{s}} \frac{dt}{2|\mathbf{p}_{ab}^*||\mathbf{p}_{cd}^*|} d\phi^* = \frac{1}{16\pi^2} \frac{dt}{2|\mathbf{p}_{ab}^*|\sqrt{s}} d\phi^*. \qquad (1.195)$$

Notice that ϕ^* in the formula above stands for the azimuthal angle around $\mathbf{p}_a^*$ in the ab rest frame, so this formula remains valid (modulo $\phi^* \to \phi$) in any reference frame where $\mathbf{p}_a$ and $\mathbf{p}_a$ are collinear.

Suggested Readings

The PDG review on kinematics, Chap. 47 of Ref. [4], offers a complete summary of the most important results for two-body phase-space. Beware that the definition of $d\Phi_n$ may differ by a factor of $(2\pi)^4$ due to the normalisation of the four-momentum conserving δ function.

Problem 1.36 Explicate the three-body phase-space measure $d\Phi_3$ in the centre-of-mass frame by using the known expression of $d\Phi_2$.

Discussion

The phase-space measure $d\Phi_n$ for an arbitrary particle multiplicity n can be constructed with the recursive formula:

$$d\Phi_n(P;\ p_1, \ldots, p_n) = d\Phi_j(q;\ p_1, \ldots, p_j)\, d\Phi_{n-j+1}(P;\ q, p_{j+1}, \ldots, p_n)\, \frac{dq^2}{2\pi} \tag{1.196}$$

In order to prove it, we introduce twice the identity into Eq. (1.184) in a suitable form, which, for sake of clarity, will be introduced between square brackets.

$$(2\pi)^4 \delta(P - \sum_{i=1}^{n} p_i) \prod_{i=1}^{n} \frac{d\mathbf{p}_i}{(2\pi)^3 2E_{\mathbf{p}_i}} =$$

$$= \left[(2\pi)^4 \delta(q - \sum_{i=1}^{j} p_i) \frac{d^4 q}{(2\pi)^4} \right] \times (2\pi)^4 \delta(P - q - \sum_{i=j+1}^{n} p_i) \prod_{i=1}^{n} \frac{d\mathbf{p}_i}{(2\pi)^3 2E_{\mathbf{p}_i}} =$$

$$= d\Phi_j(q;\ p_1, \ldots, p_j) \times (2\pi)^4 \delta(P - q - \sum_{i=j+1}^{n} p_i) \times$$

$$\times \prod_{i=j+1}^{n} \frac{d\mathbf{p}_i}{(2\pi)^3 2E_{\mathbf{p}_i}} \frac{d^4 q}{(2\pi)^4} \left[\delta(q^2 - (p_1 + \ldots + p_j)^2)\, dq^2 \right] =$$

$$= d\Phi_j(q; \ p_1, \ldots, p_j)(2\pi)^4 \delta(P - q - \sum_{i=j+1}^{n} p_i) \prod_{i=j+1}^{n} \frac{d\mathbf{p}_i}{(2\pi)^3 2E_{\mathbf{p}_i}} \frac{d^3\mathbf{q}}{(2\pi)^3 2E_{\mathbf{q}}} \frac{dq^2}{2\pi} =$$

$$= d\Phi_j(q; \ p_1, \ldots, p_j) \, d\Phi_j(P; \ q, p_{j+1}, \ldots, p_n) \frac{dq^2}{2\pi} \tag{1.197}$$

where we have made use of Eq. (1.16) to transform $\delta(q^2 - \mu^2) \, d^4q$ into $d^3\mathbf{q}/2E_{\mathbf{q}}$.

Solution

We can apply Eq. (1.196) to the two-body phase-space (1.188) and obtain:

$$d\Phi_3(P; p_1, p_2, p_3) = \frac{1}{16\pi^2} \frac{|\mathbf{p}_3|}{\sqrt{s}} \, d\Omega_3 \cdot \frac{1}{16\pi^2} \frac{|\mathbf{p}_1^*|}{m_{12}} \, d\Omega_1^* \cdot \frac{dm_{12}^2}{2\pi} =$$

$$= \frac{1}{(2\pi)^5} \frac{1}{8\sqrt{s}} |\mathbf{p}_1^*||\mathbf{p}_3| \, dm_{12} \, d\Omega_1^* \, d\Omega_3 \tag{1.198}$$

To conclude, one needs to express $|\mathbf{p}_1^*|$ and $|\mathbf{p}_3|$ as a function of m_{12}. This can be done by means of Eq. (1.91), with the replacements $\sqrt{s} \to m_{12}$ for the former, and $m_1 \to m_{12}, m_2 \to m_3$ for the latter.

Problem 1.37 Determine the three-body phase-space measure $d\Phi_3$ as a function of the centre-of-mass energies of two of the particles, after integrating over all angles.

Solution

Let's start from Eq. (1.184) and specialise the four-momenta of the three particles in their centre-of-mass frame:

$$d\Phi_3(P; \ p_1, p_2, p_3) = (2\pi)^{-5} \delta(\sqrt{s} - E_{\mathbf{p}_1, m_1}^* - E_{\mathbf{p}_2, m_2}^* - E_{\mathbf{p}_3, m_3}^*) \delta(\mathbf{p}_1^* + \mathbf{p}_2^* + \mathbf{p}_3^*) \times$$

$$\times \frac{d^3\mathbf{p}_1^*}{2E_{\mathbf{p}_1, m_1}^*} \frac{d^3\mathbf{p}_2^*}{2E_{\mathbf{p}_2, m_2}^*} \frac{d^3\mathbf{p}_3^*}{2E_{\mathbf{p}_3, m_3}^*}$$

$$\int \ldots d^3\mathbf{p}_3^* = \delta\left(\sqrt{s} - \sqrt{|\mathbf{p}_1^*|^2 + m_1^2} - \sqrt{|\mathbf{p}_2^*|^2 + m_2^2} - \sqrt{|\mathbf{p}_1^* + \mathbf{p}_2^*|^2 + m_3^2}\right) \times$$

$$\times \frac{(2\pi)^{-5}}{8E_1^* E_2^* E_3^*} |\mathbf{p}_1^*|^2 \, d|\mathbf{p}_1^*| \, d\Omega_1^* \, |\mathbf{p}_2^*|^2 \, d|\mathbf{p}_2^*| \, d\Omega_2^*$$

$$\int \ldots d\Omega_1^* \, d\phi_2^* = \delta\left(\sqrt{s} - \sqrt{|\mathbf{p}_1^*|^2 + m_1^2} - \sqrt{|\mathbf{p}_2^*|^2 + m_2^2} - \sqrt{|\mathbf{p}_1^* + \mathbf{p}_2^*|^2 + m_3^2}\right) \times$$

$$\times \frac{2(2\pi)^{-3} |\mathbf{p}_1^*|^2 |\mathbf{p}_2^*|^2}{8E_1^* E_2^* E_3^*} \, d|\mathbf{p}_1^*| \, d|\mathbf{p}_2^*| \, d\cos\theta_{12}^*$$

$$\int \ldots d\cos\theta_{12}^* = \frac{2(2\pi)^{-3}|\mathbf{p}_1^*|^2|\mathbf{p}_2^*|^2}{8E_1^*E_2^*E_3^*} \frac{d|\mathbf{p}_1^*|\,d|\mathbf{p}_2^*|}{\left|\frac{|\mathbf{p}_1^*||\mathbf{p}_2^*|}{E_3^*}\right|} = \frac{1}{4(2\pi)^3} \frac{|\mathbf{p}_1^*||\mathbf{p}_2^*|}{E_1^*E_2^*} d|\mathbf{p}_1^*|\,d|\mathbf{p}_2^*|$$

$$= \frac{dE_1^*\,dE_2^*}{4(2\pi)^3}, \tag{1.199}$$

where θ_{12} is the angle between $\mathbf{p}_2^*$ and $\mathbf{p}_1^*$, and where we have used the fact that $dE_\mathbf{p}/d|\mathbf{p}| = |\mathbf{p}|/E_\mathbf{p} = \beta_\mathbf{p}$. Finally, one notices that $P = p_1 + p_2 + p_3$ implies that:

$$m_{ij}^2 = (P - p_k)^2 = s - 2E_k^*\sqrt{s} + m_k^2 \Rightarrow dm_{ij}^2 = -2\sqrt{s}\,dE_k^* \quad \text{for} \quad i \neq j \neq k. \tag{1.200}$$

Hence, the phase-space volume $\langle d\Phi_3 \rangle$ expressed in terms of the m_{ij}^2 becomes:

$$\langle d\Phi_3 \rangle = \frac{dm_{12}^2\,dm_{23}^2}{16(2\pi)^3 s} = \frac{dm_{13}^2\,dm_{23}^2}{16(2\pi)^3 s} = \frac{dm_{13}^2\,dm_{12}^2}{16(2\pi)^3 s}. \tag{1.201}$$

Equation (1.201) shows that the three-body phase-space is uniform over the invariant masses of any two-pairs of particles.

The border of the domain in the (m_{12}^2, m_{23}^2) space is, in general, a non-trivial curve. First of all, one notices that:

$$(m_i + m_j)^2 \leq m_{ij}^2 \leq (\sqrt{s} - m_k)^2, \tag{1.202}$$

where the boundaries correspond to particles (ij) being at rest in their centre-of-mass frame, and particle k be at rest in the three-body centre-of-mass frame, respectively. Without loss of generality, we can consider the first of Eq. (1.201). Equation (1.202) implies that the domain is contained within the rectangle $[(m_1 + m_2)^2, (\sqrt{s} - m_3)^2] \times [(m_2 + m_3)^2, (\sqrt{s} - m_1)^2]$, whose sides are also tangent to the domain boundary. At a given value of m_{12}, the range of m_{23} can be determined by requiring the momentum of particle (3) in the rest frame of (12) to be either parallel or antiparallel to the momentum of particle (2), all other cases giving values of m_{23} that are, respectively, larger or smaller. In this frame, the energy of particles (2) is given by Eq. (1.89), while the energy of particle (3) can be obtained by applying a boost from the three-body rest frame to the rest frame of (12):

$$E_2^{(12)} = \frac{m_{12}^2 + m_2^2 - m_1^2}{2m_{12}} \tag{1.203}$$

$$E_3^{(12)} = \gamma_{\mathbf{p}_{(12)}^*}\left(E_3^* - \beta_{\mathbf{p}_{(12)}^*}(-|\mathbf{p}_3^*|)\right) =$$

$$= \frac{s + m_{12}^2 - m_3^2}{2\sqrt{s}\,m_{12}}\left(\frac{s - m_{12}^2 + m_3^2}{2\sqrt{s}} + \frac{|\mathbf{p}_3^*|^2\,2\sqrt{s}}{s + m_{12}^2 - m_3^2}\right) =$$

$$= \frac{s + m_{12}^2 - m_3^2}{2\sqrt{s}\,m_{12}} \left(\frac{s - m_{12}^2 + m_3^2}{2\sqrt{s}} + \frac{\left[s - (m_{12} - m_3)^2 \right] \left[s - (m_{12} + m_3)^2 \right]}{2\sqrt{s}(s + m_{12}^2 - m_3^2)} \right) =$$

$$= \frac{1}{4s\,m_{12}} \left(2s^2 - 2s(m_{12}^2 + m_3^2) \right) = \frac{s - m_{12}^2 - m_3^2}{2m_{12}}. \tag{1.204}$$

The last result could have been obtained way more easily by noticing that, in the rest frame of (12), the kinematics looks like the one of a fixed-target experiment, for which Eq. (1.87) gives $m_{12}^2 + m_3^2 + 2E_3^{(12)} m_{12} = s$. Hence:

$$m_{23}^2(m_{12}) \gtrless \left(\frac{s - m_1^2 + m_2^2 - m_3^2}{2m_{12}} \right)^2 -$$

$$- \left(\frac{\sqrt{(m_{12}^2 + m_2^2 - m_1^2)^2 + 4m_2^2 m_{12}^2}}{2m_{12}} \pm \frac{\sqrt{(s - m_{12}^2 - m_3^2)^2 + 4m_3^2 m_{12}^2}}{2m_{12}} \right)^2,$$

$$\tag{1.205}$$

where the two signs correspond to the lower and upper bound, respectively.

Discussion

If the matrix element squared is uniform over m_{ij}^2, the differential decay probabilities $d\Gamma/dm_{12}^2 dm_{23}^2$ is uniform in the domain of (m_{12}^2, m_{23}^2). Such a distribution for a three-body decay is called *Dalitz plot*. Non-uniformities of the amplitude squared over either of the m_{ij}^2 variables, would lead to a non-uniform distribution of the experimental points. For example, this is the case if the decay can be mediated by an intermediate resonance: indeed, if a narrow resonance of mass m_0 and width Γ_0 is present in e.g. the (12)-channel, then:

$$|\mathcal{M}|^2 \propto \frac{1}{(m_{12}^2 - m_0^2)^2 + m_0^2 \, \Gamma_0^2} \approx \frac{\pi}{m_0 \, \Gamma_0} \delta(m_{12}^2 - m_0^2), \tag{1.206}$$

and the corresponding Dalitz plot will feature a cluster of experimental points around the line $m_{12}^2 = m_0^2$.

Problem 1.38 Prove that in the centre-of-mass of a three-body decay, the maximum value of the three-momentum is taken by the particle with the largest mass.

Solution

The centre-of-mass momentum of particle k is given by Eq. (1.91):

$$|\mathbf{p}_k^*|^2 = \frac{\left(s - (m_{ij} + m_k)^2\right)\left(s - (m_{ij} - m_k)^2\right)}{4s}, \tag{1.207}$$

where m_{ij} is the invariant mass of the (ij) pair and s is the mass squared of the mother particle. Modulo some positive constants, the derivative of $|\mathbf{p}_k^*|^2$ with respect to m_{12} is:

$$\frac{\partial |\mathbf{p}_k^*|^2}{\partial m_{ij}} \propto -m_{ij}\left(s - (m_{ij} - m_k)^2\right) - m_{ij}\left(s - (m_{ij} + m_k)^2\right) -$$
$$- m_k\left(s - (m_{ij} - m_k)^2\right) + m_k\left(s - (m_{ij} + m_k)^2\right). \tag{1.208}$$

The last row of Eq. (1.208) is always negative since $(m_{ij} - m_k)^2 \leq (m_{ij} + m_k)^2$. Hence, $|\mathbf{p}_k^*|$ is a decreasing function of m_{ij}, so that the maximum value corresponds to the minimum value of m_{ij}, i.e. $(m_i + m_j)$. Therefore:

$$|\mathbf{p}_k^*|_{\text{max}}^2 = \frac{\left(s - (m_i + m_j + m_k)^2\right)\left(s - (m_i + m_j - m_k)^2\right)}{4s}, \tag{1.209}$$

and for any pair of indices i, k we have:

$$|\mathbf{p}_k^*|_{\text{max}} \geq |\mathbf{p}_i^*|_{\text{max}} \quad \Leftrightarrow \quad s - (m_i + m_j - m_k)^2 \geq s - (m_j + m_k - m_i)^2$$
$$m_i + m_j - m_k \leq m_j + m_k - m_i,$$
$$2(m_i - m_k) \leq 0, \qquad m_k \geq m_i. \tag{1.210}$$

Therefore, the particle that can take the maximum centre-of-mass momentum is also the one with the largest mass. Experimentally, this value corresponds to the end-point of the $|\mathbf{p}_i^*|$ distribution over several decays.

Problem 1.39 Consider the β-decay $A \to A' e^- \bar{\nu}_e$, where A and A' are two heavy nuclei and $m_A - m_{A'} - m_e = Q$ is large compared to m_e. Assume that the non-relativistic matrix element squared $|\mathcal{M}_{\text{NR}}|^2$ for this transition is approximately constant. Show that the decay width is proportional to Q^5.

Solution

Let's denote the three momenta of the decay particles by $\mathbf{p}_p$, $\mathbf{p}_e$ and $\mathbf{p}_\nu$. Since the non-relativistic matrix element is a pure constant, it is more convenient to write Eq. (1.186) using the non-relativistic normalisation

$$\langle \mathbf{p} | \mathbf{p}' \rangle = (2\pi)^3 \delta^3 (\mathbf{p} - \mathbf{p}'). \tag{1.211}$$

With this convention, the differential decay width becomes:

$$
\begin{aligned}
d\Gamma &= |\mathscr{M}_{\mathrm{NR}}|^2 (2\pi)^4 \delta^3 (\mathbf{p}_{A'} + \mathbf{p}_e + \mathbf{p}_\nu) \delta(m_A - E_{A'} - E_e - E_\nu) \frac{d^3 \mathbf{p}_{A'}}{(2\pi)^3} \frac{d^3 \mathbf{p}_e}{(2\pi)^3} \frac{d^3 \mathbf{p}_\nu}{(2\pi)^3} = \\
&= \frac{|\mathscr{M}_{\mathrm{NR}}|^2}{(2\pi)^5} \delta \left(m_A - \sqrt{|\mathbf{p}_e + \mathbf{p}_\nu|^2 + m_{A'}^2} - \sqrt{|\mathbf{p}_e|^2 + m_e^2} - |\mathbf{p}_\nu| \right) \times \\
&\quad \times |\mathbf{p}_e|^2 \, d|\mathbf{p}_e| \, d\Omega_e \, |\mathbf{p}_\nu|^2 \, d|\mathbf{p}_\nu| \, d\Omega_\nu = \\
&= \frac{4|\mathscr{M}_{\mathrm{NR}}|^2}{(2\pi)^3} |\mathbf{p}_e|^2 \left(\underbrace{m_A - m_{A'} - m_e}_{Q} - T_e \right)^2 d|\mathbf{p}_e| \tag{1.212}
\end{aligned}
$$

Let's see in more details the approximations that went into Eq. (1.212). Firstly, it was assumed that the electron energy is much smaller than the proton mass, which is indeed a good approximation since $T_e^{\max} = Q$; this approximation allows to neglect the recoil energy taken by the nucleus. The other assumption is that the neutrino mass is negligible, which is perfectly fine here. Equation (1.212) can be also written as:

$$\frac{1}{|\mathbf{p}_e|} \sqrt{\frac{d\Gamma}{d|\mathbf{p}_e|}} = Q - T_e. \tag{1.213}$$

The left-hand side of Eq. (1.213) is the so-called *Kurie plot*, which is a linear function of the electron energy with an end-point related to the Q-value of the reaction. Coming back to Eq. (1.212), we can integrate over the electron momenta. Since the intagrand grows like $|\mathbf{p}_e|^2$, the integral is dominated by the high-energy part of the electron spectrum. If we then make the approximation $m_e = 0$, the integration is trivial, giving:

$$
\begin{aligned}
\Gamma &\approx \frac{|\mathscr{M}_{\mathrm{NR}}|^2}{2\pi^3} \int_0^Q d|\mathbf{p}_e| \, |\mathbf{p}_e|^2 \left(Q^2 - 2Q|\mathbf{p}_e| + |\mathbf{p}_e|^2 \right) = \frac{|\mathscr{M}_{\mathrm{NR}}|^2}{2\pi^3} Q^5 \left(\frac{1}{3} - \frac{1}{2} + \frac{1}{5} \right) = \\
&= \frac{|\mathscr{M}_{\mathrm{NR}}|^2 Q^5}{60\pi^3}. \tag{1.214}
\end{aligned}
$$

Equation (1.214) shows that the total decay width is proportional to the fifth power of the Q-value, a property know as *Sargent rule*.

Discussion

One may wonder whether this result could have been obtained by plugging in the phase-space measure of Eq. (1.199), corrected for the non-relativistic normalisation (1.211). Indeed, one can notice that the two measures are different, since

Eq. (1.199) is linear in the two energies (after introducing the non-relativistic normalisation), while Eq. (1.212) is quadratic in $|\mathbf{p}|$. Indeed, looking back at Eq. (1.199), one can see that the integration over the polar angle between the two momenta θ_{12}^* lied on the assumption that there exists only one such angle so that the energy conservation is satisfied for a given value of $|\mathbf{p}_1^*|$ and $|\mathbf{p}_2^*|$. However, if the third particle is much heavier than the maximum energy available to the two lighter particles, so that it can be considered at rest for what concerns the energy balance, the equation of conservation of energy becomes nearly independent of θ_{12}^*, and the integration of the last delta function is not valid anymore. Therefore, if $m_3 \gg m_{1,2}$, and the matrix element is constant, then the two light particles momenta are uncorrelated in direction and fully anti-correlated in modulus, since $E_1 + E_2 = m - E_3 \approx m - m_3 = Q$: the phase-space measure is therefore given by the product of the two particle phase-spaces, which are just the number of states inside a spheric layer of radius $|\mathbf{p}|$.

Suggested Readings

The reader is addressed to Sect. 7.3 of Ref. [16] for another derivation of the Kurie plot of β-decays.

Problem 1.40 A heavy nucleus of mass M^*, initially at rest in the laboratory frame, decays to the ground state of mass M by emitting a photon. Determine the photon energy E_γ. Discuss a possible technique to suppress the energy shift due to the nuclear recoil.

Solution

Let's denote the photon energy in the laboratory by E_γ and the energy gap between the two nuclear levels by ε, such that $M^* = M + \varepsilon$. In the assumption $\varepsilon/M \ll 1$, which is generally the case since the nuclear transitions produce photons with energy of order $0.1 \div 1$ MeV, whereas nuclear masses are at least three orders of magnitude larger, we have:

$$M + \varepsilon = E_\gamma + \sqrt{M^2 + E_\gamma^2} \approx E_\gamma + M + \frac{E_\gamma^2}{2M}, \qquad (1.215)$$

$$E_\gamma = \varepsilon - \frac{E_\gamma^2}{2M} \approx \varepsilon - \frac{\varepsilon^2}{2M} = \varepsilon\left(1 - \frac{\varepsilon}{2M}\right).$$

Hence, the photon energy is smaller than the energy gap by a fraction $\varepsilon/2M \ll 1$ due to the nuclear recoil.

Discussion

If the width of the excited level is much smaller than the recoil energy taken by the nucleus, the emitted photon won't be anymore at the resonance. However, when the nucleus is bound inside a crystal, the crystal lattice behaves as a collective object, resulting in a much larger effective mass M and thus no recoil energy will be taken away: the emitted photon is then capable of inducing the inverse reaction $\gamma + M \rightarrow M^*$ with an enhanced cross section. This behaviour is called *Mössbauer effect* and is vastly employed in spectroscopy.

Problem 1.41 Consider the decay of a particle of mass M into three particles of identical mass m, assumed to be non-relativistic in the rest frame of the decaying particle. The phase-space for this decay can be represented inside an equilateral triangle centred around the origin of a cartesian coordinate system (x, y), such that the kinetic energy of each particle in units of the Q-value, $|\mathbf{p}_i|^2/2mQ$, is equal to the distances of the point from the sides. Show that the allowed kinematic configurations live inside a circle of equation $x^2 + y^2 - \frac{1}{9} \leq 0$.

Solution

Consider an equilateral triangle of unit height centred around the origin of a cartesian coordinate system (x, y). Let the vertices of the triangle be located at

$$\mathbf{r}_1 = \left(0, +\frac{2}{3}\right), \quad \mathbf{r}_2 = \left(-\frac{\sqrt{3}}{3}, -\frac{1}{3}\right), \quad \mathbf{r}_3 = \left(+\frac{\sqrt{3}}{3}, -\frac{1}{3}\right) \quad (1.216)$$

as shown in Fig. 1.13. The sides of the triangle are then defined by the three straight line equations:

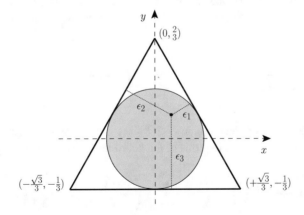

Fig. 1.13 The phase-space for a three-body decay into non-relativistic particles of energy ε_i and same mass m, represented as an equilateral triangle centred around the origin of a cartesian coordinate system (x, y), such that the kinetic energy of each particle in units of the Q-value, $|\mathbf{p}_i|^2/2mQ$, is equal to the distances of the point from the sides. Only the shaded circle is however compatible with the three-momentum conservation

$$\begin{cases} +\sqrt{3}x + y - \frac{2}{3} = 0 \\ -\sqrt{3}x + y - \frac{2}{3} = 0 \\ y + \frac{1}{3} = 0 \end{cases} \qquad (1.217)$$

The distances of a given point (x_0, y_0) inside the triangle from the three sides are therefore given by:

$$\begin{cases} \varepsilon_1 = \frac{|+\sqrt{3}x_0 + y_0 - \frac{2}{3}|}{2} = -\frac{\sqrt{3}}{2}x_0 - \frac{1}{2}y_0 + \frac{1}{3} \\ \varepsilon_2 = \frac{|-\sqrt{3}x_0 + y_0 + \frac{2}{3}|}{2} = \frac{\sqrt{3}}{2}x_0 - \frac{1}{2}y_0 + \frac{1}{3} \qquad \Rightarrow \quad \varepsilon_1 + \varepsilon_2 + \varepsilon_3 = 1, \quad (1.218) \\ \varepsilon_3 = y_0 + \frac{1}{3} \end{cases}$$

which shows that an equilateral triangle of unit height is the geometric locus of points for which the sum of the distances from the three sides is equal to unity.

Let's denote the adimensional quantity $|\mathbf{p}_i|^2/2m\,Q \equiv \varepsilon_i$, $i = 1, 2, 3$, so that:

$$\frac{|\mathbf{p}_1|^2}{2m} + \frac{|\mathbf{p}_2|^2}{2m} + \frac{|\mathbf{p}_3|^2}{2m} = M - 3m = Q \quad \Leftrightarrow \quad \varepsilon_1 + \varepsilon_2 + \varepsilon_3 = 1 \qquad (1.219)$$

Therefore, the phase-space points live inside an equilateral triangle such that the distances of its internal points from the sides are given by ε_i. However, conservation of momentum implies that $\mathbf{p}_1 + \mathbf{p}_2 + \mathbf{p}_3 = 0$, so that Eq. (1.219) can be also written as:

$$2\varepsilon_1 + 2\varepsilon_2 + 2\sqrt{\varepsilon_1 \varepsilon_2}\cos\theta_{12} = 1, \qquad (1.220)$$

where θ_{12} is the angle between two of the particles. This condition, being more restrictive than Eq. (1.219), places additional constraints on the allowed phase-space points. In particular:

$$\cos^2\theta_{12} = \left(\frac{1 - 2\varepsilon_1 - 2\varepsilon_2}{2\sqrt{\varepsilon_1 \varepsilon_2}}\right)^2 \le 1 \quad \Rightarrow \quad \varepsilon_1^2 + \varepsilon_2^2 + \varepsilon_1\varepsilon_2 - \varepsilon_1 - \varepsilon_2 + \frac{1}{4} \le 0,$$
$$(1.221)$$

By inserting any pair of equations in (1.218), say the first and the last, we can now express the inequality above in terms of the coordinates (x, y), namely:

$$\left(-\frac{\sqrt{3}}{2}x - \frac{1}{2}y + \frac{1}{3}\right)^2 + \left(y + \frac{1}{3}\right)^2 + \left(-\frac{\sqrt{3}}{2}x - \frac{1}{2}y + \frac{1}{3}\right)\left(y + \frac{1}{3}\right) -$$
$$-\left(-\frac{\sqrt{3}}{2}x - \frac{1}{2}y + \frac{1}{3}\right) - \left(y + \frac{1}{3}\right) + \frac{1}{4} \le 0 \qquad (1.222)$$

After some straightforward algebra, a large number of cancellations occurs and one gets:

$$x^2 + y^2 - \frac{1}{9} \leq 0, \tag{1.223}$$

which describes a circle of radius $1/3$ centred around the origin and tangent to the three sides of the triangle in their middle points. This is shown by the solid circle in Fig. 1.13.

Discussion

As discussed in Problem 1.37, under some suitable assumptions on the matrix-element, in a three-body decay the phase-space points are evenly distributed in the (E_1, E_2) plane, where E_1 and E_2 are the energies of any two particles. The (x, y) variables introduced in this exercise transform a tiny square of area $dE_1 \, dE_2$ into a parallelogram *uniformly* in E_1 and E_2, so that the (x, y) points will be still evenly distributed. The representation of decay events in this plane has been very popular in the early years of particle physics as a tool to infer the properties of the interaction responsible for the decay. A famous example is provided by the study of the decays of a long-lived particle, originally called τ and later-on identified as the charged kaon, into three pions, namely $K^+ \rightarrow \pi^+ \pi^+ \pi^-$.

Suggested Readings

More details on the so-called θ-τ puzzle for the three-charged pion decay of charged kaons, can be found in Chap. 3 of Ref. [12]. In particular, one can find there interesting considerations on how different matrix elements would affect the distribution of events in the (x, y) plane.

Problem 1.42 The *thrust* is an event-shape variable which, in the centre-of-mass frame of a n-particle event is defined as

$$T(\mathbf{p}_1, \ldots, \mathbf{p}_n) = \max_{\mathbf{n}} \frac{\sum_i |\mathbf{p}_i \cdot \mathbf{n}|}{\sum_i |\mathbf{p}_i|}. \tag{1.224}$$

with $|\mathbf{n}| = 1$. Consider a three-particle decay and assume the three particles to be massless. Prove that $T = \max\{x_1, x_2, x_3\}$, where $x_i = 2|\mathbf{p}_i|/\sqrt{s}$ and $\sqrt{s}$ is the centre-of-mass energy.

Solution

Let's denote the three-momenta of the involved particles by $\mathbf{p}_i$, such that $\sum \mathbf{p}_i = 0$. Let $\hat{\mathbf{n}}$ be the direction that maximises T. By definition, we must have:

$$0 = \delta T(\hat{\mathbf{n}}) = \frac{\left[\theta_1\,\mathbf{p}_1 + \theta_2\,\mathbf{p}_2 - \theta_3\,(\mathbf{p}_1 + \mathbf{p}_2)\right]\cdot\delta\hat{\mathbf{n}}}{\sqrt{s}}, \tag{1.225}$$

where $\theta_i = \pm 1$ are signs such that the individual scalar products are all positive. For any choice of the signs θ_1, the vector inside the square brackets is proportional to one of the particle momentum. Since $\hat{\mathbf{n}}\cdot\delta\hat{\mathbf{n}} = 0$, it also follows that $\hat{\mathbf{n}}$ has to be parallel to either $\mathbf{p}_1$, or $\mathbf{p}_2$, or $\mathbf{p}_3$, so that:

$$T = \max\{T(\mathbf{e}_1), T(\mathbf{e}_2), T(\mathbf{e}_3)\}, \tag{1.226}$$

where $\mathbf{e}_i$ are the directions of the three momenta. For example, for $\hat{\mathbf{n}} = \mathbf{e}_1$, one has:

$$T(\mathbf{e}_1) = \frac{(|\mathbf{p}_1| + \mathbf{e}_1\cdot\mathbf{p}_2 + ||\mathbf{p}_1| + \mathbf{e}_1\cdot\mathbf{p}_2|)}{\sqrt{s}} =$$

$$= \begin{cases} +\frac{2|\mathbf{p}_1|}{\sqrt{s}} = x_1 & \text{if } \mathbf{p}_1\cdot\mathbf{p}_2 < 0, \ \mathbf{p}_1\cdot\mathbf{p}_3 < 0 \\ -\frac{2\mathbf{e}_1\cdot\mathbf{p}_2}{\sqrt{s}} \le x_2 & \text{if } \mathbf{p}_1\cdot\mathbf{p}_2 < 0, \ \mathbf{p}_1\cdot\mathbf{p}_3 > 0 \\ -\frac{2\mathbf{e}_1\cdot\mathbf{p}_3}{\sqrt{s}} \le x_3 & \text{if } \mathbf{p}_1\cdot\mathbf{p}_2 > 0 \end{cases} \tag{1.227}$$

Similarly, we find:

$$T(\mathbf{e}_2) \begin{cases} \le x_1 & \text{if } \mathbf{p}_1\cdot\mathbf{p}_2 < 0, \ \mathbf{p}_2\cdot\mathbf{p}_3 > 0 \\ = x_2 & \text{if } \mathbf{p}_1\cdot\mathbf{p}_2 < 0, \ \mathbf{p}_2\cdot\mathbf{p}_3 < 0 \\ \le x_3 & \text{if } \mathbf{p}_1\cdot\mathbf{p}_2 > 0 \end{cases} \tag{1.228}$$

$$T(\mathbf{e}_3) \begin{cases} \le x_1 & \text{if } \mathbf{p}_3\cdot\mathbf{p}_1 < 0, \ \mathbf{p}_3\cdot\mathbf{p}_2 > 0 \\ \le x_2 & \text{if } \mathbf{p}_3\cdot\mathbf{p}_1 > 0 \\ = x_3 & \text{if } \mathbf{p}_3\cdot\mathbf{p}_1 < 0, \ \mathbf{p}_3\cdot\mathbf{p}_2 < 0 \end{cases} \tag{1.229}$$

Then, we notice that the angles $\alpha_{12}, \alpha_{23}, \alpha_{13}$ between three vectors $\mathbf{p}_1, \mathbf{p}_2, \mathbf{p}_3$ such that $\sum_i \mathbf{p}_i = 0$ satisfy $\alpha_{12} + \alpha_{23} + \alpha_{13} = 2\pi$ and $\alpha_{ij} < \pi$; therefore, if $\alpha_{ij} < \pi/2$, then the other two opening angles must be larger than $\pi/2$. In this latter case, it follows that

$$|\mathbf{p}_k| = \sqrt{|\mathbf{p}_i|^2 + |\mathbf{p}_j|^2 + 2\,\mathbf{p}_j\cdot\mathbf{p}_j} \ge \max\{|\mathbf{p}_i|, |\mathbf{p}_j|\}, \tag{1.230}$$

i.e. the vector that "recoils" against the two vectors with opening angle smaller than $\pi/2$ has also the largest modulus among the three. If we now consider all possible cases, we obtain:

$$
\begin{cases}
T(\mathbf{e}_1) \leq x_3, \; T(\mathbf{e}_2) \leq x_3, \; T(\mathbf{e}_3) = x_3 > x_1, x_2 & \text{if } \mathbf{p}_1 \cdot \mathbf{p}_2 > 0 \\
T(\mathbf{e}_1) \leq x_2, \; T(\mathbf{e}_2) = x_2 > x_1, x_3, \; T(\mathbf{e}_3) \leq x_2 & \text{if } \mathbf{p}_1 \cdot \mathbf{p}_3 > 0 \\
T(\mathbf{e}_1) = x_1 > x_2, x_3, \; T(\mathbf{e}_2) \leq x_1, \; T(\mathbf{e}_3) \leq x_1 & \text{if } \mathbf{p}_2 \cdot \mathbf{p}_3 > 0 \\
T(\mathbf{e}_1) = x_1, \; T(\mathbf{e}_2) = x_2, \; T(\mathbf{e}_3) = x_3 & \text{if } \mathbf{p}_i \cdot \mathbf{p}_j < 0 \;\; \forall i, j
\end{cases}
\tag{1.231}
$$

For all the above cases (which exhaust all the possibilities), it always holds that:

$$
T = \max\{x_1, x_2, x_3\}.
\tag{1.232}
$$

Suggested Readings

More informations on the thrust can be found in Ref. [17].

Problem 1.43 Event-shape variables are widely used to describe the structure of the hadronic events and to test perturbative chromo-dynamics (pQCD). Consider the three event-shape variables:

$$
T = \max_{\mathbf{n}} \frac{\sum_i |\mathbf{p}_i \cdot \mathbf{n}|}{\sum_i |\mathbf{p}_i|}
\tag{1.233}
$$

$$
C = 3(\lambda_1\lambda_2 + \lambda_1\lambda_3 + \lambda_2\lambda_3), \quad \lambda_k \text{ eigenvalues of } \Theta = \frac{\sum_i \frac{\mathbf{p}_i \mathbf{p}_i^{\mathrm{T}}}{|\mathbf{p}_i|}}{\sum_i |\mathbf{p}_i|}
\tag{1.234}
$$

$$
S = \frac{3}{2} \min_{\mathbf{n}} \frac{\sum_i (\mathbf{p}_i \times \mathbf{n})^2}{\sum_i \mathbf{p}_i^2}.
\tag{1.235}
$$

1. Show that T and the C are infrared-safe observables.
2. Show that the S is not.

Discussion

An observable $I_n(p_1, \ldots, p_n)$, which is a function of the four-momenta p_i of an arbitrary number of partons, is said to be *infrared-safe* (IR-safe) if:

$$
\begin{aligned}
I_{n+1}(p_1, \ldots, p_n, 0) &= I_n(p_1, \ldots, p_n) \\
I_{n+1}(p_1, \ldots, \lambda p_n, (1-\lambda)p_n) &= I(p_1, \ldots, p_n)
\end{aligned}
\tag{1.236}
$$

for any particle n. This is equivalent to requiring that the observable does not distinguish between configurations related to each other by a soft gluon emission or by the collinear splitting of a parton. For QCD, the *Kinoshita–Lee–Nauenberg theorem* ensures that inclusive-enough observables are IR-safe, see e.g. Sect. 3.5 of Ref. [18].

Solution

In order to prove that T is IR-safe, we need to verify the conditions (1.236) for an arbitrary set of four-momenta:

$$T_{n+1}(p_1, \ldots, p_n, 0) = \max_{\mathbf{n}} \frac{\sum_i^n |\mathbf{p}_i \cdot \mathbf{n}| + |0 \cdot \mathbf{n}|}{\sum_{i=1}^n |\mathbf{p}_i| + 0} = T_n(p_1, \ldots, p_n)$$

$$T_{n+1}(p_1, \ldots, \lambda p_n, (1-\lambda)p_n) = \max_{\mathbf{n}} \frac{\sum_i^{n-1} |\mathbf{p}_i \cdot \mathbf{n}| + \lambda |\mathbf{p}_n \cdot \mathbf{n}| + (1-\lambda)|\mathbf{p}_n \cdot \mathbf{n}|}{\sum_{i=1}^{n-1} |\mathbf{p}_i| + \lambda |\mathbf{p}_n| + (1-\lambda)|\mathbf{p}_{n+1}|} =$$

$$= T_n(p_1, \ldots, p_n) \tag{1.237}$$

The variable C is proportional to the second invariant of the symmetric tensor Θ. Indeed:

$$0 = \det(\Theta - \lambda \mathbb{1}) = \lambda^3 - \mathrm{Tr}\{\Theta\}\lambda^2 + \underbrace{\frac{1}{2}\left[(\mathrm{Tr}\{\Theta\})^2 - \mathrm{Tr}\{\Theta^2\}\right]}_{(\lambda_1\lambda_2 + \lambda_2\lambda_3 + \lambda_1\lambda_3) = \frac{C}{3}} - \det\Theta \tag{1.238}$$

where λ_i are the eigenvalues of Θ. To study the IR properties of C, it suffices to verify that the Θ tensor satisfies the conditions (1.236):

$$\Theta_{n+1}(p_1, \ldots, p_n, 0) = \frac{\sum_i^n \frac{\mathbf{p}_i \mathbf{p}_i^T}{|\mathbf{p}|} + 0}{\sum_{i=1}^n |\mathbf{p}_i| + 0} = \Theta_n(p_1, \ldots, p_n)$$

$$\Theta_{n+1}(p_1, \ldots, \lambda p_n, (1-\lambda)p_n) = \frac{\sum_i^{n-1} \frac{\mathbf{p}_i \mathbf{p}_i^T}{|\mathbf{p}_i|} + \frac{\lambda^2 \mathbf{p}_n \mathbf{p}_n^T}{\lambda |\mathbf{p}_n|} + \frac{(1-\lambda)^2 \mathbf{p}_n \mathbf{p}_n^T}{(1-\lambda)|\mathbf{p}_n|}}{\sum_{i=1}^n |\mathbf{p}_i| + \lambda |\mathbf{p}_n| + (1-\lambda)|\mathbf{p}_n|} =$$

$$= \Theta_n(p_1, \ldots, p_n) \tag{1.239}$$

Finally, we notice that S can be also written as:

$$S = \frac{3}{2} \min_{\mathbf{n}} \frac{\sum_i (\mathbf{p}_i \times \mathbf{n})^2}{\sum \mathbf{p}_i^2} = \frac{3}{2} \min_{\mathbf{n}} \frac{\sum_i \mathbf{p}_{\perp i}^2}{\sum_i \mathbf{p}_i^2}, \tag{1.240}$$

where $\mathbf{p}_\perp$ is the momentum component orthogonal to the direction $\mathbf{n}$. We now introduce an auxiliary tensor W defined as:

$$W = \frac{\sum_i \mathbf{p}_i \mathbf{p}_i^T}{\sum_i \mathbf{p}_i^2} = W^T. \tag{1.241}$$

For any vector $\mathbf{n}$, it holds:

$$\mathbf{n}^T W \mathbf{n} = \frac{\sum_i \mathbf{p}_{\parallel i}^2}{\sum_i \mathbf{p}_i^2} = 1 - \frac{\sum_i \mathbf{p}_{\perp i}^2}{\sum_i \mathbf{p}_i^2}. \tag{1.242}$$

Hence, minimising $S(\mathbf{n})$ is equivalent to maximise the quadratic form $\mathbf{n}^T W \mathbf{n}$ subject to the constraint $\mathbf{n}^T \mathbf{n} = 1$:

$$0 = \nabla_{\mathbf{n}} \left[\mathbf{n}^T W \mathbf{n} - \lambda \left(1 - \mathbf{n}^T \mathbf{n} \right) \right] \quad \Leftrightarrow \quad W \mathbf{n} = \lambda \mathbf{n}. \tag{1.243}$$

This implies that $\mathbf{n}$ is the eigenvector of the W tensor corresponding to the largest of its three eigenvalues $\lambda_1 \geq \lambda_2 \geq \lambda_3$:

$$S = \frac{3}{2} \left(1 - \lambda_1 \right). \tag{1.244}$$

To study the IR properties of S, we need to test whether W satisfies the conditions (1.236):

$$W_{n+1}(p_1, \ldots, p_n, 0) = \frac{\sum_i^n \mathbf{p}_i \mathbf{p}_i^T + 0}{\sum_{i=1}^n |\mathbf{p}_i^2| + 0} = W_n(p_1, \ldots, p_n) \tag{1.245}$$

$$W_{n+1}(p_1, \ldots, \lambda p_n, (1-\lambda)p_n) = \frac{\sum_i^{n-1} \mathbf{p}_i \mathbf{p}_i^T + \lambda^2 \mathbf{p}_n \mathbf{p}_n^T + (1-\lambda)^2 \mathbf{p}_n \mathbf{p}_n^T}{\sum_{i=1}^n |\mathbf{p}_i^2| + \lambda^2 |\mathbf{p}_n^2| + (1-\lambda)^2 |\mathbf{p}_n^2|}$$

$$\neq W_n(p_1, \ldots, p_n). \tag{1.246}$$

The last inequality is a consequence of the non-linear dependence of W on the particle momenta. Another way to see that S is not collinear-safe is to consider the case where $\mathbf{n}$ is aligned with one of the particle momenta $\mathbf{p}$: in this case, a collinear splitting of that particle would not change the numerator ($\mathbf{n} \times \mathbf{p} = 0$), but it would change the denominator due to the quadratic dependence on the momenta.

We conclude by mentioning that, differently from the sphericity, the event-shape variable:

$$S' = \left(\frac{4}{\pi} \right)^2 \min_{\mathbf{n}} \left(\frac{\sum_i |\mathbf{p}_i \times \mathbf{n}|}{\sum_i |\mathbf{p}_i|} \right), \tag{1.247}$$

called *sphericity*, is instead IR-safe.

Suggested Readings

See Refs. [17, 18] for more details on IR properties of observables relevant for experimental tests of pQCD.

Problem 1.44 Consider the parton-level reaction $e^+ e^- \to q \bar{q} g$, where all final-state partons are assumed to be massless. Define the adimensional parameters $x_i = 2 p_i P / s$ with $p_i = q, \bar{q}, g$ and $P = \sum_i p_i$. Show that the allowed values for the x_i parameters are located inside a triangle in the plane $(x_q, x_{\bar{q}})$. Consider now the metric $y_{ij} \equiv 2 p_i p_j / s$. An event is said to contain three jets if $\min_{i \neq j} y_{ij} > y$ for some value of the jet parameter y. Determine the geometric locus of three-jet events in the $(x_q, x_{\bar{q}})$ plane as a function of y.

Discussion

In this problem, the concept of *jet* is introduced. A jet is defined by a jet-finding algorithm that determines how to cluster together an ensemble of particles in an iterative procedure based on a definition of distance between particles. Metrics based on the R_{ij} distance of Eq. (1.167) are quite popular at hadron colliders, while the Lorentz-invariant metric of the exercise, also known ad JADE distance, are more popular at lepton colliders. Observables related to a jet-finding algorithm should be predictable within a given theory, if one wishes to compare its predictions to the experiment. A problem which often occurs in theories of massless particles like QCD and QED is connected with the divergences associated with soft or collinear splitting of the massless particles, see Problem 1.47. These divergences can be consistently categorised and re-absorbed for IR-safe observables, see Problem 1.43. The replacement of parton-level observables with jet-based ones allows to recover a well-posedness necessary to carry out the perturbative calculations. A jet algorithm can be studied in an experiment by using as input the experimental signature of the theoretical particles.

Solution

If $y > \frac{1}{3}$, then the event contains exactly two jets. This is easy to prove since:

$$y_{12} + y_{13} + y_{23} = \sum_{i \neq j} \frac{2 p_i p_j}{s} = \frac{(\sum_i p_i)^2}{s} = 1, \qquad (1.248)$$

hence at least one of the three measures has to be smaller than $1/3$. The algorithm will then cluster the two "closest" partons, leaving two resolved jets. We then focus on the more interesting case $0 \leq y \leq \frac{1}{3}$, where three-jet configurations can arise. It is also interesting to notice that for three partons:

$$\min_{i \neq j} \left\{ \frac{(p_i + p_j)^2}{s} \right\} = \min_k \left\{ \frac{s - 2 p_k P}{s} \right\} = \min_k \{1 - x_k\} = 1 - \max_k \{x_k\} = 1 - T, \qquad (1.249)$$

where T is the thrust parameter introduced in Problem 1.42. The three x_i parameters are not independent since they satisfy the relation:

$$x_q + x_{\bar{q}} + x_g = \frac{2(q + \bar{q} + g)P}{s} = 2. \qquad (1.250)$$

Each event can be parametrised by two energy fractions x_q and $x_{\bar{q}}$. In the $(x_q, x_{\bar{q}})$ plane, the allowed points lie in the geometric locus defined by:

$$\begin{cases} x_q \le 1 \\ x_{\bar{q}} \le 1 \\ x_q + x_{\bar{q}} \ge 1. \end{cases} \qquad (1.251)$$

which identifies a rectangular triangle in the $(x_q, x_{\bar{q}})$ plane. The top-side (left) of this triangle corresponds to $x_{\bar{q}} = 1$ ($x_q = 1$), i.e. a collinear emission of the gluon by the quark (antiquark). The top-right vertex of the triangle corresponds instead to a soft gluon emission, since $x_q + x_{\bar{q}} \to 2$ implies $x_g \to 0$ as for Eq. (1.250). For a three-jet event one has instead:

$$\min_{k=1,2,3} \{1 - x_k\} > y \;\Rightarrow\; 1 - x_{q,\bar{q},g} > y, \qquad (1.252)$$

The three-jet events are therefore associated with points located inside a triangle similar to Eq. (1.251), but with shorted sides defined by:

$$\begin{cases} 2y \le x_q \le 1 - y \\ 2y \le x_{\bar{q}} \le 1 - y \\ x_q + x_{\bar{q}} \ge 1 + y. \end{cases} \qquad (1.253)$$

Figure 1.14 shows the triangle for a generic value of y. Notice that for $y > 0$, the sides of the larger triangle are excluded from the three-jet parameter space. Since all

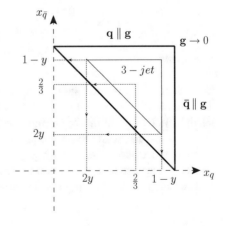

Fig. 1.14 The phase-space representation of a $q\bar{q}g$ event in terms of the Lorentz-invariants variables $x_i = 2p_iP/s$ computed in the centre-of-mass frame. The domain corresponding to 3-jet events according to the JADE algorithm with parameter y is the one delimited by the light lines

the divergence related to a soft or collinear splitting are located on the top and right side of the triangle, the three-jet phase-space is free from IR divergences.

Suggested Readings

A detailed calculation of the LO three-jet cross section with the JADE algorithm can be found in Ref. [18].

Bando n. 13705/2010

Problem 1.45 The reaction $K^- p \to \pi^0 \Lambda^0$ is studied in a fixed-target experiment. Determine the beam energy for which the Λ^0 baryon can be produced at rest in the laboratory frame.

Solution

Let's denote the four-momenta of the involved particles by p_K, p_p, p_π and p_Λ. In the laboratory frame, they can be written as:

$$p_K = (E_K, \mathbf{p}_K), \qquad p_p = (m_p, \mathbf{0}), \qquad p_\Lambda = (m_\Lambda, \mathbf{0}), \qquad p_\pi = (E_\pi, \mathbf{p}_K). \quad (1.254)$$

Conservation of energy and momentum implies that $p_K + p_p = p_\pi + p_\Lambda$. Since nothing is known about the π^0 kinematics after the scattering, we get rid of it by squaring the π^0 four-momentum:

$$p_\pi = p_K + p_\pi - p_\Lambda,$$
$$m_\pi^2 = m_K^2 + m_p^2 + m_\Lambda^2 + 2E_K m_p - 2E_K m_\Lambda - 2m_p m_\Lambda,$$
$$2E_K (m_\Lambda - m_p) = m_K^2 + m_p^2 + m_\Lambda^2 - 2m_p m_\Lambda - m_\pi^2,$$
$$E_K = \frac{(m_\Lambda - m_p)^2 + m_K^2 - m_\pi^2}{2(m_\Lambda - m_p)}. \quad (1.255)$$

Discussion

Equation (1.255) has the same form of the centre-of-mass energy in a two-body decay with $\sqrt{s} = m_\Lambda - m_p$, see Eq. (1.89). Indeed, we could have guessed this result by noticing that this scattering reaction is kinematically identical to the decay $\Lambda^0 \to p\, K^-\, \pi^0$, with the proton at rest, which is analogous to a two-body decay where only a centre-of-mass energy $\sqrt{s} = m_\Lambda - m_p$ is available for the two mesons.

Notice that not all scatterings at energy E_K produce Λ's at rest in the laboratory. Indeed, this particular configuration corresponds to a Λ emitted forward ($\theta^* = 0$)

in the centre-of-mass frame: for any other angle θ^*, a different kinematics in the laboratory will arise.

Bando n. 1N/R3/SUB/2005

Problem 1.46 A proton beam of energy $E_1 = 20$ GeV collides head-on against another beam of protons of energy $E_2 = 5$ GeV. Determine:

1. the centre-of-mass energy of a binary pp collision;
2. the velocity of the centre-of-mass in the laboratory frame;
3. the angle in the laboratory of a relativistic particle produced at 90° in the centre-of-mass;
4. the energy that a proton beam must have in order to generate the same centre-of-mass energy in a fixed-target collision.

Solution

1. Let's denote the four-momenta of the colliding protons by $p_i = (E_i, \mathbf{p}_i)$. The centre-of-mass energy is given by the square-root of the Mandelstam variable:

$$s = (p_1 + p_2)^2 = 2m_p^2 + 2(E_1 E_2 + |\mathbf{p}_1||\mathbf{p}_2|) =$$

$$= 2m_p^2 + 2E_1 E_2 \left(1 + \sqrt{1 - \left(\frac{m_p}{E_1}\right)^2}\sqrt{1 - \left(\frac{m_p}{E_2}\right)^2}\right). \qquad (1.256)$$

The proton mass $m_p = 0.938$ GeV is small compared to either of the two energies, so that a first-order expansion of the square roots is accurate enough for most purposes, giving $\sqrt{s} \approx 20$ GeV. A complete calculation yields $\sqrt{s} = 19.95$ GeV.

2. The velocity of the centre-of-mass frame is the velocity of a "particle" of four-momentum $p_1 + p_2$, see Eq. (1.96), hence:

$$\beta = \frac{\left|\sqrt{E_1^2 - m_p^2} - \sqrt{E_2^2 - m_p^2}\right|}{E_1 + E_2} = \frac{15.07 \text{ GeV}}{25 \text{ GeV}} = 0.603. \qquad (1.257)$$

3. A relativistic particle with centre-of-mass velocity $\beta^* \approx 1$ and polar angle $\theta^* = \pi/2$ with respect to the beam direction, will emerge at an angle θ in the laboratory frame given by Eq. (1.44). If we chose the direction of the x axis as aligned with the most energetic proton, then, the boost parameter to the laboratory frame is given by $-\beta$. We can therefore use Eqs. (1.44) and (1.257) to get:

$$\tan\theta = \frac{\sin(\pi/2)}{\gamma\left(\cos(\pi/2) + \beta\right)} = \frac{\sqrt{s}}{\sqrt{s}\,\beta\,\gamma} = \frac{\sqrt{s}}{|\mathbf{p}_2| - |\mathbf{p}_1|} = \frac{19.95 \text{ GeV}}{15.07 \text{ GeV}} = 1.32,$$

$$\theta = \text{atan}(1.32) = 52.9°. \qquad (1.258)$$

4. For a proton collision against a fixed-target at the same centre-of-mass energy, the beam momentum E_{fix} must satisfy:

$$\sqrt{2m_p^2 + 2m_p E_{\text{fix}}} = \sqrt{s}, \qquad 2m_p (m_p + E_{\text{fix}}) = s,$$

$$E_{\text{fix}} = \frac{s - 2m_p^2}{2m_p} = \frac{(19.95 \text{ GeV})^2 - (0.938 \text{ GeV})^2}{2 \cdot 0.938 \text{ GeV}} = 211 \text{ GeV}, \qquad (1.259)$$

which is more than one order of magnitude larger than the highest proton energy for head-on collisions.

Bando n. 1N/R3/SUB/2005, Bando n. 13153/2009

Problem 1.47 Can a photon decay to an electron-positron pair in vacuum? Can a particle radiate a photon in vacuum? Consider a photon conversion in the following two cases: $\gamma \to e^+ e^-$ in the electromagnetic field of a heavy nucleus of mass m_N, initially at rest, and $\gamma \to e^+ e^-$ in the neighbourhood of an electron, also at rest. For both cases, determine the threshold energy of the photon such that the reaction can take place.

Discussion

The four-momentum p of a particle is either *time-like* ($p^2 > 0$) or *light-like* ($p^2 = 0$). In both cases, the sum of two such four-vectors is always time-like. Indeed, if at least one of the particles is massive, say particle a, one can make a boost to its rest frame and compute explicitly the invariant:

$$(p_a + p_b)^2 = p_a^2 + p_b^2 + 2p_a p_b = m_a^2 + m_b^2 + 2m_a E_b^{(a)} \geq$$
$$\geq m_a^2 + m_b^2 + 2m_a m_b = (m_a + m_b)^2 > 0. \qquad (1.260)$$

If instead both particles are massless, then in any frame $(p_a + p_b)^2 = 2E_a E_b (1 - \cos \theta_{ab}) \geq 0$. By iteratively clustering pairs of four-momenta, it then follows that

$$(p_a + p_b + p_c + \ldots)^2 \geq (m_a + m_b + m_c + \ldots)^2. \qquad (1.261)$$

The right-hand side of this inequality is the tightest lower bound and corresponds to a configuration where all particles are at rest in their centre-of-mass frame. For a reaction $a + b \to c + d + \ldots$, with b at rest in the laboratory frame, we then have:

$$s = (p_a + p_b)^2 = (p_c + p_d + \ldots)^2,$$
$$m_a^2 + m_b^2 + 2E_a m_b = (p_c + p_d + \ldots)^2 \geq (m_c + m_d + \ldots)^2. \qquad (1.262)$$

Since the laboratory energy of E_a is a linear function of s, its minimum value corresponds to the minimum possible value of s, which is given by the left-hand side of Eq. (1.261). The threshold condition for a fixed-target scattering corresponds to a beam energy such that:

$$E_a^{thr} = \frac{(m_c + m_d + \ldots)^2 - m_a^2 - m_b^2}{2m_b}. \tag{1.263}$$

Solution

Let's consider the reaction $\gamma \to e^+ e^-$, where p_γ, p_{e^+}, and p_{e^-} are the four-momenta of the tree particles. Conservation of energy and momentum implies:

$$p_\gamma = p_{e^+} + p_{e^-}, \quad p_\gamma^2 = 2m_e^2 + 2p_{e^+}p_{e^-}, \quad 0 = m_e^2 + p_{e^+}p_{e^-}. \tag{1.264}$$

In the rest frame of the electron, the Lorentz-invariant product $p_{e^+}p_{e^-}$ is $E'_{e^+} m_e > 0$, hence the last equality in Eq. (1.264) cannot be satisfied. Notice that this is not the case if m_e were zero: in this latter case, either a collinear splitting with $\mathbf{p}_{e^-}$ parallel to $\mathbf{p}_{e^+}$, or the emission of an infinitely soft electron or positron, would allow Eq. (1.264) to be still fulfilled. Let's now consider the reaction $e^- \to e^- \gamma$. Like in the previous case, one gets the condition:

$$p_{e^-} = p'_{e^-} + p_\gamma, \quad p_{e^-}^2 = p'^2_{e^-} + p_\gamma^2 + 2p_\gamma p'_{e^-}, \quad 0 = p_\gamma p'_{e^-} \tag{1.265}$$

In the rest frame of the final electron, $p_\gamma p'_{e^-} = E'_\gamma m_e > 0$. Notice that the only possibility for Eq. (1.265) to be satisfied is again through a collinear splitting in the limit $m_e \to 0$, or through an infinitely soft photon emission.

As seen from Eq. (1.264), photon conversion in vacuum is not allowed, whereas photon conversion in the presence of a spectator particle is kinematically allowed. Let's consider the reaction $\gamma X \to e^+ e^- X$, where X can be either an electron or a heavy nucleus at rest in the laboratory frame. The reaction can occur if the photon energy is above the threshold for e^+e^- production, namely:

$$E_\gamma^{thr} = \begin{cases} \frac{(2m_e+m_e)^2 - m_e^2}{2m_e} = 4\,m_e & \text{electron} \\ \frac{(2m_e+m_N)^2 - m_N^2}{2m_N} = \frac{m_N}{2}\left[\frac{4m_e}{m_N} + \mathcal{O}\left(\frac{m_e}{m_N}\right)\right] \approx 2\,m_e & \text{nucleus} \end{cases} \tag{1.266}$$

Thus, photon conversion in the electromagnetic field of a heavy nucleus requires a factor of two smaller threshold energy compared to the conversion in the electron field.

Problem 1.48 A photon converts to $\gamma \to e^+e^-$ in the electromagnetic field of a heavy nucleus of mass m_N. Calculate the minimum momentum transfer to the nucleus $|\mathbf{q}|$ when the photon has an initial energy $E_\gamma = 1$ GeV. Estimate an order of magnitude for the opening angle between the two leptons in the laboratory frame.

Discussion

The process by which a photon converts in matter is known as *Bethe–Hadler*, after the authors who first computed its theoretical cross section. It can be heuristically explained like follows. A photon has four-momentum $p_\gamma = (E_\gamma, E_\gamma\, \mathbf{e}_x)$, which gives $p_\gamma^2 = E_\gamma^2 - E_\gamma^2 = 0$. In order to transmute into a pair of massive electrons, the photon needs to "acquire" a mass. This can be achieved by transfering a quantity $\mathbf{q}$ of its momentum to the nearby nucleus with no energy loss, provided that the total four-momentum is conserved. In this way: $p_\gamma \to p_X = (E, E - |\mathbf{q}|)$ and p_X^2 is now positive. Exchanging three-momentum with a negligible energy loss is indeed possible if $m_N \gg E_\gamma$.

Solution

Conservation of energy-momentum implies that $p_\gamma + p_N = p_{e^+} + p_{e^-} + p_N'$. Let's define:

$$p_X = p_{e^+} + p_{e^-}, \qquad q = p_\gamma - p_X = p_N' - p_N = \left(\sqrt{|\mathbf{q}|^2 + m_N^2} - m_N, \mathbf{q}\right) \quad (1.267)$$

If the energy of the photon is much smaller than the nuclear mass m_N, then the exchanged momentum $|\mathbf{q}|$ is also much smaller than m_N, so that:

$$q \approx \left(\frac{|\mathbf{q}|^2}{2m_N}, \mathbf{q}\right). \qquad (1.268)$$

From the momentum conservation equation $p_X = p_\gamma - q$, it follows:

$$\frac{|\mathbf{q}|^4}{4m_N^2} + \left(1 + \frac{E_\gamma}{m_N}\right) |\mathbf{q}|^2 - 2E_\gamma |\mathbf{q}| \cos\theta + m_X^2 \approx$$
$$|\mathbf{q}|^2 - 2E_\gamma |\mathbf{q}| \cos\theta + m_X^2 = 0, \qquad (1.269)$$

where θ is the angle of $\mathbf{q}$ with respect to the photon direction. Terms suppressed by powers of m_N have been neglected. From Descartes' rule of sign, we see that Eq. (1.269) admits a valid solution only if $\cos\theta > 0$. Explicitly:

$$|\mathbf{q}|_\pm = E_\gamma \cos\theta \pm \sqrt{E_\gamma^2 \cos^2\theta - m_X^2}. \qquad (1.270)$$

It can be easily verified that $\partial|\mathbf{q}|_\pm/\partial m_X^2 \lessgtr 0$ and $\partial|\mathbf{q}|_\pm/\partial\cos\theta \gtrless 0$, so that the minimum momentum transfer corresponds to a nucleus recoil parallel to the incoming photon momentum and to the minimum possible invariant mass of the e^+e^- pair, which is given by $m_X^{\min} = 2m_e$, see Eq. (1.261). Hence:

$$|\mathbf{q}|_{\min} = E_\gamma \left(1 - \sqrt{1 - \frac{4m_e^2}{E_\gamma^2}} \right) \approx \frac{2m_e^2}{E_\gamma} \approx 0.5 \text{ KeV.} \qquad (1.271)$$

We now estimate the typical size of the opening angle ϕ between the electron and positron. If $m_X = 2m_e$, the velocity of the two particles in their rest frame is zero as for Eq. (1.92): any boost from the reference frame to the laboratory frame will maintain the two momenta parallel to each other, hence $\phi = 0$. The largest opening angle for a two-body decay as a function of the velocity β of the mother particle (X in our case) and on the velocity β^* of the daughter particles in their c.o.m frame is given by Eq. (1.65):

$$\tan(\phi_{\max}) = \frac{2\gamma\beta\beta^*}{\gamma^2\beta^2 - \beta^{*2}} = \frac{2|\mathbf{p}_X|\sqrt{m_X^2 - 4m_e^2}}{|\mathbf{p}_X|^2 - (m_X^2 - 4m_e^2)}. \qquad (1.272)$$

Since pair-production is an electromagnetic process where, at leading order, a virtual photon is exchanged between the nucleus and a virtual electron, one can expect the differential cross section to feature the typical $\sim 1/|\mathbf{q}|^4$ behaviour, so that the bulk of the conversions will have $|\mathbf{q}| \gtrsim |\mathbf{q}|_{\min}$, or equivalently $m_X = f \cdot 2m_e$, with f of order one. Since $|\mathbf{q}|_{\min} \ll E_\gamma$, for small transfered momenta, the momentum $|\mathbf{p}_X|$ of the e^+e^- pair is $|\mathbf{p}_X| \approx E_\gamma$. Equation (1.273) then gives

$$\phi \sim \frac{|\phi_{\max} - \phi_{\min}|}{2} \approx \frac{m_e}{E_\gamma}\sqrt{f^2 - 1} = \mathcal{O}(\text{mrad}). \qquad (1.273)$$

The possibility to measure the opening angle, and thus to determine the decay plane of lepton pair, opens the possibility to use the Bethe–Hadler process as a polarimeter of the incoming photon, since the orientation of the decay plane is correlated with the polarisation of the photon.

Suggested Readings

Reference [19] provides a complete review of the photo-production mechanism of lepton-antilepton pairs. The idea of using the double photon-conversion for the measurement of the CP properties of the Higgs boson in the $H \to \gamma\gamma$ channel has been investigated by the phenomenological work of Ref. [20].

Bando n. 13153/2009

Problem 1.49 The strong reaction $p\,p \to X\,K^+\,K^-$ is studied in laboratory, where X denotes an unknown particle. Determine the values of the electric charge, strangeness, and baryon number of X. Let the mass of X be twice the proton mass. Determine the minimal energy in the laboratory frame necessary to produce X in a fixed target experiment.

Solution

Since the strong interaction conserves the electric charge (Q), the strangeness (S), and the baryon number (B), it follows that $Q_X = +2$, $S_X = 0$, and $B_X = 2$. See Chap. 5 for more details. If $m_X = 2m_p$, the threshold energy for a fixed-target collision is given by Eq. (1.263):

$$E_p^{\text{thr}} = \frac{(2m_p + 2m_K)^2 - 2m_p^2}{2m_p} = m_p \left[1 + 4\left(\frac{m_K}{m_p}\right) + 2\left(\frac{m_K}{m_p}\right)^2\right] = 3.43 \text{ GeV},$$

$$(1.274)$$

where we have used the PDG values $m_p = 0.938$ GeV and $m_K = 0.494$ GeV [4].

Problem 1.50 Determine the threshold energy for the antiproton production $p\,p \to p\,p\,p\,\bar{p}$, when using a proton beam on a liquid-hydrogen target. Consider now the case that the target consists of a heavy material, so that the target proton is actually a bounded nucleon. Assume a Fermi energy $E_F = 30$ MeV. By how much does the threshold energy get reduced by the nuclear motion?

Solution

Let's first consider the case where the target consists of liquid hydrogen. The thermal motion, being characterised by an energy $k_B T \approx 25$ meV $(T/300 \text{ K})$, is totally negligible, so that the target proton can be safely assumed at rest. The threshold energy is therefore given by Eq. (1.263):

$$E_p^{\text{thr}} = \frac{(4m_p)^2 - 2m_p^2}{2m_p} = 7m_p = 6.56 \text{ GeV},$$

$$(1.275)$$

or, in terms of kinetic energy, $T_p^{\text{thr}} = 6m_p = 5.6$ GeV.

If the target proton is bounded inside a nucleus, its momentum can be as large as the Fermi momentum $|\mathbf{p}_F|$. Since $E_F \ll m_p$, the bounded nucleon can be treated as classical. The most favourable kinematical configuration corresponds to a nucleon of momentum $|\mathbf{p}_F|$ moving against the incoming proton. The threshold condition of Eq. (1.262) gets modified to:

$$2m_p^2 + 2E_p^{\text{thr}}\left(m_p + \frac{|\mathbf{p}_F|^2}{2m_p}\right) + 2\sqrt{(E_p^{\text{thr}})^2 - m_p^2}\,|\mathbf{p}_F| = (4m_p)^2.$$

$$(1.276)$$

Introducing the adimensional parameter $\varepsilon = |\mathbf{p}_F|/m_p = \sqrt{2E_F/m_p} \approx 0.23$, we can simplify the above equation by neglecting terms of $\mathcal{O}(\varepsilon^2)$:

$$(E_p^{thr})^2 - 14m_p \left(1 + \frac{\varepsilon^2}{2}\right) + (7m_p)^2 + \varepsilon^2 m_p^2 = 0$$

$$E_p^{thr} = 7m_p \left(1 + \frac{\varepsilon^2}{2}\right) - \sqrt{(7m_p)^2(1 + \varepsilon^2) - (7m_p)^2 - \varepsilon^2 m_p^2} =$$

$$= 7m_p \left(1 + \frac{\varepsilon^2}{2}\right) - \sqrt{48 \, m_p^2 \, \varepsilon^2} = 7m_p \left(1 + \frac{E_F}{m_p}\right) - \sqrt{96 \, m_p \, E_F} = 5.24 \text{ GeV},$$

$$(1.277)$$

corresponding to a proton kinetic energy $T_p^{thr} = 4.3$ GeV, i.e. about 1.3 GeV less than that necessary with a hydrogen target.

Suggested Readings

This reaction was first produced in laboratory by a E. Segrè et al. at the AGS accelerator at Berkeley. The reader is encouraged to study the discovery paper [21]. A guided discussion on the experimental set up can be also found in Ref. [12].

Bando n. 18211/2016

Problem 1.51 Given the Boltzman constant $k_B = 8.6 \times 10^{-5}$ eV K^{-1}, estimate the typical wavelength of the cosmic background radiation. Which part of the EM spectrum does it belong to?

Solution

The cosmic microwave background (CMB) features a black-body spectrum with temperature $T \approx 2.7$ K. The energy density per unit of frequency is given by Planck's law

$$\frac{dE}{d\nu} = \frac{8\pi h}{c^3} \frac{\nu^3}{\exp(h\nu/k_B T) - 1}.$$

$$(1.278)$$

The peak frequency mode has an energy of about $3k_B T$, so that:

$$3k_B T = h\bar{\nu} = \frac{hc}{\bar{\lambda}}, \qquad \bar{\lambda} = \frac{hc}{3k_B T} =$$

$$= \frac{6.6 \times 10^{-34} \text{ J s} \cdot 3 \times 10^8 \text{ m s}^{-1}}{3 \cdot 8.6 \times 10^{-5} \text{ K}^{-1} \cdot 1.6 \times 10^{-19} \text{ J} \cdot 2.7 \text{ K}} \approx 1.7 \text{ mm},$$

$$(1.279)$$

corresponding to a frequency $\bar{\nu} = 160$ GHz. The typical radiation spectrum is therefore located in the microwave domain.

Bando n. 13705/2010

Problem 1.52 A proton of the cosmic radiation can produce π^0 by interacting with the cosmic microwave background (CMB) at 3 K. You can assume that the CMB photons are uniformly distributed in space with an energy density $\rho_\gamma = 0.38\,\mathrm{eV/cm^3}$, and have an average energy $E_\gamma = 0.7$ meV. What is the proton energy threshold? At the energy threshold, what fraction of the proton energy gets lost, and what is the minimum angle between the two photons in the π^0 decay? Let us assume that the photo-absorption cross section above threshold is $\sigma_{\gamma p} = 200\ \mu$barn. Provide an estimate of the proton mean free path.

Discussion

The existence of a threshold energy for the inelastic reaction $p\ \gamma_{\mathrm{CMB}} \to p\ \pi^0$, where γ_{CMB} is a photon of the cosmic microwave radiation, implies a suppression of the cosmic ray spectrum above that threshold, known as the *GZK cut-off*.

Solution

The lowest proton energy threshold corresponds to a CMB photon of energy E_γ moving against the proton. For this configuration, one has:

$$m_p^2 + 2m_p\,m_\pi + m_\pi^2 = m_p^2 + 2E_p^{\mathrm{thr}} E_\gamma + 2|\mathbf{p}_p|E_\gamma. \qquad (1.280)$$

Since $E_p^{\mathrm{thr}}/m_p \sim m_{\pi^0}/E_\gamma \gg 1$, it can be safely assumed that $|\mathbf{p}_p| \approx E_p$. Therefore:

$$
\begin{aligned}
E_p^{\mathrm{thr}} &= m_{\pi^0}\frac{2m_p + m_{\pi^0}}{4E_\gamma} = 0.135\ \mathrm{GeV}\left(\frac{(2\cdot 0.938 + 0.135)\ \mathrm{GeV}}{4\cdot 7\times 10^{-13}\ \mathrm{GeV}}\right) = \\
&= 0.9\times 10^{18}\ \mathrm{eV}.
\end{aligned}
\qquad (1.281)
$$

At the threshold, both the proton and the neutral pion are produced at rest in their centre-of-mass frame. The gamma factor of the centre-of-mass is given by $\gamma = (E_p^{\mathrm{thr}} + E_\gamma)/(m_p + m_{\pi^0}) \approx E_p^{\mathrm{thr}}/(m_p + m_{\pi^0})$. The relative energy loss suffered by the proton after the photon absorbtion is therefore given by:

$$E_p' = \gamma\,m_p = \frac{m_p}{m_p + m_{\pi^0}}E_p^{\mathrm{thr}} \quad\Rightarrow\quad \frac{E_p^{\mathrm{thr}} - E_p'}{E_p^{\mathrm{thr}}} = \frac{m_{\pi^0}}{m_p + m_{\pi^0}} \approx 0.12. \qquad (1.282)$$

The minimum open angle between the two photons from the π^0 decay is given by Eq. (1.61):

$$\phi_{\min} = \frac{2}{\gamma} = \frac{2(m_p + m_{\pi^0})}{E_p^{\mathrm{thr}}} \approx 2\times 10^{-19}\ \mathrm{rad}. \qquad (1.283)$$

Given a process with cross section σ, the interaction length λ is defined as the inverse of the probability of interaction per unit length and per incoming particle, see Eq. (1.291). Since both the proton and the CMB photon are moving in opposite direction at speed c, the relative velocity is $v_{\text{rel}} = 2c$, see Eq. (1.20). By definition of cross section, the number of interactions per unit time and unit volume is given by:

$$\frac{dN_{\gamma p}}{dt \, d^2\mathbf{A} \, dx} = \rho_p \rho_\gamma \, v_{\text{rel}} \cdot \sigma_{\gamma p}, \quad \frac{dN_{\gamma p}}{dt} = \left(c \, \rho_p \, d^2\mathbf{A} \, dx\right) \cdot 2 \, \rho_\gamma \, \sigma_{\gamma p} = d\Phi_p \, dx \cdot 2 \, \rho_\gamma \, \sigma_{\gamma p},$$

$$\frac{dN_{\gamma p}}{(d\Phi_p \, dt) \, dx} = 2 \, \rho_\gamma \, \sigma_{\gamma p} \quad \Rightarrow \quad \lambda = \frac{1}{2 \, \rho_\gamma \, \sigma_{\gamma p}} = \frac{1}{2 \frac{0.38 \text{ eVcm}^{-3}}{0.7 \times 10^{-3} \text{ eV}} \cdot 2 \times 10^{-28} \text{ cm}^2} =$$

$$= 5 \times 10^{24} \text{ cm} = 1.5 \times 10^6 \text{ parsec}, \tag{1.284}$$

Here, $d\Phi_p$ in the intermediate calculations is the flux of incoming proton entering the infinitesimal volume $d^2\mathbf{A} \, dx$, so that $dN_{py}/d\Phi_p \, dt$ is the number of interactions *per* incoming proton.

1.3 Cross Section

The cross section σ of a scattering process $a \, b \rightarrow X$, where a is a moving projectile and b is at rest, is defined as the number of reactions X measured per unit time, per unit scattering center, and per unit of incoming flux density:

$$\sigma = \frac{1}{J_a} \frac{dN_X}{dt \, dN_b}, \tag{1.285}$$

where dN_b is the number of scattering centers in an infinitesimal volume $d^3\mathbf{r}$ irradiated by the flux density J_a. The latter is defined as the number of particles a crossing per unit time and unit area the normal surface at the position of the volume $d^3\mathbf{r}$:

$$J_a = n_a |\mathbf{v}_a|, \tag{1.286}$$

where n_a and $|\mathbf{v}_a|$ are the particle density ($[n] = \text{m}^{-3}$) and velocity, respectively. The dimension of σ is therefore $[\sigma] = \text{cm}^2$. The definition (1.285) can be generalised to an arbitrary kinematics of the involved particles, so that the distinction between projectile and target can be ultimately ignored. Firstly, we introduce the particle density of the target, n_b, and write:

$$\sigma = \frac{\left(dN_X/dt \, d^3\mathbf{r}\right) \cdot d^3\mathbf{r}}{\left(n_b \cdot d^3\mathbf{r}\right) n_a |\mathbf{v}_a|} = \frac{\left(dN_X/dt \, d^3\mathbf{r}\right)}{n_b \, n_a \, |\mathbf{v}_a|} \tag{1.287}$$

where we have introduced at the numerator the number of reactions X observed per unit time and unit volume. The quantity $n_b \, n_a \, |\mathbf{v}_a|$ at the denominator can be written in terms of the density and velocity in the generic reference frame by means of the relative velocity of Eq. (1.20), so that:

$$\sigma = \frac{\left(dN_X/dt\, d^3\mathbf{r}\right)}{n_a\, n_b\sqrt{(\mathbf{v}_a - \mathbf{v}_b)^2 - (\mathbf{v}_a \times \mathbf{v}_b)^2/c^2}} = \frac{\left(dN_X/dt\, d^3\mathbf{r}\right)}{n_a\, n_b\, v_{\mathrm{rel}}}, \tag{1.288}$$

see Problem 1.11. Under a generic boost the numerator of Eq. (1.288) is invariant, since dN_X is a pure number and $d^4x \equiv dt\, d^3\mathbf{r}$ is invariant under transformations of the Poincaré group, see Problem 1.2. The same holds for the combination $n_a\, n_b\, v_{\mathrm{rel}}$, as discussed in Problem 1.3: the cross section is therefore a Lorentz-invariant.

We now come back to the early definition of Eq. (1.285) and introduce the useful concept of *interaction length*. Let's consider a uniform flux density $J_a = \Phi_a/A$, with Φ_a being the total flux across the surface A ($[\Phi] = \mathrm{s}^{-1}$) irradiating a target of thickness δx and uniform particle density n_b. From Eq. (1.285) we have:

$$\frac{dN_X}{dt} = \frac{\Phi_a}{A} \cdot (n_b \cdot A \cdot \delta x) \cdot \sigma = \underbrace{\Phi_a(n_b\, \delta x)}_{\mathscr{L}} \cdot \sigma. \tag{1.289}$$

The quantity $n_b\, \delta x$ is the surface density of the target, and once multiplied by the flux, it gives the *luminosity* $\mathscr{L}$ of a fixed target experiment. Notice that $[\mathscr{L}] = \mathrm{cm}^{-2}\, \mathrm{s}^{-1}$. For different beam structure, the luminosity can be still defined from Eq. (1.288) as the coefficient of proportionality between the total event rate and the cross section, see Sect. 3.3.

Dividing both sides of Eq. (1.290) by the incoming flux Φ_a, the probability of interaction per incoming particle is obtained:

$$\frac{1}{\Phi_a}\frac{dN_X}{dt} = \text{probability of interaction per particle} = (n_b\sigma) \cdot \delta x, \tag{1.290}$$

so that $n_b\, \sigma$ is the probability of interaction per particle and per unit length, and its inverse

$$\lambda \equiv \frac{1}{n_b\, \sigma}, \tag{1.291}$$

is called *interaction length* of the process under consideration. Since λ^{-1} is a probability of interaction per incoming particle and per unit length, in the presence of multiple exclusive processes of interaction between a and b, the probability of any such interaction is given by the sum of all probabilities, and the total interaction length is therefore given by the inverse of the sum of all inverse interaction lengths.

The differential cross section for a scattering $p_1\, p_2 \rightarrow p_3\, \cdots\, p_n$ can be calculated theoretically using the formula:

$$d\sigma(p_1, p_2; p_3, \ldots, p_n) = \frac{1}{4I(p_1, p_2)}|\mathscr{M}(p_1, \ldots, p_n)|^2\, d\Phi_n(p_1 + p_2; p_3, \ldots, p_n) \tag{1.292}$$

where the incoming and outgoing particles are asymptotic states of the free Hamiltonian, i.e. eigenstates of the linear momentum and spin, and:

- $I(p_1, p_2)$ is the invariant discussed in Problem 1.3, and accounts for the incoming particle flux. Indeed, the factor $4I = (2E_1)(2E_2)v_{rel}$ is just but the product of the two beam densities times their relative velocity as in Eq. (1.288). For a free particle, the space density is given by the squared norm of the wave function $|\psi_{\mathbf{p}}(\mathbf{r})|^2 = 2E_{\mathbf{p}}$. The choice of wave function normalisation is convention as long as it is used consistently. For example, another popular normalisation is the non-relativistic normalisation of Eq. (1.211);
- $\mathcal{M}$ is the relativistic scattering amplitude, see Eq. (1.186);
- $d\Phi_n$ is the relativistic phase-space measure of Eq. (1.184).

Spin indices are omitted in Eq. (1.293) to simplify the notation, although one should always remember that the cross section depends in general on the spin vectors r_i of the scattered particles. When the incoming particles are unpolarised and spin is not observed, one can replace the matrix element squared in Eq. (1.293) by

$$\overline{\sum}|\mathcal{M}| \equiv \frac{\sum_{r_1,\dots,r_n}|\mathcal{M}|^2}{(2S_1 + 1)(2S_2 + 1)}, \tag{1.293}$$

which has often the virtue of greatly simplifying the calculations.

Problems

Problem 1.53 Prove that the differential cross section for a $2 \to 2$ scattering in the centre-of-mass frame can be written as:

$$\frac{d\sigma}{d\Omega^*} = \frac{1}{64\pi^2 s}\frac{|\mathbf{p}_f^*|}{|\mathbf{p}_i^*|}|\mathcal{M}|^2, \tag{1.294}$$

where $|\mathbf{p}_i^*|$ and $|\mathbf{p}_f^*|$ are the centre-of-mass momenta before and after the scattering.

Solution

The relativistic invariant I at the denominator of Eq. (1.293) can be also written as:

$$I^2 = (p_1 p_2)^2 + m_1^2 m_2^2 = \frac{(s - m_1^2 - m_2^2)^2 + 4m_1^2 m_2^2}{4} = s|\mathbf{p}^*|^2, \tag{1.295}$$

where the identity has been proved in Eq. (1.91). By making use of the two-body phase-space measure expressed in terms of the centre-of-mass solid angle, see Eq. (1.188), we get:

$$d\sigma = \frac{1}{4|\mathbf{p}_i^*|\sqrt{s}}|\mathcal{M}|^2\frac{1}{16\pi^2}\frac{|\mathbf{p}_f^*|}{\sqrt{s}}d\Omega^* \quad \Rightarrow \quad \frac{d\sigma}{d\Omega^*} = \frac{1}{64\pi^2 s}\frac{|\mathbf{p}_f^*|}{|\mathbf{p}_i^*|}|\mathcal{M}|^2, \tag{1.296}$$

If the scattering amplitude is normalised using a non-relativistic normalisation, then the above expression becomes:

$$
\frac{1}{64\pi^2 s} \frac{|\mathbf{p}_f^*|}{|\mathbf{p}_i^*|} |\mathcal{M}|^2 = \frac{1}{64\pi^2} \frac{(2E_1^*)\,(2E_2^*)\,(2E_3^*)\,(2E_4^*)}{(E_1^*+E_2^*)(E_3^*+E_4^*)} \frac{|\mathbf{p}_f^*|}{|\mathbf{p}_i^*|} |\mathcal{M}_{NR}|^2 =
$$

$$
= \frac{1}{4\pi^2} \frac{1}{\left(\frac{1}{E_1^*}+\frac{1}{E_2^*}\right)\left(\frac{1}{E_3^*}+\frac{1}{E_4^*}\right)} \frac{|\mathbf{p}_f^*|}{|\mathbf{p}_i^*|} |\mathcal{M}_{NR}|^2 = \frac{1}{4\pi^2} \frac{|\mathbf{p}_f^*|^2}{(|\mathbf{v}_1|+|\mathbf{v}_2|)\,(|\mathbf{v}_3|+|\mathbf{v}_4|)} |\mathcal{M}_{NR}|^2
$$

$$
= \frac{1}{4\pi^2} \frac{|\mathbf{p}_f^*|^2}{v_{rel}^i\, v_{rel}^f} |\mathcal{M}_{NR}|^2, \tag{1.297}
$$

where the relative velocity has been used.

Discussion

Consider the $2 \to 2$ scattering $a\,b \to c\,d$. Suppose that both this reaction, denoted by (1), and its time-reversed $c\,d \to a\,b$, denoted by (2), can be performed in the laboratory under controlled conditions. In particular, if the two reactions are studied at the same centre-of-mass energy, and if the scattering particles are unpolarised and the measurement is inclusive with respect to the spin of the final-state particles, then Eq. (1.293) together with Eq. (1.296) implies

$$
\frac{(d\sigma_{(1)}/d\Omega^*)}{(d\sigma_{(2)}/d\Omega^*)} = \frac{(2S_c+1)(2S_d+1)}{(2S_a+1)(2S_b+1)} \left(\frac{|\mathbf{p}_{cd}^*|}{|\mathbf{p}_{ab}^*|} \right)^2, \tag{1.298}
$$

where one has to further assumed that the interaction is invariant under time-reversal so that the spin-averaged matrix element squared for (1) and (2) are identical at the same centre-of-mass angle $\cos\theta^*$. If the spin of three of the four particles is known, the spin of the fourth can be *measured* by comparing the event rates of (1) and (2). As an example, this technique was employed by Steinberger et al. [22] to measure the spin of the charged pion profiting from the reaction $\pi^+ d \to p\,p$ and its inverse $p\,p \to \pi^+ d$, which could be both obtained in fixed-target collisions.

Suggested Readings

The measurement of the pion spin from detailed balance arguments is documented in Ref. [22]. An introduction to the same topic can be found in Chap. 2 of Ref. [12].

Problem 1.54 What is the $\pi^+ p$ cross section at the peak of the $\Delta(1232)$ resonance?

Discussion

When the centre-of-mass energy E^* approaches the mass E_0 of a resonance of spin J and width Γ, the scattering cross section between two unpolarised particles with spin S_1 and S_2, detected in the channel f, is described by the *Breit–Wigner* formula:

$$\sigma_f(E^*) = \frac{4\pi}{|\mathbf{p}^*|^2} \frac{2J+1}{(2S_1+1)(2S_2+1)} \frac{\Gamma^2/4}{(E^* - E_0)^2 + \Gamma^2/4} \mathrm{BR}_i \, \mathrm{BR}_f, \qquad (1.299)$$

where $\mathrm{BR}_{i,f}$ are the branching ratio of the resonance into the initial and final state particles, respectively, and $\mathbf{p}^*$ is the centre-of-mass momentum given by Eq. (1.91).

Solution

According to the Breit–Wigner formula of Eq. (1.299), the cross section at the peak is independent of the width Γ and depends only on the resonance mass E_0:

$$\sigma_\Delta = \frac{8\pi}{|\mathbf{p}^*|^2} = \frac{32\pi \, m_\Delta^2}{\left[m_\Delta^2 - (m_p + m_\pi)^2\right]\left[m_\Delta^2 - (m_p - m_\pi)^2\right]} =$$
$$= 485 \, \mathrm{GeV}^{-2} = 188 \, \mathrm{mbarn}, \qquad (1.300)$$

where we have used the values $\mathrm{BR}_{\pi^+ p} = 1$, $J = 3/2$, $S_p = 1/2$, and $S_\pi = 0$. In the last row of Eq. (1.300), we have made use of Eq. (1.9) to convert the result into SI units. This result is in good agreement with the experimental data, see e.g. Fig. 2.11 of Ref. [16].

Suggested Readings

A concise summary of the Breit–Wigner scattering can be found in the PDG review dedicated to kinematics, Chap. 47 of Ref. [4].

Bando n. 18211/2016

Problem 1.55 How was it possible to measure the number of SM neutrino families at LEP by studying the Z^0 resonance?

Discussion

In the neighbourhood of the Z^0 mass, the cross section for $e^+ e^- \to Z^0 \to$ hadrons for unpolarised electron-positron beams features an energy dependence described by a *relativistic Breit–Wigner*:

$$\sigma_{\text{had}}(s) = \frac{12\pi}{m_Z^2} \frac{\Gamma_e \Gamma_{\text{had}}}{\Gamma_Z^2} \frac{s \Gamma_Z^2}{(s - m_Z^2)^2 + s^2 \Gamma_Z^2/m_Z^2}, \tag{1.301}$$

where Γ_Z is the width at the Z^0 mass, see Ref. [4]. In Eq. (1.301), the Breit–Wigner width grows with s. Notice that the kinematic part of Eq. (1.301) reduces to Eq. (1.299) when $s \approx m_Z^2$, Indeed:

$$\frac{s \Gamma_Z^2}{(s - m_Z^2)^2 + s^2 \Gamma_Z^2/m_Z^2} \approx \frac{m_Z^2 \Gamma_Z^2}{(\sqrt{s} - m_Z)^2(\sqrt{s} + m_Z)^2 + m_Z^2 \Gamma_Z^2} = \frac{\Gamma_Z^2/4}{(\sqrt{s} - m_Z)^2 + \Gamma_Z^2/4}. \tag{1.302}$$

Radiative corrections due to soft and collinear photon radiation from the incoming leptons distort the lineshape by inducing an asymmetric tail and by reducing the peak cross section to a value:

$$\sigma_{\text{had}}^{\text{peak}} = \frac{12\pi}{m_Z^2} \frac{\Gamma_e \Gamma_{\text{had}}}{\Gamma_Z^2}(1 - \delta_{\text{rad}}) \equiv \sigma_{\text{had}}^0(1 - \delta_{\text{rad}}). \tag{1.303}$$

In the SM, the Z^0 boson width is given by the sum of the hadronic, leptonic, and neutrino partial widths:

$$\Gamma_Z = \Gamma_{\text{had}} + N_\nu \Gamma_\nu + (3 + \delta_\tau)\Gamma_\ell, \tag{1.304}$$

where N_ν is the number of neutrino families that couple to the Z^0 and δ_τ is a phase-space corrections that accounts for the large τ mass.

Solution

There are three main methods to measure at LEP the number of light and active neutrinos, i.e. charged under the SM group.

The first method is based on subtracting the leptonic and hadronic width from the total visible width Γ_Z. This can be best implemented as fit to the lineshape of the hadronic cross section $\sigma_{\text{had}}(s)$ of Eq. (1.301) measured at different value of $\sqrt{s}$. By taking the SM values for δ_{rad}, Γ_{had}, Γ_ν, and Γ_e, the only unknown parameters in the fit are m_Z and N_ν, which can be therefore simultaneously extracted from the fit. It should be noticed that the sensitivity to Γ_ν arises from both the width of the Breit–Wigner and from the cross section value at the peak. The result published by the ALEPH Collaboration in 1989 gave a result

$$N_\nu = 3.27 \pm 0.30 \quad \text{and} \quad N_\nu < 4 \quad \text{at 98\% CL}. \tag{1.305}$$

The lineshape-based analysis relies on model assumptions for the partial widths and on the lineshape itself. Part of these assumptions can be relieved by using additional observables measured at the Z^0 peak. The ratio between the hadronic and leptonic decay widths, $R_\ell^0 = \Gamma_{\text{had}}/\Gamma_\ell$, is an observable that could be measured to large accuracy at LEP. By assuming universality of the charged lepton coupling to the Z^0 boson, the ratios R_e^0, R_μ^0, and R_τ^0 can be assumed identical, modulo the mass-related corrections for τ leptons, encoded in a correction factor $R_\tau^0/R_\ell^0 \equiv 1 + \delta_\tau$. By using the hadronic cross section at the peak, σ_{had}^0, one can therefore express the ratio $R_{\text{inv}}^0 \equiv \Gamma_{\text{inv}}/\Gamma_\ell$, as a function of m_Z, σ_{had}^0, and R_ℓ^0 as:

$$\sigma_{\text{had}}^0 = \frac{12\pi}{m_Z^2} \frac{\Gamma_e \, \Gamma_{\text{had}}}{\Gamma_Z^2} = \frac{12\pi}{m_Z^2} \frac{R_\ell^0}{(R_{\text{inv}}^0 + 3 + \delta_\tau + R_\ell^0)^2},$$

$$R_{\text{inv}}^0 = \left(\frac{12\pi R_\ell^0}{m_Z^2 \sigma_{\text{had}}^0} \right)^{\frac{1}{2}} - R_\ell^0 - (3 + \delta_\tau). \tag{1.306}$$

If one further assumes that the total width arises from neutrinos, the value of R_{inv}^0 thus obtained can be related to the number of active light neutrinos as:

$$R_{\text{inv}}^0 = \frac{N_\nu \, \Gamma_\nu^{\text{SM}}}{\Gamma_\ell^{\text{SM}}} \quad \Rightarrow \quad N_\nu = R_{\text{inv}}^0 \left(\frac{\Gamma_\ell}{\Gamma_\nu} \right)_{\text{SM}} \tag{1.307}$$

The combined result from the four LEP experiments is $N_\nu = 2.984 \pm 0.008$ [23].

The third method is based on a direct measurement of the invisible width Γ_{inv}^0 from the cross section of the process $e^+ e^- \to \nu \bar{\nu} \gamma$, where the final-state radiation (FSR) photon is required to trigger the event. The combined LEP measurement gave a result of $N_\nu = 2.92 \pm 0.05$, compatible with the method of Eq. (1.307), although plagued by larger uncertainties.

Suggested Readings

The measurement of the number of light neutrino families from the Z^0 lineshape is documented in Ref. [24]. The details on the combined LEP measurement of N_ν can be found in Ref. [23]. See also the dedicated PDG review on this subject [4] for more details.

Problem 1.56 A neutral narrow resonance X of mass M, natural width Γ, and spin J, is produced in proton-proton collisions at a centre-of-mass energy $\sqrt{s}$ and detected through its decay to a pair of photons. Two partonic channels contribute to the resonance production: $q\bar{q} \to X$ and $gg \to X$. Write down the LO cross section for $pp \to X \to \gamma\gamma$ in terms of the resonance parameters and of the proton PDF.

Solution

At LO in perturbation theory, the production amplitude is described by the s-channel diagrams $q\bar{q} \to X$ and $gg \to X$, followed by the disintegration of X into a pair of photons. For values of the partonic centre-of-mass energy $\hat{s}$ in the neighbourhood of

M^2, the partonic cross section is described by the relativistic Breit–Wigner similar to Eq. (1.301). In high-energy hadron-hadron interactions, the cross section can be factorised into a short-distance component, characterised by an energy scale $\hat{s}$, and a long-distance component that accounts for the parton distribution inside the protons, described by the gluon, quark, and antiquark PDF's:

$$\sigma = \int dx_1 \, dx_2 \left[\frac{\sum_{ij} f_i(x_1, \hat{s}) f_j(x_2, \hat{s})}{1 + \delta_{ij}} \right] \hat{\sigma}(\hat{s} = x_1 x_2 s),$$

$$\text{with} \quad \hat{\sigma} = \frac{16\pi}{M^2} \frac{(2J+1)}{\rho_i \rho_j} \frac{\Gamma_{\gamma\gamma} \, \Gamma_{ij}}{\Gamma^2} \frac{M^2 \Gamma^2}{(\hat{s} - M^2)^2 + M^2 \Gamma^2} (1 + \delta_{ij}) \qquad (1.308)$$

Here, $f_i(x, \hat{s})$ is the PDF of parton i evaluated at the momentum fraction x and at a factorisation scale $\hat{s}$, see e.g. Ref. [18]. All of the short-distance physics is encoded in the partonic cross section $\hat{\sigma}$, which is parametrised as in Eq. (1.301). For unpolarised initial-states, one has to average over the possible configurations of quantum numbers as described by the appropriate density matrix, resulting in a dilution factor ρ_i and ρ_j. The symmetry fractor $(1 + \delta_{ij})$ accounts for the undistinguishability of the initial-state particles. An heuristic motivation for this extra factor will be provided later. Notice that the energy-dependence of the width has been suppressed because we can work under the assumption that $\Gamma \ll M$, so that $\hat{s} \sim M^2$. Under this *narrow-width* assumption, we can further approximate the Breit–Wigner by a delta function:

$$\frac{1}{(\hat{s} - M^2)^2 + M^2 \Gamma^2} \approx \frac{\pi}{M \Gamma} \delta(\hat{s} - M^2). \qquad (1.309)$$

The factor of π at the numerator ensures the proper normalisation, as one can readily verify by integrating both sides of Eq. (1.309) over $\hat{s}$. The total cross section thus become:

$$\sigma = \frac{16\pi^2}{M\Gamma} \frac{(2J+1)}{\rho_i \rho_j} (1 + \delta_{ij}) \Gamma_{\gamma\gamma} \, \Gamma_{ij} \int dx_1 \, dx_2 \left[\frac{\sum_{ij} f_i(x_1, \hat{s}) f_j(x_2, \hat{s})}{1 + \delta_{ij}} \right] \delta(x_1 x_2 s - M^2)$$

$$= \frac{(2J+1)}{M\Gamma s} \Gamma_{\gamma\gamma} \, \Gamma_{ij} \sum_{ij} \underbrace{\frac{16\pi^2}{\rho_i \rho_j} \int_{\tau \equiv \frac{M^2}{s}}^{1} \frac{dx}{x} f_i(x, \hat{s}) f_j\left(\frac{\tau}{x}, \hat{s}\right)}_{=C_{ij}(\tau)}$$

$$\equiv \frac{(2J+1)}{M\Gamma s} \Gamma_{\gamma\gamma} \sum_{ij} C_{ij}\left(\frac{M^2}{s}\right) \Gamma_{ij} \qquad (1.310)$$

The long-distance physics in fully encoded into the *parton luminosity* factor $C_{ij}(\tau)$. Specialising Eq. (1.310) to the case of interest, we have:

$$\sigma = \frac{(2J+1)}{M\Gamma s} \Gamma_{\gamma\gamma} \times \begin{cases} \Gamma_{gg} \times \frac{\pi^2}{8} \int_{\tau}^{1} \frac{dx}{x} g(x) \, g\left(\frac{\tau}{x}\right) & g\,g \to X \\ \Gamma_{q\bar{q}} \times \frac{4\pi^2}{9} \int_{\tau}^{1} \frac{dx}{x} \left[q(x) \, \bar{q}\left(\frac{\tau}{x}\right) + \bar{q}(x) \, q\left(\frac{\tau}{x}\right) \right] & q\,\bar{q} \to X \end{cases}$$

$$(1.311)$$

Table 1.1 Numerical value of the parton luminosity factors $C(\tau)$ evaluated at $\tau = M^2/s$ with $M = 750$ GeV and $\sqrt{s} = 13$ TeV, and factorisation scale $\mu^2 = M^2$, using the MSTW2008NLO PDF set

$C_{d\bar{d}}$	$C_{u\bar{u}}$	$C_{s\bar{s}}$	$C_{c\bar{c}}$	$C_{b\bar{b}}$	C_{gg}
627	1054	83	36	15	2137

To derive this result, we have used the fact that the sum over the gluon PDF's is symmetric under $i \leftrightarrow j$, and that the density factors are

$$\rho = \begin{cases} \frac{1}{2 \cdot 2} \frac{1}{N_A^2} = \frac{1}{4 \cdot 8^2} = \frac{1}{256} & g\,g \\ \frac{1}{2 \cdot 2} \frac{1}{N_C^2} = \frac{1}{4 \cdot 3^2} = \frac{1}{36} & q\,\bar{q} \end{cases} \tag{1.312}$$

For example, at a resonance mass $M = 750$ GeV and at a centre-of-mass energy $\sqrt{s} = 13$ TeV, the parton luminosities using the MSTW2008NLO PDF set are reported in Table 1.1. In one further assumes that one production channel dominates over the others (for example, due to the largeness of $C_{ij}(\tau)$ or of the partial width Γ_{ij}), then, by measuring σ, M, and Γ, one can constrain the product of the branching ratios into photons and into the channel ij in the combination:

$$\mathrm{BR}(X \to \gamma\,\gamma) \cdot \mathrm{BR}(X \to ij) = \frac{\sigma\,s}{(2J+1)\,C_{ij}} \frac{M}{\Gamma}. \tag{1.313}$$

Discussion

We now give an heuristic motivation for the symmetry factor $(1 + \delta_{ij})$ appearing in the Breit–Wigner of Eq. (1.308). Consider for example the case of a spin-0 particle produced in the s-channel by massless particles a, and decaying to a pair of massless particles b. The cross section can be computed directly from Eq. (1.293) using the spin-average of Eq. (1.293):

$$\sigma = \sum_{r_1,r_2} \sum_{r_3,r_4} \frac{1}{\rho_a^2} \frac{1}{4|\mathbf{p}^*|\sqrt{s}} \int d\Phi_2 |\mathcal{M}_a|^2 \underbrace{\frac{1}{(s-M^2)^2 + \Gamma^2 M^2}}_{\text{propagator}} |\mathcal{M}_b|^2 =$$

$$= \frac{1}{2M^2\rho_a^2} 2M \underbrace{\frac{1}{2M} \sum_{r_3,r_4} \int d\Phi_2 |\mathcal{M}_b|^2}_{\Gamma_b} \left[\frac{1}{(s-M^2)^2 + \Gamma^2 M^2} \right] \frac{2M}{\Phi_2} \underbrace{\frac{1}{2M} \sum_{r_1,r_2} |\mathcal{M}_a|^2 \Phi_2}_{\Gamma_a}$$

$$= \frac{2\,(\Phi_2)^{-1}}{M^2} \frac{\mathrm{BR}_a\,\mathrm{BR}_b}{\rho_a^2} \frac{M^2\Gamma^2}{(s-M^2)^2 + \Gamma^2 M^2}, \tag{1.314}$$

where we have approximated $|\mathbf{p}^*| = M/2$ and $\sqrt{s} = M$. The two-body phase-space for massless particles is given by Eq. (1.189). If the particles are distinguishable, one has to integrate over the full solid angle, giving $\Phi_2 = 1/8\pi$, and the usual factor of 16π at the numerator, whereas identical particles give a factor of two smaller phase-space, hence the symmetry factor $(1 + \delta_{ij})$ of Eq. (1.308).

Suggested Readings

A putative new high-mass resonance decaying to a pair of photons at a mass of about 750 GeV was reported by both the ATLAS and CMS Collaborations in December 2016 [25, 26]. The analysis of additional data, however, disproved the excess, which was then attributed to a mere statistical fluctuation of the background, dominated by prompt diphoton production, see e.g. Ref. [27]. A lot of theoretical speculation was stimulated by the early observation, see e.g. Ref. [28] for further details.

Bando n. 18211/2016

Problem 1.57 An excited state of ^{57}Fe decays by emitting a 14.4 keV photon ($t_{1/2} = 68$ ns). Determine the FWHM of the energy distribution of the emitted photon.

Solution

The spectral width Γ and decay time τ are related by the relation $\Gamma = \hbar/\tau$, see Eq. (1.187), or, in natural units, $\Gamma = \tau^{-1}$. By using Eq. (1.8), we can express $t_{1/2}$ in keV^{-1}, giving:

$$\Gamma = \tau^{-1} = \frac{\ln 2}{t_{1/2}} = 6.7 \times 10^{-12} \text{ keV}, \tag{1.315}$$

If a Breit–Wigner distribution as in Eq. (1.299) is assumed, the full-width-at-half-maximum is given by $\Gamma_{\text{FWHM}} = \Gamma = 0.7 \times 10^{-11}$ keV. The relative width Γ_{FWHM}/E is therefore of order 10^{-12}, and thus too small to be measured in spectroscopy.

Discussion

The natural width of spectral lines is generally very small. In hot media, the energy resolution is dominated by other effects, like collisional and Doppler broadening.

Problem 1.58 In classical mechanics, the differential cross section $d\sigma/d\Omega$ of a particle of mass m and momentum $\mathbf{p}$ scattered by a central potential is given by the formula:

$$\frac{d\sigma}{d\Omega} = \frac{b}{\sin\theta} \left| \frac{db}{d\theta} \right|, \tag{1.316}$$

where $d\Omega = d\phi \, d\cos\theta$ is the solid angle parametrised by the polar and azimuthal angles with respect to $\mathbf{p}$, and $b = b(\theta, |\mathbf{p}|)$ is the so-called impact parameter, i.e. the

Fig. 1.15 Cartoon showing
the classical scattering of a
particle against a central
potential

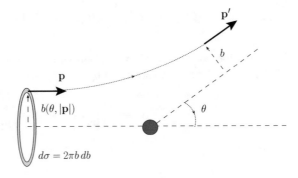

distance from the polar axis at time $t = -\infty$ of those particles that get scattered at
an angle θ at time $t = +\infty$. Use this formula to derive the Rutherford cross section
for a particle of charge e scattered by a heavy nucleus with charge $Z e$.

Discussion

It is easy to verify that Eq. (1.316) gives the correct result. Let's imagine that the
source of the potential is located at the origin of the reference frame. Those particles
that at time $t = +\infty$ are scattered at a polar angle θ have momentum $\mathbf{p}' = |\mathbf{p}|\, \mathbf{e}_\theta$.
Since the potential is central, angular momentum with respect to the origin is con-
served: all particles with impact parameter b will end up at the same polar angle θ,
hence it's possible to define a function $b = b(\theta, |\mathbf{p}|)$, see Fig. 1.15. By definition,
$d\sigma = d^2\mathbf{b} = b\, db\, d\phi$ is the infinitesimal cross section for scattering at the polar
angle centred around (θ, ϕ), hence:

$$\frac{d\sigma}{d\phi} = b\, db \quad \Rightarrow \quad \frac{d\sigma}{d\phi\, d\cos\theta} = \frac{b}{\sin\theta}\left|\frac{db}{d\theta}\right| \tag{1.317}$$

If the classical source is replaced by a pointlike source of finite mass M, Eq. (1.317)
still holds provided that the polar angle θ is measured in the centre-of-mass frame,
and m and $|\mathbf{p}|$ gets replaced by the reduced mass $\mu = m M/(M + m)$ and centre-of-
mass momentum, respectively.

Solution

We need to compute $b(\theta, |\mathbf{p}|)$ for the Coulomb potential. To this purpose, it is con-
venient to exploit the time-invariance of $\mathbf{L} = \mathbf{r} \times \mathbf{p}$. In the orbital plane, with
the x-axis parallel to the initial momentum, the momentum after the scattering is
$\mathbf{p}' = |\mathbf{p}|(\cos\theta, \sin\theta)$. The angular momentum for a particle with impact parameter
b is given by

$$|\mathbf{L}| = -m|\mathbf{r}|^2 \frac{d\theta}{dt} = |\mathbf{p}|b \quad \Rightarrow \quad dt = -\frac{m|\mathbf{r}|^2}{|\mathbf{p}|b} d\theta. \tag{1.318}$$

The Coulomb potential $V(\mathbf{r})$ and its Fourier transform $\tilde{V}(\mathbf{q})$ for a point-like source of charge Ze at the origin are given by

$$V(\mathbf{r}) = \frac{Ze}{4\pi|\mathbf{r}|}, \qquad \tilde{V}(\mathbf{q}) = \frac{Ze}{|\mathbf{q}|^2}, \tag{1.319}$$

so that

$$\frac{dp_y}{dt} = -\partial_z V(\mathbf{r}) = \frac{Ze^2}{4\pi|\mathbf{r}|^2} \sin\theta(t),$$

$$dp_y = \frac{Ze^2}{4\pi|\mathbf{r}|^2} \sin\theta \left(-\frac{m|\mathbf{r}|^2}{|\mathbf{p}|b} \right) d\theta = -\frac{mZe^2}{4\pi|\mathbf{p}|b} \sin\theta \, d\theta$$

$$|\mathbf{p}|\sin\theta = \int_\pi^\theta dp_y = \frac{mZe^2}{4\pi|\mathbf{p}|b} \int_\theta^\pi d\theta' \sin\theta' = \frac{mZe^2}{4\pi|\mathbf{p}|b} (1 + \cos\theta), \tag{1.320}$$

from which we get the relation:

$$b(\theta, |\mathbf{p}|) = \frac{mZe^2}{4\pi|\mathbf{p}|^2} \frac{1 + \cos\theta}{\sin\theta} = \frac{mZe^2}{4\pi|\mathbf{p}|^2 \tan\frac{\theta}{2}}. \tag{1.321}$$

Using Eq. (1.317), we thus obtain:

$$\frac{d\sigma}{d\Omega} = \frac{1}{\sin\theta} \frac{mZe^2}{4\pi|\mathbf{p}|^2 \tan\frac{\theta}{2}} \left| \frac{mZe^2}{4\pi|\mathbf{p}|^2} \left(-\frac{1}{\tan^2\frac{\theta}{2}} \right) \frac{1}{2\cos^2\frac{\theta}{2}} \right| =$$

$$= \frac{1}{4} \left(\frac{mZe^2}{4\pi|\mathbf{p}|^2} \right)^2 \frac{1}{\sin^4\frac{\theta}{2}} = \left[\frac{Z\alpha(\hbar c)}{4T\sin^2\frac{\theta}{2}} \right]^2 \tag{1.322}$$

where $T = |\mathbf{p}|^2/2m$ is the kinetic energy of the particle and $\alpha = e^2/(4\pi\hbar c)$, which reproduces the well-known non-relativistic Rutherford's formula. See Problem 1.59 for its relativistic generalisation. In natural units, Rutherford's formula becomes:

$$\frac{d\sigma}{d\Omega} = \left[\frac{Z\alpha}{4T\sin^2\frac{\theta}{2}} \right]^2, \tag{1.323}$$

and the conversion into MKS units can be done by either remembering that $\alpha \to (1/137) \cdot 197$ MeV fm, or by using the conversion GeV$^{-2} \to$ mbarn of Eq. (1.9).

The above result, which has been obtained according to the laws of classical mechanics, coincides with the (non-relativistic) quantum-mechanical expectation from the exchange of a virtual photon. Indeed, the amplitude squared for the scattering is $|\mathscr{M}_{NR}|^2 = |\tilde{V}(\mathbf{q})|^2 = Z^2 e^4/|\mathbf{q}|^4$, where $\mathbf{q}$ is the four-momentum exchange. Since that the reaction is elastic, the velocity and centre-of-mass momenta are identical before and after the scattering, and Eq. (1.297) gives:

$$\frac{d\sigma}{d\Omega} = \frac{1}{4\pi^2} \frac{|\mathbf{p}^*|^2}{v_{rel}^{*2}} \frac{Z^2 e^4}{|\mathbf{q}|^4}. \tag{1.324}$$

For the Rutherford scattering, $|\mathbf{q}| = 2|\mathbf{p}| \sin\theta/2$, $v_{rel} = |\mathbf{v}|$, and the centre-of-mass frame coincides with the laboratory frame if $m \ll M$, M being the mass of the source generating the potential. Hence, Eq. (1.324) can be written as:

$$\frac{d\sigma}{d\Omega} = \frac{4}{(4\pi)^2} \frac{Z^2 e^4 |\mathbf{p}^*|^2}{16|\mathbf{v}|^2|\mathbf{p}^*|^4 \sin^4 \frac{\theta}{2}} = \left[\frac{Z\alpha}{4T \sin^2 \frac{\theta}{2}}\right]^2, \tag{1.325}$$

which agrees with Eq. (1.323).

Problem 1.59 The differential cross section of a spin-1/2 particle of mass m scattering against a point-like heavy particle is described by the *Mott formula*:

$$\frac{d\sigma}{d\Omega} = \left[\frac{\alpha}{2|\mathbf{p}|^2 \sin^2 \frac{\theta}{2}}\right]^2 \left(m^2 + |\mathbf{p}|^2 \cos^2 \frac{\theta}{2}\right), \tag{1.326}$$

where $\mathbf{p}$ is the electron momentum.

- Show that the formula reduces to the Rutherford cross section in the classical limit $|\mathbf{p}| \ll m$.
- What is the value of the total cross section σ?
- Show that in the ultra-relativistic limit the formula reduces to:

$$\frac{d\sigma}{d\Omega} \approx \frac{\alpha^2 \cos^2 \frac{\theta}{2}}{4E^2 \sin^4 \frac{\theta}{2}}, \tag{1.327}$$

where E is the total relativistic energy. What is the reason behind the $\cos^2 \theta/2$ dependence of the cross section, which is not present in Rutherford's formula?

Discussion

The well-known Rutherford's formula applies to the non-relativistic scatettering between two charged particles, a moving projectile and a steady target, the latter being much heavier than the former so that the scattering occurs without any energy transfer, see Fig. 1.16. Indeed, the very same formula can be obtained by using the classical picture of a point-like charged particle interacting with the static electric field of the target, see Problem 1.58.

Fig. 1.16 Cartoon of the scattering of an electron off a much heavier particle of mass M

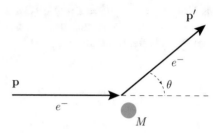

Solution

In the limit $|\mathbf{p}| \ll m$, the formula of Eq. (1.326) reduces to:

$$\frac{d\sigma}{d\Omega} \approx \left[\frac{\alpha}{2(mv)^2 \sin^2 \frac{\theta}{2}}\right]^2 m^2 = \left[\frac{\alpha}{4T \sin^2 \frac{\theta}{2}}\right]^2, \qquad (1.328)$$

where T is the kinetic energy, $T \equiv E - m$.

The total cross section is obtained by integrating the differential cross section over the full solid angle. For $\theta \in [0, \varepsilon]$, $\varepsilon \ll 1$, the integral goes like:

$$\int_0^\varepsilon d\theta \, \sin\theta \frac{1}{\sin^4 \frac{\theta}{2}} \sim \int_0^\varepsilon d\theta \frac{\theta}{\theta^4} = \int_0^\varepsilon \frac{d\theta}{\theta^3} = +\infty. \qquad (1.329)$$

This is a consequence of the electromagnetic interaction range being infinite.

In the ultrarelativistic limit $|\mathbf{p}| \approx E \gg m$, the Mott formula reduced to:

$$\frac{d\sigma}{d\Omega} \approx \left[\frac{\alpha \cos \frac{\theta}{2}}{2E \sin^2 \frac{\theta}{2}}\right]^2 = \frac{\alpha^2 \cos^2 \frac{\theta}{2}}{4E^2 \sin^4 \frac{\theta}{2}}. \qquad (1.330)$$

The $\cos\theta/2$ term at the numerator arises from the spin-$1/2$ nature of the electron: since the electromagnetic interaction does not change the chirality, an initial relativistic fermion of a given chirality will also have a fixed helicity h as discussed in Problems 1.6 and 1.8. A left-handed electron with $h = -1/2$ cannot be scattered exactly backward, since this configuration would result in a spin-flipped state, thus changing the J_z component of the state along the scattering axis.

Problem 1.60 Two ion beams, one composed of $^4_2\text{He}^+$ ions and the other of $^A_3\text{Li}^+$ ions of unknown mass number A, collide against a fixed gold target. A velocity selector filters ions of the same initial velocity $|\mathbf{v}| = 0.1\,c$. The two beams provide the same integrated flux. A particle detector, located at a polar angle θ with respect to the beam direction, counts the number of scattered ions. The ratio between the number of countings for the two beams is $R = N_{He}/N_{Li} = 3.0 \pm 0.1$. What is the most probable mass number A of the lithium ions?

Solution

Given that the target nuclei are much heavier than any of the two ions, and that the ion velocity is small enough to treat the scattering particles as classical, we can use Rutherford's formula (1.323). Under the conditions assumed by the problem, deviations between the two event counts can only arise from differences in the mass, velocity, or charge of the two beams, i.e.:

$$R = \frac{N_{He}}{N_{Li}} = \left(\frac{m_{Li}}{m_{He}} \frac{|\mathbf{v}_{Li}|^2}{|\mathbf{v}_{He}|^2} \frac{Z_{He}}{Z_{Li}} \right)^2 = \left(\frac{A}{4} \right)^2, \quad A = \sqrt{R} \cdot 4 = 6.93. \quad (1.331)$$

By propagating the error on the ratio, we find

$$\frac{\delta A}{A} = \frac{1}{2} \frac{\delta R}{R}, \quad (1.332)$$

from which we estimate $A = (6.93 \pm 0.11)$, which is compatible with the closest integer to the 1σ level. The most probable value for the mass number of the lithium ions is therefore $A = 7$, which corresponds indeed to a stable isotope.

Problem 1.61 A beam of alpha particles of energy $T = 0.1$ GeV collides against a fixed target of aluminium (density $\rho = 2.7$ gcm^{-3}, molar mass $A = 27$ g/mol) of thickness of $d = 1$ cm. The beam flux at the target is $\Phi = 10^9$ s^{-1}. A scintillating detector is placed at an angle $\theta = 30°$ from the beam axis, and $L = 1$ m away from the target. The active surface of the detector has a cross section of 1 cm $\times$ 1 cm as seen from the target. Estimate the counting rate measured by the detector.

Solution

The instantaneous luminosity $\mathscr{L}$ of the experimental set-up is given by Eq. (1.290):

$$\mathscr{L} = \Phi \frac{\rho N_A}{A} d = 10^9 \text{ s}^{-1} \cdot \frac{2.7 \text{ g cm}^{-3} \cdot 6.02 \times 10^{23} \text{ mol}^{-1}}{27 \text{ g mol}^{-1}} \cdot 1 \text{ cm}$$

$$\approx 6 \times 10^{31} \text{ cm}^{-2}\text{s}^{-1}. \quad (1.333)$$

Using the Rutherford's cross section formula (1.323), we get:

$$\frac{d\sigma}{d\Omega}(\theta) = \frac{Z_\alpha^2 Z_{Al}^2 \alpha^2}{16 T^2 \sin^4 \frac{\theta}{2}} = \frac{2^2 \cdot 13^2 \cdot (1/137)^2}{16 \cdot 0.1^2 \cdot \sin^4(\frac{30 \cdot \pi/180}{2})} = 50.2 \text{ GeV}^{-2}/\text{std} =$$

$$(1.334)$$

$$= 1.95 \cdot 10^{-26} \text{ cm}^2/\text{std}. \quad (1.335)$$

The last equality in Eq. (1.334) comes from the conversion $\text{GeV}^{-2} \rightarrow \text{cm}^2$, see Eq. (1.9). The integrated cross section in a small solid angle $\Delta\Omega = (1\text{ cm} \times 1\text{ cm})/1\text{ m}^2 = 10^{-4}$ std is therefore given by:

$$\sigma \approx \frac{d\sigma}{d\Omega}(\theta) \cdot \Delta\Omega = 1.95 \times 10^{-30}\text{ cm}^2. \tag{1.336}$$

Using the instantaneous luminosity of Eq. (1.333), we therefore expect an average rate of

$$\frac{dN}{dt} = \mathscr{L}\,\sigma \approx 120\text{ Hz}. \tag{1.337}$$

Problem 1.62 A beam made of π^+ and K^+ with momentum $|\mathbf{p}| = 5\,\text{GeV}$ is directed towards a bubble chamber filled with liquid hydrogen. Events where an energetic δ-ray is produced are selected and the energy of the emitted electron is measured with negligible uncertainty. Determine the minimum kinetic energy T of the δ-ray such that the incoming particle can be unambiguously identified as a pion. Estimate the probability of such events if the chamber has a total length $L = 1$ m.

Solution

The discrimination between pions and kaons is possible because the maximum energy transfer at fixed momentum $|\mathbf{p}|$ decreases with the mass of the incoming particle, see Problem 1.26. Indeed, according to Eq. (1.141), we have

$$T_{\max} = \frac{2\,m_e|\mathbf{p}|^2}{m_e^2 + M^2 + 2m_e\sqrt{|\mathbf{p}|^2 + M^2}} = \begin{cases} 1.05\text{ GeV} & \pi^+ \\ 103\text{ MeV} & K^+ \end{cases} \tag{1.338}$$

Therefore, any δ-ray with energy above 103 MeV can be assumed to originate from a pion scattering.

To compute the probability of a δ-ray emission with energy $T_{\max}^K \leq T \leq T_{\max}^\pi$, we integrate the differential cross section. This is best done by expressing the differential cross section as a function of the four-momentum transfer q^2, giving:

$$q^2 = -|\mathbf{q}^*|^2 = -4|\mathbf{p}^*|^2 \sin^2\frac{\theta^*}{2} \quad \Rightarrow \quad d\Omega^* = \frac{d\phi\,dq^2}{2|\mathbf{p}^*|^2}. \tag{1.339}$$

Inserting this last expression into Eq. (1.324), we get:

$$\frac{d\sigma}{dq^2} = \frac{1}{4\pi^2}\frac{|\mathbf{p}^*|^2}{v_{\text{rel}}^{*2}}\frac{Z^2 e^4}{q^4}\frac{\pi}{|\mathbf{p}^*|^2} = \frac{4\pi}{v_{\text{rel}}^{*2}}\frac{Z^2\alpha^2}{q^4}, \tag{1.340}$$

where v_{rel} is the relative velocity in the centre-of-mass frame frame. The latter can be expressed in terms of the velocity $\mathbf{v}$ of the beam particles in the laboratory frame through Eq. (1.52):

$$v^*_{rel} = |\mathbf{v}|(1 + \beta^*_e \beta^*_\pi),$$

(1.341)

where β^*_e and β^*_π are the velocities in the centre-of-mass frame. The latter can be computed from Eqs. (1.91) to (1.89). Since the pion mass is much larger than the electron mass, we can approximate:

$$\beta^*_e \approx \frac{s - m^2_\pi}{s - m^2_\pi} \approx 1, \qquad \beta^*_\pi \approx \frac{s - m^2_\pi}{s + m^2_\pi} \approx \frac{m_e E}{m^2_\pi + m_e E} = 0.12.$$

(1.342)

Furthermore, if we denote the four-momentum of the initial (final) electron by k (k'), it follows that

$$k' = k + q, \qquad m^2_e = m^2_e + 2m_e q_0 + q^2, \qquad q_0 = \frac{-q^2}{2m_e}.$$

(1.343)

Since $q_0 = T$, we can easily transform Eq. (1.340) into a cross section differential in the kinetic energy of the recoiling electron:

$$\frac{d\sigma}{dT} = \frac{2\pi\, Z^2\alpha^2}{v^{*2}_{rel}\, m_e} \frac{1}{T^2}.$$

(1.344)

The total cross section for T in the range $[T^K_{max}, T^\pi_{max}]$ is therefore given by:

$$\sigma = \int_{T^K_{max}}^{T^\pi_{max}} \frac{d\sigma}{dT} = \frac{2\pi Z^2\alpha^2}{v^{*2}_{rel}\, m_e} \left(\frac{1}{T^K_{max}} - \frac{1}{T^\pi_{max}} \right) =$$

$$= \frac{2\pi \cdot (\frac{1}{137})^2 (197)^2 \text{ MeV}^2\, 10^{-26} \text{ cm}^2}{0.511 \text{ MeV}} \frac{\left(1 + m^2_\pi/|\mathbf{p}|^2\right)^2}{(1 + 0.12)^2} \left(\frac{\text{MeV}^{-1}}{103} - \frac{\text{MeV}^{-1}}{1.05 \times 10^3} \right)$$

$$= 1.8 \times 10^{-27} \text{ cm}^2.$$

(1.345)

The probability p of such an emission across a length $L = 1$ m can be estimated by introducing the interaction length of Eq. (1.291):

$$p \approx \frac{L}{\lambda} = L \cdot \left(\sigma \frac{\rho N_A}{A} \right) =$$

$$10^2 \text{ cm} \cdot 1.8 \times 10^{-27} \text{ cm}^2 \cdot \frac{0.06 \text{ g cm}^{-3} \cdot 6.02 \times 10^{23} \text{ mol}^{-1}}{1 \text{ g mol}^{-1}} = 6.5 \times 10^{-3}$$

(1.346)

where we have used the value $\rho = 0.06$ g cm^{-3} for the mass density of liquid hydrogen. Notice that this probability comes out to be small, thus justifying the use

Fig. 1.17 Probability of
single collisions in which
released electrons have an
energy E or larger (left scale)
and practical range of
electrons in Ar/C H_4 (P10) at
NTP (dot-dashed curve, right
scale). Taken from Ref. [4]

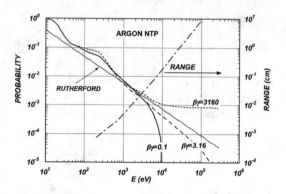

of Eq. (1.346) to estimate the probability of interaction (otherwise we should have
taken the cumulative of the exponential distribution, see Problem 2.14).

Discussion

According to Eq. (1.344), the probability of emitting an electron of energy equal or
larger than T follows an approximate T^{-1} law, see e.g. Fig. 1.17 taken from Ref. [4].
This property is relevant in gaseous detectors which detect the passage of a particle
by the ionisation trail left behind by the passage of a charged particle, see Chap. 2.
The emission of energetic electrons, whose range scales like T^2 (see Problem 2.3),
sets an intrinsic limitation to the position accuracy achievable by such detectors.

Suggested Readings

This problem is inspired from a similar exercise that can found in Ref. [16], Chap. 11.
The reader is recommended to study in detail that chapter to find more information
on the subject.

Bando n. 13705/2010

Problem 1.63 A neutron beam passes through a chamber of length $d = 1$ m which
can be either empty or filled with hydrogen at 20 °C and 760 mmHg. The neutrons
are detected by a counter located at the end of the cylinder. By using the same beam,
5×10^6 countings are measured when the chamber is empty, and 4.6×10^6 when the
chamber is filled with H_2. Estimate the neutron-proton cross section and its statistical
uncertainty.

Solution

Let's denote the detection efficiency of the detector by ε. The number of countings
for the two set-up is given by Eq. (1.290), namely:

$$\begin{cases} N_1 = N_0 \, \varepsilon \, (1 - 2nd\sigma) \\ N_2 = N_0 \, \varepsilon \end{cases} \tag{1.347}$$

where N_0 is the total number of neutrons entering the chamber, n is the molecular density of the gas when the chamber is filled, and σ is the total neutron-proton cross section. The factor of two in front of the gas density accounts for the fact that each hydrogen molecule contains two protons. The molecular density n can be obtained from the law of ideal gases $PV = \mathcal{N} RT$, with $\mathcal{N} = N/N_A$, from which:

$$n = \frac{N}{V} = \frac{P N_A}{RT} = \frac{760 \cdot 133 \, \text{Pa} \cdot 6.02 \times 10^{23} \, \text{mol}^{-1}}{8.314 \, \text{J mol}^{-1} \, \text{K}^{-1} \cdot 293 \, \text{K}} = 2.5 \times 10^{19} \, \text{cm}^{-3}. \tag{1.348}$$

Taking the ratio between the two countings, the unknown efficiency ε cancels, giving:

$$\frac{N_1}{N_2} = 1 - 2nd\sigma, \quad \sigma = \left(\frac{N_2 - N_1}{N_1}\right) \frac{1}{2nd} =$$
$$= \left(\frac{5.0 \times 10^6 - 4.6 \times 10^6}{5.0 \times 10^6}\right) \frac{1}{2 \cdot 2.5 \times 10^{19} \, \text{cm}^{-3} \cdot 10^2 \, \text{cm}} = 16 \, \text{barn}. \tag{1.349}$$

The event counts are uncorrelated and large enough so that the Gaussian statistics applies, see Problem 4.11. By standard error propagation, see Sect. 4.2, it follows that:

$$(\delta\sigma)^2 \propto \left(\frac{N_2}{N_1^2}\right)^2 (\delta N_1)^2 + \left(\frac{1}{N_1}\right)^2 (\delta N_2)^2 = \left(\frac{N_2}{N_1}\right)^2 \left(\frac{1}{N_1} + \frac{1}{N_2}\right),$$
$$\frac{\delta\sigma}{\sigma} = \sqrt{\frac{N_2}{N_1}} \frac{\sqrt{N_1 + N_2}}{N_2 - N_1} = 0.8 \times 10^{-2}, \tag{1.350}$$

hence: $\sigma = (16 \pm 1)$ barn.

Bando n. 1N/R3/SUB/2005

Problem 1.64 A neutrino beam of mean energy $\langle E_\nu \rangle = 20$ GeV is produced from the decay of charged pions. Estimate:

- the energy of the pion beam (assumed monochromatic) that has generated the neutrino beam;
- the divergence of the neutrino beam at the far-end detector located at a distance $d = 100$ km downstream of the beampipe;
- the order-of-magnitude for the neutrino-nucleon cross section;
- the mean free path of the neutrinos in a detector with the density of water;
- the ratio between the cross section on protons and on electrons.

Solution

From Problem 1.19 and (1.109), we know that the energy of the neutrino in the laboratory frame is uniformly distributed in the range $E_\nu^*(\gamma \pm \sqrt{\gamma^2 - 1})$, where $E_\nu^* = (m_\pi^2 - m_\mu^2)/2m_\pi$ is the centre-of-mass energy of the neutrino in the most probable decay $\pi^+ \to \mu^+ \nu_\mu$, and $\gamma = E_\pi/m_\pi$. The mean energy is therefore:

$$\langle E_\nu \rangle = \gamma E_\nu^* = \left(\frac{E_\pi}{m_\pi}\right) \frac{m_\pi^2 - m_\mu^2}{2m_\pi} \quad \Rightarrow \quad E_\pi = \frac{2 \langle E_\nu \rangle}{1 - (m_\mu/m_\pi)^2} = 96 \text{ GeV}. \tag{1.351}$$

For a massless neutrino, the polar angle in the laboratory frame can range up to π radians. However, for large boosts, the probability of decaying at an angle $\theta \gg 1/\gamma$ is negligible. Since $\gamma \approx 670 \gg 1$, we can make use of Eqs. (1.73) and (1.76) to obtain:

$$\langle \theta \rangle \approx 2.3 \times 10^{-3}, \qquad \sigma_\theta \approx 4.9 \times 10^{-3}, \qquad \theta_{90\%} \approx 4.5 \times 10^{-3}. \tag{1.352}$$

Taking the 90% quantile as an estimator of the beam divergence of the far-end detector located at $d = 100$ km, we get a beam spread of about 450 m along the transverse coordinate.

The charged-current (CC) neutrino cross section on an isoscalar target in the deep inelastic scattering (DIS) regime, appropriate for this value of the neutrino energy, can be computed by using the effective Fermi Lagrangian of Eq. (2.82), giving:

$$\sigma_{\nu d} = \frac{G_F^2 s}{4\pi^2}, \qquad \sigma_{\nu \bar{u}} = \frac{G_F^2 s}{4\pi^2} \frac{(1 + \cos\theta)^2}{4}$$

$$\Rightarrow \quad \sigma_{\nu N} = \frac{1}{2} \left(\sigma_{\nu p} + \sigma_{\nu n}\right) = \frac{G_F^2}{\pi} M E_\nu \left[Q + \frac{\bar{Q}}{3}\right], \tag{1.353}$$

where $Q = \int dx\, x\, q(x)$ ($\bar{Q} = \int dx\, x\, \bar{q}(x)$) is the average momentum carried by quarks (antiquarks) inside the proton, and M indicates the average nucleon mass. See e.g. Ref. [4] for more informations. For anti-neutrinos, one needs to swap $Q \leftrightarrow \bar{Q}$. Taking $M = 0.938$ GeV, and using Eq. (1.9) to convert the result into SI, we get:

$$\sigma_{\nu N} = \frac{(1.166 \text{ GeV}^{-2})^2}{\pi} \cdot 0.938 \text{ GeV} \cdot E_\nu \left[Q + \frac{\bar{Q}}{3}\right]$$

$$\approx 1.56 \times 10^{-38} \text{ cm}^2 \left(\frac{E_\nu}{\text{GeV}}\right) \left[Q + \frac{\bar{Q}}{3}\right]. \tag{1.354}$$

The mean free-path for neutrinos in a medium of density $\rho = 1$ g cm^{-3} is given by Eq. (1.291). From DIS experiments, one measures $\bar{Q} \ll Q \approx 0.5$, see e.g. Problem 5.20. Thus, we get:

$$\lambda = \frac{A}{\sigma \, \rho \, N_A} = \frac{18 \text{ g mol}^{-1}}{1.56 \times 10^{-38} \text{ cm}^2 \cdot 20 \cdot 0.5 \cdot 1 \text{ g cm}^{-3} \cdot 6.02 \times 10^{23} \text{ mol}^{-1}}$$
$$\approx 2 \times 10^{14} \text{ m}. \tag{1.355}$$

The same effective theory predicts the total cross section of the CC interaction $\nu_\mu \, e^- \to \mu^- \, \nu_e$ to be as the first of Eq. (1.353):

$$\sigma_{ve} = \frac{G_F^2 \, s}{\pi} = \frac{2 \, G_F^2}{\pi} m_e \, E_\nu \quad \Rightarrow \quad \frac{\sigma_{ve}}{\sigma_{\nu N}} = \frac{2 \, m_e}{M \left[Q + \bar{Q}/3 \right]} \approx 2 \times 10^{-3}. \tag{1.356}$$

Suggested Readings

For more details on the interaction of high-energy neutrino with matter, the reader is addressed to Chap. 8 of Ref. [12]. A compendium of useful formulas can be also found in Sect. 49 of Ref. [4].

Appendix 1

We report here a simple computer program in Python which implements Newton's method for finding the roots of a real-valued function f, specialised here to the case where f is the first derivative of $\tan(\phi)$ in Eq. (1.67) with respect to $x \equiv \cos \theta^*$.

After initialising the program with the values of β, β_1^*, and β_2^*, the roots of f are searched for by iteratively incrementing the variable x as:

$$x_{i-1} \to x_i = x_{i-1} - \frac{f'(x_{i-1})}{f(x_{i-1})}$$

starting from an initial value x_0. The loop stops when the desired accuracy is attained, i.e. $\Delta x_i / x_{i-1} < \varepsilon$, or the maximum number of iterations is exceeded.

A critical point of such a method applied to the case of interest arises from the fact that $x \in [-1, 1]$, while the intermediate values x_i may occasionally fall outside of this range. When this happens, one can try tuning the starting value x_0 until the convergence is attained.

```python
import math
class Newton:
    def __init__(self, b0, b1, b2):
        self.a = b0/math.sqrt(1-b0*b0)*(b1 + b2)/(b1*b2)
        self.b = -b0*b0
        self.c = b0*(b1-b2)/(b1*b2)
        self.d = (b0*b0/(b1*b2)-1+b0*b0)
        self.err = -1 #-1=NOT CONVERGED, 0=OK, 1=ERROR
    # The first derivative of tan(phi) [=f in Newton's method]
    def f_prime(self, x):
        # need to define a=self.a, b=self.b, ...
        val = a*(b*x*x*x-(d+2*b)*x-c)/math.sqrt(1-x*x)/math.pow(b*x*x+c*x+d,2)
        return val
    # The second derivative of tan(phi) [=f' in Newton's method]
    def f_second(self, x):
        # need to define a=self.a, b=self.b, ...
        val0 = a/math.pow(1-x*x,1.5)/math.pow(b*x*x + c*x +d,3)
        val1 = (3*b*x*x - (d+2*b))*(1-x*x)*(b*x*x + c*x + d)
        val2 = (b*x*x*x-(d+2*b)*x-c)*(-5*b*x*x*x-3*c*x*x-(d-2*b)*x+2*c)
        return val0*(val1-val2)
    # Do up~to it_max iterations starting from x_start until accuracy<res
    def iterate(self, x_start=0.0, it_max=10, res=0.01):
        x_min = x_start
        n_it = 0
        for it in xrange(it_max):
            n_it += 1
            if abs(x_min)>1:
          self.err = 1
          break
            f0 = self.f_prime(x_min)
            f1 = self.f_second(x_min)
            delta = -f0/f1
            x_min += delta
            if(abs(delta)<res):
                self.err = 0
                break
        return (x_min, n_it)
# Run from command line: $ python newton.py
newton = Newton(b0=0.8, b1=0.3, b2=0.5)
result = newton.iterate(x_start=0.0, it_max=10, res=0.01)
print"Result:", result[0]
```

Appendix 2

The computer program below illustrates the generation of toy MC events where
an unpolarised resonance of mass M decays into a pair of massless particles. This
routine profits from a number of built-in functions available in ROOT that implement
a good deal of four-vectors algebra.

```python
import ROOT
import math

# the random number generator
ran = ROOT.TRandom3()

# M = resonance\index{Resonance} mass, ntoys = number of MC\index{Monte Carlo}
 events, beta = boost vector\index{Boost}
def toys(M=80., ntoys=1000, beta=ROOT.TVector3(0., 0., 0.1)):

    # this is the same for all events
    E_cm = M/2.

    for ntoy in xrange(ntoys):

        # cos(theta*) is uniform in [-1,1], phi in [-pi,pi]
        cos = ran.Uniform(-1,1)
        phi = ran.Uniform(-math.pi, +math.pi)

        # the 3-momentum in the centre-of-mass frame
        v1_cm = ROOT.TVector3()
        v1_cm.SetPx(E_cm*cos)
        v1_cm.SetPy(E_cm*math.sqrt(1-cos*cos)*math.sin(phi))
        v1_cm.SetPz(E_cm*math.sqrt(1-cos*cos)*math.cos(phi))
        v2_cm = -v1_cm

        # the 4-vectors in the centre-of-mass frame
        p1_cm = ROOT.TLorentzVector(v1_cm, E_cm)
        p2_cm = ROOT.TLorentzVector(v2_cm, E_cm)

        # apply a boost to the lab frame
        p1_cm.Boost(beta)
        p2_cm.Boost(beta)

        # compute mT and pT
        pT_1 = math.sqrt(p1_cm.Py()*p1_cm.Py() + p1_cm.Pz()*p1_cm.Pz())
        pT_2 = math.sqrt(p2_cm.Py()*p2_cm.Py() + p2_cm.Pz()*p2_cm.Pz())
        mT2 = 2*pT_1*pT_2-2*p1_cm.Py()*p2_cm.Py()-2*p1_cm.Pz()*p2_cm.Pz()
        mT = math.sqrt(mT2)
        print "Result:", mT, pT_1
```

References

1. J.D. Jackson, *Classical Electrodynamics*, 3rd edn. (Wiley, New York, 1999)
2. M. Peskin, D.V. Schroeder, *An Introduction to Quantum Field Theory*, Advanced Book Program (Westview Press, Boulder, 1995)
3. A. Di Giacomo, *Lezioni di Fisica teorica* (Edizioni ETS, Pisa, 1992)
4. C. Patrignani et al., Particle data group. Chin. Phys. C **40**, 100001 (2016)
5. H.L. Andreson et al., Phys. Rev. **85**, 936 (1952). https://doi.org/10.1103/PhysRev.85.936
6. J. Smith et al., Phys. Rev. Lett. **50**, 1738 (1983). https://doi.org/10.1103/PhysRevLett.50.1738
7. CMS Collaboration, CMS-PAS-TOP-15-002 (2015). https://cds.cern.ch/record/2053086
8. K. Agashe et al., Phys. Rev. D **88**(5), 057701 (2013). https://doi.org/10.1103/PhysRevD.88.057701

9. CMS Collaboration, J. High Energy Phys. **04**, 124 (2015). https://doi.org/10.1007/JHEP04(2015)124
10. W.R. Leo, *Techniques for Nuclear and Particle Physics Experiments*, 2nd edn. (Springer, Berlin, 1993)
11. L. Leprince-Ringuet, J. Crussard, Compt. Rend. **219**, 618–620 (1944). J. Phys. Radium **7**, 65–69 (1946)
12. R. Cahn, G. Goldhaber, *The Experimental Foundations of Particle Physics*, 2nd edn. (Cambridge University Press, Cambridge, 1989)
13. NA7 Collaboration, Phys. Lett. B **138**, 454–458 (1984). https://doi.org/10.1016/0370-2693(84)91938-5
14. CMS Collaboration, Phys. Lett. B **751**, 143–163 (2015). https://doi.org/10.1016/j.physletb.2015.10.004
15. G. Salam, Eur. Phys. J. C **67**, 637686 (2010). https://doi.org/10.1140/epjc/s10052-010-1314-6
16. D.H. Perkins, *Introduction to Hig Energy Physics*, 4th edn. (Cambridge University Press, Cambridge, 2000)
17. R.K. Ellis, W.J. Stirling, B.R. Webber, *QCD and Collider Physics* (Cambridge Monographs, Cambridge, 1996)
18. G. Dissertori, I.G. Knowles, M. Schmelling, *Quantum Chromodynamics: High Energy Experiments and Theory* (Oxford University Press, Oxford, 2002)
19. Y.-S. Tsai, Rev. Mod. Phys. **46**, 816 (1974). https://doi.org/10.1103/RevModPhys.46.815
20. F. Bishara et al., J. High Energy Phys. **04**, 084 (2014). https://doi.org/10.1007/JHEP04(2014)084
21. O. Chamberlain et al., Phys. Rev. **100**, 947 (1955). https://doi.org/10.1103/PhysRev.100.947
22. R. Durbin et al., Phys. Rev. **83**, 646 (1951). https://doi.org/10.1103/PhysRev.83.646
23. The ALEPH Collaboration, the DELPHI Collaboration, the L3 Collaboration, the OPAL Collaboration, the SLD Collaboration, the LEP Electroweak Working Group, the SLD electroweak, heavy flavour groups. Phys. Rept. **427**, 257–454 (2006). https://doi.org/10.1016/j.physrep.2005.12.006
24. The ALEPH Collaboration, Phys. Lett. B **231**, 519–529 (1989). https://doi.org/10.1016/0370-2693(89)90704-1
25. CMS Collaboration, Phys. Rev. Lett. **117**, 051802 (2016). https://doi.org/10.1103/PhysRevLett.117.051802
26. ATLAS Collaboration, J. High Energy Phys. **09**, 1 (2016). https://doi.org/10.1007/JHEP09(2016)001
27. CMS Collaboration, Phys. Lett. B **768**, 57–80 (2017). https://doi.org/10.1016/j.physletb.2017.02.010
28. R. Francsechini et al., J. High Energy Phys. **03**, 144 (2016). https://doi.org/10.1007/JHEP(2016)144

Chapter 2
Particle Detectors

Abstract The subject of the second chapter is the interaction of particles with matter. The first section discusses the mechanism by which various types of particles interact with different media. Particular emphasis is given to the concept of energy loss and range in matter. The second section focuses on the experimental techniques for particle identification. The third section is dedicated to the functioning of particle detectors.

2.1 Passage of Particles Through Matter

The kinematics of a particle moving through matter is affected by the interaction with the medium, which can be traced back to one or multiple incoherent collisions with the scattering centres, or to coherent effects that involve the medium as a whole. When the interaction is elastic, the particle transfers to the medium part of its energy or momentum at each collision. This is the case of the energy loss by electron collision, multiple scattering, Compton scattering. Inelastic reactions absorb or transmute the particle into something else, and can also give rise to new forms of radiation or leave behind excited states. This is for example the case of photon conversion, bremsstrahlung, neutron capture, charged-current neutrino interactions. Depending on the particle type, on its energy, and on the properties of the medium, one mechanism usually dominates over the others.

Energy Loss by Collision

Moderately relativistic charged particles lose energy mostly by the interaction with the electromagnetic field of atoms (electron collision). In the $10^{-1} \lesssim \beta\gamma \lesssim 10^{3}$ regime, the rate of energy loss per unit of traversed length, dE/dx, depends almost exclusively on the particle velocity β and on the properties of the medium. The formula describing the average rate of energy loss, or *linear stopping power*, is called *Bethe formula* and is given by:

$$-\frac{dE}{dx} = 2\pi N_A \rho \frac{Z}{A} \frac{\alpha^2 (\hbar c)^2}{m_e c^2} \frac{z^2}{\beta^2} \left[\ln \frac{2 m_e c^2 \gamma^2 \beta^2 W_{\max}}{I^2} - \beta^2 - \delta(\beta) - 2\frac{C(I, \beta)}{Z} \right], \quad (2.1)$$

© Springer International Publishing AG 2018
L. Bianchini, *Selected Exercises in Particle and Nuclear Physics*,
UNITEXT for Physics, https://doi.org/10.1007/978-3-319-70494-4_2

with:

N_A, α Avogadro number (6.02×10^{23} mol^{-1}) and fine structure constant ($\alpha \approx 1/137$)

ρ, A, Z mass density (g cm^{-3}), atomic weight (g mol^{-1}), and atomic number of the material

z, β, γ electric charge in units of e, velocity, and gamma-factor of the incident particle

W_{max} maximum energy transfer in a binary collision (see Problem 1.26 for its derivation)

I mean excitation potential of the material, given by the approximate formula $I \approx 16 \cdot Z^{0.9}$ eV.

δ, C *density* and *shell* corrections factors, see Problem 2.2. For their parametrisation, the reader is addressed to more advanced textbooks on the topic.

The units of dE/dx deserve a few more words. It is quite common to express the energy loss as a *mass stopping power*, i.e. in units of MeV g^{-1} cm^2 rather than in MeV cm^{-1}. This is motivated by the fact that the energy loss by collision is proportional to the density of scattering centers, i.e. $Z N_A \rho/A$. Since Z/A is quite uniform across different materials, the energy loss per unit of surface density is less dependent on the medium.

As shown by Eq. (2.1), for a fixed medium the energy loss by collision depends only on the particle velocity β and on its charge z. The functional form features a fast rise as β approaches 0 due to the β^{-2} factor, it approaches a global minimum at around $\beta \approx 0.94 \div 0.97$, or γ between about 3 and 4, and then rises logarithmically with γ. Particle sitting on the minimum and on the plateau of their dE/dx curve are characterised by a rather uniform and close-to-minal energy loss, and for this reason they are said to be minimum ionising particles (MIP). Using the approximation $m_e/M \ll 1$ in Eq. (1.141) for the maximum energy transfer, and neglecting both shell and density corrections, which are however relevant for large γ, we can arrive an approximate formula:

$$-\frac{dE}{dx} \approx \left(0.307 \text{ MeV mol}^{-1} \text{ cm}^2\right) \rho \frac{Z}{A} \frac{z^2}{\beta^2} \left[\ln \frac{2 m_e c^2 \gamma^2 \beta^2}{I} - \beta^2\right]. \quad (2.2)$$

Considering that for most of the elements and their compounds $Z/A \approx 0.5$ g^{-1} mol, and given that the term within square brackets is slowly varying with γ between 10 and 15, when dealing with particles of sufficiently large initial energy, one can often use an average value:

$$-\frac{dE}{dx} \approx 1.5 \div 2.0 \left(\frac{\rho}{\text{g cm}^{-3}}\right) \frac{z^2}{\beta^2} \text{ MeV/cm}. \quad (2.3)$$

Figure 2.1 shows the function at the right-hand side of Eq. (2.2) for $Z/A = 0.5$ g^{-1} mol and for two extreme values of the ionisation potential I. The num-

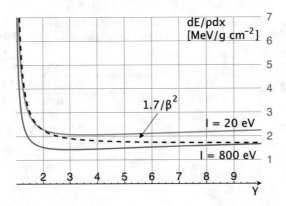

Fig. 2.1 The approximate Bethe formula of Eq. (2.2) as a function of the γ factor of the incident particle and for two extreme values of the mean excitation potential I. The global minimum at around $\gamma = 3 \div 4$ and the logarithmic growth are evident from the curves. For values of γ above the argmin of the function, the mass stopping power is in the ballpark of 2 MeV g^{-1} cm^2

ber of electrons extracted from their orbitals per unit length by the interaction with a MIP can be crudely estimated from Eq. (2.3) to be

$$\frac{dN_e}{dx} \approx \frac{1}{I}\frac{dE}{dx}. \tag{2.4}$$

For example, for a typical ionisation potential $I = 20$ eV and a water-like mass density, a unit-charge MIP produces about 10^5 electrons/cm.

For electrons and positrons moving inside matter, a formula similar to Eq. (2.1) holds. A few modifications have to be introduced, however, to account for the smaller mass and for the identity of the incident electron with the electrons that it ionizes, see e.g. Sect. 2.4 of Ref. [1] or Sect. 33.4 of Ref. [2]. In particular, one needs to replace $W_{\max}$ by $m_e c^2 (\gamma - 2)/2$ and $2 m_e c^2$ by $m_e c^2$ in the argument of the logarithm, and add a number of extra β-dependent terms inside the square brackets of Eq. (2.1), giving:

$$-\frac{dE}{dx} = \frac{(0.307 \text{ MeV mol}^{-1} \text{ cm}^2)}{2} \rho \frac{Z}{A} \frac{z^2}{\beta^2} \times$$

$$\times \left[\ln \frac{m_e c^2 \gamma^2 \beta^2 (m_e c^2 (\gamma - 1))}{2 I^2} + \frac{1}{\gamma^2} - \frac{2\gamma - 1}{\gamma^2} \ln 2 + \frac{1}{8}\left(\frac{\gamma - 1}{\gamma}\right)^2 - \delta(\beta) \right] \tag{2.5}$$

Numerically, it turns out that the stopping power for heavy ions and electrons and positrons with the same velocity β are rather similar, indeed they are consistent with each other to within about 15% up to γ factors of about 100, after which energy loss by radiation prevails anyway. The relative difference between Eqs. (2.1) and (2.8) for a few illustrative values of γ is reported in Table 2.1.

Table 2.1 Relative difference between the energy loss by collision dE/dx for ions and electrons at the same velocity and for a value of the mean excitation potential $I = 20$ (800) eV. The density correction δ is neglected

γ	β	$\frac{dE_{ion}/dx - dE_e/dx}{dE_{ion}/dx}$
1.01	0.140	11.3% (17.1%)
1.1	0.417	10.7% (16.3%)
2	0.866	8.48% (12.5%)
4	0.968	8.10% (11.4%)
10	0.995	9.06% (12.2%)
100	0.999	12.3% (17.5%)

The ionisation charge and the residual atomic excitation produced by the passage of a charged particle can be detected through various methods and thus yield a measurement of the particle position or energy. For example, this is the working principle of gaseous detectors like proportional chambers, drift and streamer tubes, RPC, liquid noble-gas detectors, etc. Semiconductor materials are also largely employed in experiments. When a charged particle moves inside a semiconductor, a number of electron-hole pairs are produced by the electrons being excited from the valence to the conductive band. One strength of these materials relies on their small band gap energy, a few eV infact, yielding a large number of signal carriers. Through appropriate doping and polarisation of the semiconductor, these electron-hole pairs can drift across the medium without significant recombination, to be finally collected for signal generation.

Other materials have the property of converting a fraction of the energy lost by a moving charged particle in the form of molecular or electronic excitation of long-lived states, that subsequently decay by emitting photons of characteristic wavelength (fluorescence). Because of such property, these materials are called *scintillators*, and the emitted radiation is called *scintillation light*. A key property of the scintillation mechanism is that the medium is transparent to its own light over distances large enough that the photons can be efficiently collected. The total light output per unit length is approximately proportional to the stopping power, a property which can also allow one to measure the total particle energy for fully absorbed particles. Scintillators can be classified into two families: *organic*, for which the scintillation mechanism relies on the fluorescence of organic molecules (e.g. plastic, organic crystals), and *inorganic*, for which the fluorescence originates from the band structure of the crystal (possibly activated by the introduction of suitable inpurities), or from electron-ion or ion-ion recombination. A broader overview on the field can be found in e.g. Chap. 7 of Ref. [1].

Table 2.2 shows the mean energy loss necessary to produce one signal carrier, which can be either a ion-electron pair, an electron-hole pair, or a scintillation photon, depending on the excitation mechanism. As shown in the table, the largest signal yields are provided by semiconductors, followed by the best scintillators and by ionisation in noble gases. Some of the most popular scintillators materials in HEP are actually characterised by relatively low light yield.

Table 2.2 Mean energy loss necessary to produce one signal carrier, listed in increasing order. For scintillators, the mean energy is defined as the inverse of the light yield (LY) in $[\gamma/MeV]$, a quantity that is commonly used to quantify the brightness of the scintillator. The values are taken from Ref. [1, 2] and, since they usually depend on the ambiental conditions, they should be considered more as an order-of-magnitude estimate. For more precise values, the reader is addressed to the technical literature

Material	Excitation	Mean excitation energy ε [eV]
Ge (77 K)	Electron-hole	3.0
Si	Electron-hole	3.6
CsI (Tl)	Scintill. γ	12
NaI (Tl)	Scintill. γ	22
Xe	Electron-ion	22
Isobutane	Ionisation	23
Ar	Electron-ion	26
CO_2	Ionisation	33
LISO (Ce)	Scintill. γ	35
He	Electron-ion	41
Plastic	Scintill. γ	100
BGO	Scintill. γ	300
PbWO	Scintill. γ	5000

Multiple Scattering

Multiple scattering (MS) through small angles refers to the ensemble of incoherent elastic collisions against the nuclear fields that charged particles undergo when crossing a piece of material. Their collective effect it to randomise the direction of the incoming particle with no significant energy loss. More informations on the subject can be found in Ref. [2]. The probability of multiple scattering through small angles is large because of the $\sin^{-4}\theta/2$ dependence of the Rutherford cross section (see Problem 1.58). However, there is also some finite probability that the scattering occurs at large angles, with subsequent emission of a knocked-out electron, or δ-ray (see Problem 1.62 for how to estimate such a probability). The quantity that characterises multiple scattering through small angles is the mean square angle per unit length Θ_s^2, which in the standard theory is given by:

$$\Theta_s^2 = \left(\frac{E_s}{\beta c|\mathbf{p}|}\right)^2 \frac{1}{X_0}, \quad \text{with} \quad E_s = \sqrt{\frac{4\pi}{\alpha}} m_e c^2 = 21 \text{ MeV}. \quad (2.6)$$

Notice that the quantity E_s is the same that enters the definition of the *Molier radius* for the lateral width of an electromagnetic shower, see Problem 2.34. The effect of MS inside a medium of length L and radiation length X_0 is to randomise the position and direction of a charged particle at the exit of the medium. Considering their projections onto a plane, the displacement y and angle θ_y are described by the joint p.d.f:

$$p(y, \theta_y \mid L) = \frac{2\sqrt{3}}{\pi} \frac{1}{\Theta_s^2 L^2} \exp\left[-\frac{4}{\Theta_s^2}\left(\theta_y^2 - \frac{3\,y\,\theta_y}{L} + \frac{3\,y^2}{L^2}\right)\right] \qquad (2.7)$$

From Eq. (2.7), one can easily compute the standard deviation of θ_y and y, and their correlation:

$$\sqrt{\langle\theta_y^2\rangle} = z\sqrt{\frac{\Theta_s^2 L}{2}} = z\frac{14.8\ \mathrm{MeV}}{\beta c|\mathbf{p}|}\sqrt{\frac{L}{X_0}} \qquad (2.8)$$

$$\sqrt{\langle y^2\rangle} = \sqrt{\langle\theta_y^2\rangle}\frac{L}{\sqrt{3}} = z\frac{8.54\ \mathrm{MeV}}{\beta c|\mathbf{p}|}\sqrt{\frac{L}{X_0}}\,L$$

$$\frac{\langle y\,\theta_y\rangle}{\sqrt{\langle\theta_y^2\rangle\langle y^2\rangle}} = \frac{\sqrt{3}}{2}. \qquad (2.9)$$

A more accurate treatment of MS modifies the first of Eq. (2.8) to the well-known formula:

$$\sqrt{\langle\theta_y^2\rangle} = z\frac{0.0136\ \mathrm{GeV}}{\beta c|\mathbf{p}|}\sqrt{\frac{L}{X_0}}\left[1 + 0.038\ln\left(\frac{L}{X_0}\right)\right], \qquad (2.10)$$

See Ref. [2] for further informations.

Energy Loss by Bremsstrahlung

For energies above a material-dependent threshold known as *critical energy* (E_c), energy loss by radiation in the electromagnetic field of the atoms (*bremsstrahlung*) prevails. An approximate parametrisation for the critical energy for electrons and positrons is provided by the formula

$$E_c = \frac{800\ \mathrm{MeV}}{(Z + 1.2)}. \qquad (2.11)$$

In the bremsstrahlung-dominated regime, the energy loss per unit length is approximately proportional to the energy itself:

$$-\frac{dE}{dx} = \frac{E}{X_0}, \qquad (2.12)$$

where X_0, called *radiation length*, is approximately independent of E. In units of mass per unit area, the radiation length is provided by the approximate expression:

$$X_0 = \frac{(m_e c^2)^2 A}{4\,Z\,(Z + 1)\,N_A\,\alpha^3(\hbar c)^2\left[\ln(183 Z^{-1/3}) - f(Z)\right]}$$

$$\approx \frac{716\,A\ \mathrm{g\ cm^{-2}}}{Z(Z + 1)\ln(287\sqrt{Z})} \approx 180\frac{A}{Z^2}\ \mathrm{g\ cm^{-2}}, \qquad (2.13)$$

Table 2.3 Radiation lengths for some materials that can be commonly found in particle physics experiments, listed by decreasing order or X_0 [cm]. From Ref. [1]

Material	X_0 [g/cm^2]	X_0 [cm]
Air	36	300×10^2
Scintill.	44	42
H_2O	36	36
Si	21.9	9.4
NaI	9.5	2.6
Fe	13.8	1.8
BGO	8.0	1.1
Pb	6.4	0.56

where A is the mass number in units of g mol^{-1}. More informations on $f(Z)$ can be found in dedicated textbook, see e.g. Ref. [1]. Notice that both the nucleus and the atomic electrons contribute to this $\mathcal{O}(\alpha^3)$ process: the former through a charge $Z e$ (hence the term $\sim Z^2$), the latter through Z incoherent scatterings of strength e (hence the term $\sim Z$). The last of Eq. (2.13) is a further approximation that helps remembering the order-of-magnitude of X_0 and its dependence on the atomic and mass number. The radiation length for a few representative materials commonly found in particle physics experiments are reported in Table 2.3. The energy loss by radiation is the dominant mechanism of energy degradation for ultra-relativistic charged particles. Notice that the radiation length X_0 is proportional to the mass squared of the charged ion (m_e in Eq. (2.13)). The next-to-lightest charged particle is the muon with a mass nearly 200 times larger than m_e. The threshold at which energy loss by radiation starts to be comparable to energy loss by collision is therefore much higher.

Energy Loss by Coherent Radiation: Cherenkov and Transition Radiation

If $\beta > 1/n(\omega)$, $n(\omega)$ being the refraction index of the medium at the frequency ω, the particle emits energy in the form of *Cherenkov radiation* of wavelength $\lambda = 2\pi c/\omega$. The energy loss per unit length is given by:

$$-\frac{dE}{dx} = z^2 \frac{\alpha \hbar}{c} \int d\omega \, \omega \sin^2 \theta_c(\omega), \qquad (2.14)$$

where θ_c is the angle of the shock-wave direction with respect to the particle direction, which satisfies the relation:

$$\cos \theta_c = \frac{1}{\beta n}. \qquad (2.15)$$

Although the Cherenkov spectrum is continuous, photodetectors have a limited range of sensitivity which depends of the quantum efficiency of the photocathode. In order to estimate the number of photons to which the detector will be sensitive, we can integrate Eq. (2.13) over the relevant spectrum to yield:

$$\frac{dN_\gamma}{dx\,d\lambda} = \frac{2\pi\,\alpha\,z^2}{\lambda^2}\sin^2\theta_c(\lambda) \quad\Rightarrow\quad \frac{dN_\gamma}{dx} \approx 2\pi\alpha z^2\,\langle\sin^2\theta_c\rangle\,\frac{\lambda_2 - \lambda_1}{\lambda_1\lambda_2} \tag{2.16}$$

$$\frac{N_\gamma}{L} \approx \frac{1.15\times 10^3}{(\hat{\lambda}/400\text{ nm})}z^2\,\langle\sin^2\theta_c\rangle\,\frac{\Delta\lambda}{\hat{\lambda}}, \tag{2.17}$$

where $\hat{\lambda} = \sqrt{\lambda_1\lambda_2}$ and the mean value of $\sin^2\theta_c$ is used, which is appropriate if n is slowing varying. For example, with a photodetector sensitive in the range 300 to 500 nm, this gives $N_\gamma \lesssim 500$ photons/cm for a particle with $z = 1$, to be compared with the about 10^5 electrons/cm electrons released by a MIP from collision loss, see Eq. (2.7). When coupled to a photodetector, the geometric and quantum efficiency of the photocathode further reduce the photo-electrons (p.e.) output. Equation 2.16 for $z = 1$ can be written as:

$$N_{p.e.} = L\,N_0\,\langle\sin^2\theta_c\rangle \tag{2.18}$$

where N_0 is the so-called *Cherenkov detector quality factor*, which is of order 100 cm^{-1} for realistic photodetectors sensitive in the visible-UV range: practical counters in experiments feature values of the quality factor ranging between 30 and 180 cm^{-1} [2].

Detectors based on the detection of Cherenkov radiation can be used for measuring the total energy of the crossing particle as well as for particle identification. In the first case, one exploits the proportionality between the collected light yield and the range of the particle, which is approximately proportional to the initial particle energy, see Problem 2.3. For the second purpose, one should distinguish between *threshold* detectors, which trigger the passage of a particle with velocity above the Cherenkov threshold, and *imaging* detectors, which are instead designed to exploit the angle of emission of individual Cherenkov photons. For highly energetic particles with $\beta \approx 1$, the employment of threshold Cherenkov detectors for particle identification becomes problematic since the index of refraction needs to approach one. To this purpose, radiators with very low density, like He, CO_2, or silica aerogel, are commonly used. Indeed, the refraction index for a homogeneous medium depends on the density according to the relation:

$$n = 1 + \frac{2\pi\,f(0,\mathbf{k})}{|\mathbf{k}|^2}N, \tag{2.19}$$

where N is the density of scattering centres, $f(0,\mathbf{k})$ is the forward scattering amplitude and $\mathbf{k}$ is the wave-number vector. For example, a simple model based on a collection of damped electronic oscillators with resonant frequency ω_k and damping constant ν_k would give [3]:

$$n(\omega) = 1 + 2\pi\,r_e\,c^2\,N\sum_k \frac{f_k}{\omega_k^2 - \omega^2 - i\nu_k\omega}. \tag{2.20}$$

Table 2.4 Refraction index and γ threshold for various radiators commonly used in Cherenkov detectors. The values refer to wavelengths in the visible domain

Material	$n - 1$	γ
He (NTP)	3.3×10^{-5}	123
Air (NTP)	2.7×10^{-4}	43
CO_2 (NTP)	4.3×10^{-4}	34
$C_5 H_{12}$ (NTP)	1.7×10^{-3}	17.2
Silica aerogel	$0.007 \div 0.13$	$2.1 \div 8.5$
H_2O	0.33	1.52
Glass	$0.46 \div 0.75$	$1.22 \div 1.37$

See Tables 2.4 and 6.1 of Ref. [2] for the index of refraction of some popular radiators. In the $\beta \to 1$ regime, though, the light output becomes small as for Eq. (2.16). For example, if a threshold Cherenkov is used for particle identification in a beam of fixed momentum $\mathbf{p}$, the refraction index can be set to the inverse velocity of the slowest particle, say β_2, and then:

$$\langle \sin^2 \theta_c \rangle = 1 - \frac{\beta_2^2}{\beta_1^2} = \frac{1 - m_1^2/|\mathbf{p}|^2 - 1 + m_2^2/|\mathbf{p}|^2}{\beta_1^2} = \frac{m_2^2 - m_1^2}{|\mathbf{p}|^2 + m_1^2}, \qquad (2.21)$$

which decreases like the square of the beam momentum.

When a relativistic charged particle crosses the boundary between vacuum and a medium, a coherent radiation is emitted in the forward region $\theta \sim 1/\gamma$. The total energy radiated depends linearly on the γ factor of the particle according to the formula:

$$I = \alpha z^2 \gamma \frac{\hbar \omega_p}{3} = \left(0.07 z^2 \sqrt{\frac{\rho}{\text{g cm}^{-3}} \frac{Z}{A}} \, \text{eV} \right) \gamma, \qquad (2.22)$$

where $\omega_p = \sqrt{4\pi n_e/m} \, e$ is the *plasma frequency* of the medium [3]. Although the energy emitted per each crossing is rather small, the total yield for particles with large γ, like GeV-electrons can be enhanced by interleaving several layers of medium, as it is usually done in the so-called *transition radiation detectors* (TRD). The latter find applications as tracking devices with built-in particle-identification capability. In terms of emitted photons, the spectrum is concentrated in the region $0.1 \gamma < \omega/\omega_p < \gamma$, so that more energetic particles give rise to a harder spectrum. More informations on the subject can be found in Ref. [2].

Interaction of Photons with Matter

Photons interact with matter by three mechanisms: *photoelectric effect*, *Rayleigh and Compton scattering*, and *pair-production*. Depending on the material and on the photon energy, one mechanism at the time usually dominates over the others. The photoelectric effect consists in the absorption of the photon by an atom, with the subsequent expulsion of an electron of energy

$$E_e = h\nu - \mathscr{B}_e, \tag{2.23}$$

where $h\nu$ is the photon energy and $\mathscr{B}_e$ is the electron binding energy. Conversely, photon scattering against the atomic electrons does not destroy the photon, but modifies its energy and direction, see e.g. Problem 1.25. The scattering can either leave the atom in the ground state (coherent, or Rayleigh scattering) or kick-out the electron (incoherent, or Compton scattering), thus leaving the atom in an excited state. Pair-production is the conversion of a photon into e^+e^- in the electromagnetic field of the atom, see e.g. Problem 1.48 for the kinematics of this reaction.

At low energy, photoelectric effect prevails. As the atomic electrons are bound in discrete states, the photoelectric cross section as a function of the photon energy features a number of thresholds corresponding to the opening of new atomic level. For energies above the innermost level, the so-called K-shell, the cross section steeply falls with energy like $\sim E^{7/2}$. The K-shell threshold for high-Z elements can be crudely estimated by using the energy levels formula for the hydrogen atom:

$$E(n) = -\frac{1}{2n^2}\alpha^2 Z^2 m_e c^2. \tag{2.24}$$

From this approximations, one expects $E_K \approx 10$ keV for metals like iron (measured value 7.1 keV), and $E_K \approx 100$ keV for lead (measured value 88 keV). At lower energies, the L and M atomic levels give rise to as many new thresholds. Depending on the photon energy, the cross section changes with the atomic number of the medium. For MeV photons, it is roughly proportional to Z^β, with $\beta = 4 \div 5$. The cross section at the K-threshold is of the order of 10^3 barn in lead and about 10^6 barn in iron. See Ref. [9] for a compendium of measured values.

Above the K-threshold, the photoelectric and Compton scattering cross sections become of comparable size. The latter changes mildly with energy for photon energies up to the pair-production threshold, after which pair-production becomes dominant. For $k \equiv E/m_e \lesssim 2$, the total cross sections is approximately given by the *Klein–Nishina formula* for Z incoherent scattering centers [9]:

$$\sigma_{\text{Comp}} = Z\,\sigma_{\text{KN}} \approx Z\left(\frac{8\pi}{3}r_e^2\right)\frac{1 + 2k + 1.2k^2}{(1 + 2k)^2} \tag{2.25}$$

with $8\pi/3r_e^2 = 0.665$ barn. The low-energy limit of Eq. (2.25) gives the *Thomas cross section* for Z free electrons, while the k-dependent term reduces the cross section for increasing photon energies. In the Compton scattering, a fraction of the photon energy is transferred to the outgoing electron. The differential cross section in the recoil energy of th electron T can be obtained from the Klein–Nishina formula, giving

$$\frac{d\sigma_{\text{Comp}}}{dT} = \frac{3\,\sigma_{\text{Th}}}{8\,m_e\,c^2\,k^2}\left[2 + \frac{\left(\frac{T}{E}\right)^2}{k^2\left(1 - \frac{T}{E}\right)^2} + \frac{\frac{T}{E}}{1 - \frac{T}{E}}\left(\frac{T}{E} - \frac{2}{k}\right)\right] \tag{2.26}$$

$$\text{with}\quad 0 \le T \le \frac{2k}{1 - 2k}E,$$

see also Problem 1.25. Due to the second term within square brackets in Eq. (2.26), the differential cross section rises steeply with T up to the kinetic bound, giving rise to a characteristic peak in the electron spectrum known as *Compton peak*.

When the photon energy exceeds the e^+e^- threshold, pair-production in the nuclear and electronic fields dominates. For energies below about 10 MeV, the interaction cross section varies logarithmically with the photon energy, and then becomes almost independent of energy. Using Tsai's formula [4], we get

$$\frac{d\sigma_{\text{pair}}}{dx} = \frac{A}{X_0\,N_A}\left[1 - \frac{4}{3}x\,(1 - x)\right] \quad \Rightarrow \quad \sigma_{\text{pair}} = \frac{7}{9}\frac{A}{X_0\,N_A} \approx 7.2\,Z^2\ \text{mbarn}, \tag{2.27}$$

where x is the photon energy fraction transferred to the electron/positron, and we have used the last formula in Eq. (2.13) to approximate X_0. Notice that the appearance of macroscopic properties of the medium in the cross section, like the mass number and the Avogadro number, are fictitious, since they exactly cancel the same quantities inside X_0. The latter is conveniently introduced to show that the interaction length for e^+e^- production is indeed related to the radiation length by $\lambda_{\text{pair}} = (9/7)\,X_0$. See Ref. [2] for more details.

Neutrons

The interaction between neutrons and matter depends strongly on the neutron energy. For energies in excess of about 100 MeV, neutrons initiate a hadronic cascade, with the production of primary hadrons (e.g. pions) sharing a fair fraction of the initial neutron energy. *Fast* neutrons, i.e. from a few hundreds of keV to a few tens of MeV, slowly thermalise by elastic scattering in high-Z materials, or faster in hydrogenised materials, see Problem 1.24. Inelastic scattering, like $A(n, n')B$, $A(n, 2n')B$, can also occur in the presence of nuclear resonances. *Epithermal* neutrons, i.e. from about 0.1 eV to about 100 keV, and *thermal* neutrons, i.e. around 25 meV, undergo preferentially nuclear reactions, like radiative neutron capture $A(n, \gamma)B$, nuclear spallation $A(n, p)B$, $A(n, \alpha)B$, and nuclear fission.

Problems

Bando n. 13153/2009

Problem 2.1 Give a qualitative description of how the energy loss by ionisation of a charged particle of mass m depends on the particle momentum.

Solution

The energy loss by ionisation dE/dx of a particle of mass m and charge $z\,e$ is described by the Bethe formula of Eq. (2.1). To good approximation, it is a function of the particle velocity and charge only, namely:

$$-\frac{dE}{dx} = z^2 \, f(\beta) = z^2 f'(|\mathbf{p}|) \tag{2.28}$$

see also Problem 2.4. At a given value of m, the function $f'(|\mathbf{p}|)$ features the following qualitative behaviour:

$$f'(|\mathbf{p}|) \sim \begin{cases} a\,|\mathbf{p}|^{-2}\ln|\mathbf{p}| & |\mathbf{p}| \gtrsim (m/m_e)\,I \\ b\,|\mathbf{p}|^{-2} + c & |\mathbf{p}| \lesssim m \\ c & |\mathbf{p}| \approx 3m \div 4m \\ c + d\,\ln|\mathbf{p}| & |\mathbf{p}| \gg m \end{cases} \tag{2.29}$$

In words: it first decreases as $|\mathbf{p}|^{-2}\ln|\mathbf{p}|$ at small momenta, until the momentum reaches a few times the mass value. At this point, it plateaus and increases only logarithmically with $|\mathbf{p}|$, see Fig. 2.2.

Bando n. 5N/R3/TEC/2005

Problem 2.2 Motivate the presence of the density and shell correction terms to the Bethe formula.

Discussion

The Bethe formula describes the energy loss of a charged particle due to the elastic collisions with the atomic electrons. In this respect, it assumes that the electrons are at rest compared to the moving particle, which is nearly unaffected by each

Fig. 2.2 The Bethe formula dE/dx in arbitrary units (a.u.) as a function of $|\mathbf{p}|$, compared to its piecewise approximation in four momentum ranges

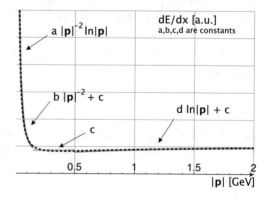

individual binary collision. A non-relativistic version of the Bethe-Bloch equation can be indeed obtained by considering the total momentum transfer that an infinitely massive moving charge has on a free electron initially at rest and located at an impact parameter b with respect to the direction of flight. The net effect is obtained by considering an ensemble of such electrons up to a maximum value of b such that the momentum transfer is above the mean ionisation energy necessary to strip the electron from its orbital. See e.g. Sect. 2.2 of Ref. [1].

Solution

The Bethe formula turns out to be accurate in the high- and low-velocity regimes only if the density δ and shell C corrections are added, as shown in Eq. (2.1). The former accounts for the polarisation of the medium by the electric field of the incident particle, which decreases the effective volume available for electron collision. As such, it tends to reduce the energy loss, and is more relevant at high-energy, see e.g. Ref. [2] for a parametrisation of δ. Conversely, if the particle velocity is comparable with the electron velocity, which is of order α, then the assumption that the electrons are at rest breaks down and a correction $C(I, \beta)$ has to be included.

Suggested Readings

The reader is addressed to Sect. 2.2 of Ref. [1] and Chap. 33 of Ref. [2] for further details on this topic.

Problem 2.3 Derive an approximate expression for the range R of a charged particle of mass m and initial energy E that loses energy by collision with the atomic electrons. How does R depends on the initial kinetic energy in the ultra-relativistic limit $E \gg m$ and in the classical limit?

Solution

The energy loss by collision is given by the Bethe formula of Eq. (2.1). The *range* of a particle is the average distance it travels before losing all of its kinetic energy and thus come to a stop. In the *continuous slowing-down approximation* (CSDA), it can be obtained by integrating the inverse linear stopping power over the full range of kinetic energy, i.e.:

$$R(E) = \int_E^m dE \, \frac{1}{dE/dx}. \tag{2.30}$$

The analytical integration of the Bethe formula is an hard task to due to the logarithmic term. However, as we have seen in the introduction Sect. 2.1, for sufficiently large initial energy, one can neglect the dependence of this term on the velocity β and use an approximate version of the type:

$$\frac{dE}{dx} = -\frac{C z^2}{\beta^2}, \tag{2.31}$$

with $C \approx 1.7$ MeV $\left[\rho/(\text{g cm}^{-3})\right]$ cm^{-1}, see Eq. (2.3). Using Eq. (2.31) in place of the full Bethe formula, the range is given by:

$$
R(E) = \int_m^E dE\, \frac{\beta^2}{Cz^2} = \frac{1}{Cz^2} \int_m^E dE \left(1 - \frac{m^2}{E^2}\right) =
$$
$$
= \frac{1}{C z^2}\left[(E-m) + m^2\left(\frac{1}{E} - \frac{1}{m}\right)\right] = \frac{1}{C z^2} \frac{(E-m)^2}{E} = \frac{m}{C z^2}\frac{(\gamma-1)^2}{\gamma}.
$$
$$(2.32)$$

Hence, we find that R/m is a function of $\gamma = E/m$:

$$
\frac{R}{m} = \frac{1}{C z^2}\frac{(\gamma-1)^2}{\gamma} = \frac{1}{C z^2}\frac{\left(\sqrt{1+(\beta\gamma)^2}-1\right)^2}{\sqrt{1+(\beta\gamma)^2}}.
$$
$$(2.33)$$

The second of Eq. (2.33) can be directly compared to Fig. 2.3, which shows the range of a heavy ion in different materials as obtained from a full integration of Eq. (2.1), as a function of $\beta\gamma$ (from Ref. [2]). A good numerical agreement is found with the approximate formula of Eq. (2.33) up to $\beta\gamma \gtrsim 1$. For smaller values of β, Eq. (2.33) underestimates the true range by a fair amount. This is a consequence of having neglected the logarithmic term in the stopping power.

Fig. 2.3 Range of heavy charged particles in liquid (bubble chamber) hydrogen, helium gas, carbon, iron, and lead. From Ref. [2]

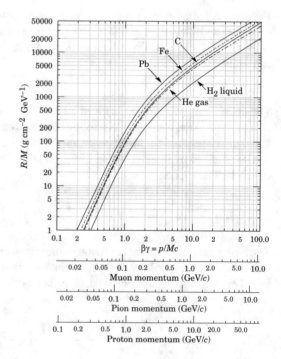

According to Eq. (2.32), the range for an ultra-relativistic particle is proportional to its energy:

$$R_{\gamma \gg 1} \approx \frac{m\,\gamma}{C\,z^2} = \frac{E}{C\,z^2}. \tag{2.34}$$

This result comes out intuitive if one considers that a particle with $\gamma \gg 1$ moves at the speed of light; the energy loss is approximately constant until the velocity drops below c. At this point, the stopping power steepens due to the β^{-2} dependence and the residual energy gets degraded in a short path, so that $R \sim E/(dE/dx|_{MIP}) \sim E$. For a non-relativistic particle, the range is instead a quadratic function of the kinetic energy T:

$$R_{NR} \approx \frac{(E-m)^2}{C\,z^2\,m} = \frac{T^2}{C\,z^2\,m}. \tag{2.35}$$

However, one should remember that for $\beta \lesssim 0.5$, the approximation of Eq. (2.31) is not valid anymore and the resulting range is underestimated. For example, for an α particle emitted in the decay of ^{210}Po with $T = 5.3$ MeV, the range in air predicted by Eq. (2.35) is about $5.3^2/(2 \cdot 10^{-3} \cdot 2^2 \cdot 4 \times 10^3) \approx 1$ cm, whereas the CSDA range from a full integration of the Bethe function gives about 5 cm [5].

Suggested Readings

A good starting point to learn more about the concept of range is Chap. 2 of Ref. [1].

Problem 2.4 Determine the relation between the stopping power dE/dx for two particles of masses m_1 and m_2, electric charges $z_1\,e$ and $z_2\,e$, and same momentum $|\mathbf{p}|$, moving through the same medium. What is the relation between the range R_1 and R_2 of the two particles under the same conditions?

Solution

The energy loss by collision is given by the Bethe formula of Eq. (2.1), which, as a function of the particle momentum, can be written as:

$$\frac{dE_i}{dx}(|\mathbf{p}|) = z_i^2\, f\left(\frac{|\mathbf{p}|}{m_i}\right), \tag{2.36}$$

so that:

$$\frac{dE_2}{dx}(|\mathbf{p}|) = z_2^2\, f\left(\frac{|\mathbf{p}|}{m_2}\right) = \frac{z_2^2}{z_1^2} z_1^2\, f\left(\frac{m_1}{m_2}\frac{|\mathbf{p}|}{m_1}\right)$$

$$= \frac{z_2^2}{z_1^2}\frac{dE_1}{dx}\left(\frac{m_1}{m_2}|\mathbf{p}|\right). \tag{2.37}$$

Owing to such scaling law, the stopping power $dE(|\mathbf{p}|)/dx$ as a function of $|\mathbf{p}|$ for different particle types are all related by a uniform scaling of the orizontal axis equal the mass ratio, and by a scaling on the vertical axis by the ratio of the squared charges.

Let's now consider the range as defined in Problem 2.3. For a given kinetic energy T, the range is given by:

$$R_i(T) = \int_T^0 dE \, \frac{1}{dE_i/dx} = \int_T^0 dE \, \frac{1}{z_i^2 \, f\left(\frac{E}{m_i}\right)}, \tag{2.38}$$

so that:

$$R_2(T) = \int_T^0 dE \, \frac{1}{z_2^2 \, f\left(\frac{E}{m_2}\right)} = \frac{z_1^2}{z_2^2} \int_T^0 dE \, \frac{1}{z_1^2 \, f\left(\frac{m_1}{m_2}\frac{E}{m_1}\right)} = \frac{z_1^2}{z_2^2}\frac{m_2}{m_1} \int_{\frac{m_1}{m_2}T}^0 dE' \, \frac{1}{z_1^2 \, f\left(\frac{E'}{m_1}\right)}$$

$$= \frac{z_1^2}{z_2^2}\frac{m_2}{m_1} R_1\left(\frac{m_1}{m_2}T\right). \tag{2.39}$$

Discussion

The simultaneous measurement of the stopping power dE/dx and of the particle momentum, or of its kinetic energy, or of its velocity, provides a tool to identify the particle type thanks to the scaling law of Eq. (2.37). The canonical example of a detector that allows for a simultaneous measurement of these quantities is the *time projection chamber* (TPC).

Suggested Readings

For an overview on the TPC, the reader is encouraged to consult the PDG review on detectors for accelerators [2]. See also Ref. [1] for the scaling law of stopping powers and ranges.

Problem 2.5 The range R of a particle is the distance over which the particle loses all of its kinetic energy. For a heavy ion, the energy loss per unit length of traversed material can be approximated by the formula

$$\frac{dE}{dx} = -\frac{C z^2}{\beta^2}, \tag{2.40}$$

where $C \approx 1.7$ MeV cm^{-1}, z is the ion charge in units of e, and β is the particle velocity.

- What kind of interaction between the ion and the material is responsible for this energy loss?
- Explain how the mass of a charged particle can be determined from the simultaneous measurement of dE/dx and of the momentum $|\mathbf{p}|$.

- Estimate the range R in water of a proton with $T = 60$ MeV.

Solution

As discussed in Sect. 2.1, heavy ions moving in matter lose energy due to elastic collision with the atomic electrons.

Since $dE/dx \sim z^2 f(\beta)$ and $|\mathbf{p}| = m\beta\gamma$, the simultaneous measurement of the two quantities allows to measure m for different ansatz on z. A comparison of the mass values thus obtained with the spectrum of known particles allows one to identify the particle type.

In order to estimate the range of a proton in water, we can use Eq. (2.35) obtained from the limit $\gamma \to 1$ in Eq. (2.32). We can obtain the same result starting from Eq. (2.40) and using the fact that $T = |\mathbf{p}|^2/2m$ for a classical particle:

$$R(T) = \int_T^0 dE \frac{1}{dE/dx} = \int_0^T dE \frac{\beta^2}{z^2 C} = \int_0^T dT' \frac{2 T'}{m_p c^2 z^2 C} = \frac{1}{m_p c^2} \frac{T^2}{z^2 C} =$$

$$= \frac{(60)^2 \text{ MeV}^2}{10^3 \text{ MeV} \cdot 1^2 \cdot 1.7 \text{ MeV cm}^{-1}} = 2.1 \text{ cm}, \qquad (2.41)$$

to be compared with a CSDA value of 3.1 cm from a full integration of the Bethe formula [5].

Bando n. 13153/2009

Problem 2.6 Discuss the characteristics of the Bragg peak and its main applications.

Solution

The energy loss of a charged ion in matter is described by the Bethe formula (2.1). Due to the dominant $1/\beta^2$ behaviour at velocities below about 0.9, the energy deposition per unit length becomes increasingly more intense as the particle velocity decreases. By tuning the initial particle energy T to attain a certain range R, the Bethe formula predicts that most of T will be infact dissipated near the end of the trajectory.

Since $E = E(\beta)$, the Bethe formula can be solved as an ordinary differential equation (ODE) in β, giving a solution $dE(x)/dx$. The latter features a peak at $x \approx R$, the so-called *Bragg peak*. Indeed, by using the approximation (2.3) and assuming the ion to be non-relativistic, the ODE can be easily solved analytically, yielding:

$$\frac{d}{dx}\left(\frac{1}{2} m \beta^2\right) = -\frac{z^2 C}{\beta^2}, \qquad m \beta \frac{d\beta}{dx} = -\frac{z^2 C}{\beta^2}, \qquad \beta^3 d\beta = -\frac{z^2 C}{m} dx,$$

$$\beta^4 - \beta_0^4 = -\frac{4 z^2 C}{m} x, \qquad \beta^2(x) = \beta_0^2 \sqrt{1 - \frac{4 z^2 C}{m \beta_0^4} x} = \beta_0^2 \sqrt{1 - \frac{x}{R}}, \qquad (2.42)$$

Fig. 2.4 Sketch of a typical
Bragg curve for protons or
heavy ions moving in a
dense medium

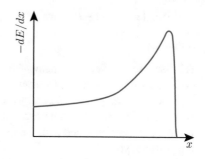

where we have used Eq. (2.35) to define the range R of the particle. From Eq. (2.42)
we therefore get:

$$\frac{dE}{dx}(x) = -\frac{z^2 C}{\beta_0^2}\frac{1}{\sqrt{1 - x/R}}. \tag{2.43}$$

The energy ΔE deposited in the interval $[\lambda R, R]$ can be easily computed from
Eq. (2.43) to give:

$$\Delta E(\lambda) = \int_{\lambda R}^{R} dx \left|\frac{dE}{dx}\right| = T\sqrt{1 - \lambda}. \tag{2.44}$$

The value of λ such that a fraction α of the initial energy is lost in the interval $[\lambda R, \ R]$
is therefore given by $\lambda = 1 - \alpha^2$. For example, 50% of the kinetic energy T is lost
in the last quarter of the particle path, and 25% in the trailing 6% of the path. A
caveat: Eq. (2.43) has been obtained under the assumption that $dE/dx \sim \beta^{-2}$. This
is a poor approximation for $\beta\gamma \lesssim 1$, and the resulting stopping power gets largely
overestimated. Furthermore, when $\beta\gamma \lesssim 0.1$, the shell corrections are relevant,
reducing significantly the stopping power, and the Bethe formula ultimately breaks
down. Overall, the Bragg curve is much less peaked than predicted by Eq. (2.43),
and infact the maximum occurs before the full range is attained, see Fig. 2.4.

   The Bragg peak finds one major application in medical physics as a tool for curing
solid tumors: the intense energy deposition in the neighbourhood of the beam range
allows to burn selected tissue depths with reduced damage to the upstream tissue.

Bando n. 13705/2010

Problem 2.7 A 2 cm-thick plastic scintillator is coupled to a photomultiplier with
gain $G = 10^6$ and detection threshold $Q_{th} = 1$ pC, such that all the scintillation
light can be assumed to be detected. A beam of particles of energy 10 GeV impinges
perpendicularly to the scintillator:

- Estimate the charge collected at the anode, if the beam is made of muons.
- If the beam is made of neutrons, estimate the minimum scattering angle on protons such that the neutron can be detected.

Discussion

Scintillators have been briefly discussed in Sect. 2.1. A scintillator is always coupled with a photomultiplier that transforms the scintillating photons in photoelectrons (p.e.). Because of the geometry of the medium and of the QM-nature of the photo-electric effect, a photomultiplier is only sensitive to a fraction ε_C of the total light output, of which only a fraction ε_Q is actually converted into p.e. By themselves, such p.e. do not usually represent an amount of charge large enough to generate a significant signal, i.e. above the electronic noise. For this reason, the primary p.e. undergo a multiplicative enhancement between the photocathode and the anode. This can be for example achieved by accelerating them with intense electric fields, so that they can initiate a chain reaction that brings to the fan exponential charge multi-plication. The enhancement factor, i.e. the total output charge per initial p.e., is the called *gain* (G) of the photomultiplier. The enhanced charge is finally read-out at the anode by a chain of amplifiers which transforms it into voltage or currents. A key point in all this procedure is that the proportionality between the initial number of p.e. and the final signal amplitude is preserved. After coupling the amplification stage to the read-out electronics, characterised by an electronic noise N_e, the relative energy resolution from a scintillator that produces n_γ Poisson-distributed photons for a particle of energy E, can be parametrised as [2]:

$$\frac{\sigma(E)}{E} = \sqrt{\frac{f_N}{n_\gamma \, \varepsilon_Q \, \varepsilon_C} + \left(\frac{N_e}{Q \, n_\gamma \, \varepsilon_Q \, \varepsilon_C}\right)^2}, \qquad (2.45)$$

where f_N is the called *excess noise factor* and arised from the amplification process. The role of the gain factor in reducing the signal uncertainty is made clear by Eq. (2.45).

Solution

A 10 GeV muon loses energy mostly by collision with the atomic electrons as dis-cussed in Sect. 2.1. In particular, it behaves as a MIP, and its mean energy loss per unit length is provided by Eq. (2.3). For a plastic scintillator, the mass density is approximately $\rho \approx 1$ g cm^{-3}. With this value, the energy loss is given by:

$$-\frac{dE}{dx} \approx 2.0 \text{ MeV g}^{-1} \text{ cm}^2 \cdot 1 \text{ g cm}^{-3} = 2 \text{ MeV cm}^{-1}. \qquad (2.46)$$

While crossing a thickness $d = 2$ cm, the total energy lost by the muon is $\Delta E = |dE/dx| \cdot d \approx 4$ MeV. The mean excitation energy for a plastic scintillator can be found in Table 2.2. Assuming $\varepsilon = 100$ eV, $\varepsilon_C = 1$, and $\varepsilon_Q = 1$, we expect to collect an average charge at the anode of about:

$$Q = G \cdot \frac{\Delta E}{\varepsilon} \cdot e = 10^6 \cdot \frac{4 \text{ MeV}}{100 \text{ eV}} \cdot 1.6 \times 10^{-19} \text{ C} = 6.4 \text{ nC}, \qquad (2.47)$$

i.e. more than three orders of magnitude larger than the threshold charge Q_{th}.

If the beam is made of neutrons, their detection proceeds through the measurement of the recoil energy of protons and other nuclei that interact with the beam particles. The threshold energy such that a recoil proton gives rise to a detectable signal is determined by the condition:

$$G \cdot \frac{T_{th}}{\varepsilon} \cdot e = Q_{th}, \quad \Rightarrow \quad T_{th} = \frac{10^{-12} \text{ C} \cdot 10^2 \text{ eV}}{10^6 \cdot 1.6 \times 10^{-19} \text{ C}} = 0.62 \text{ keV}, \qquad (2.48)$$

which is small compared to the proton mass and to the beam momentum. It is easy to show that for very small recoil energy, momentum has to be exchanged perpendicularly. Indeed, if we indicated the four-momenta of the initial (final) neutron and proton by p and k (p' and k'), and the angle that the recoiling proton forms with the beam momentum as θ_p, then:

$$p' = p + k - k',$$
$$m_n^2 = m_n^2 + 2m_p^2 - 2E_n m_p - 2(E_n E_p' - |\mathbf{p}_n||\mathbf{p}_p'| \cos \theta_p) - 2E_p' m_p,$$
$$\cos \theta_p = \frac{E_p' (E_n + m_p) - m_p (E_n + m_p)}{|\mathbf{p}_n||\mathbf{p}_p'|} = \frac{T_p (E_n + m_p)}{|\mathbf{p}_n||\mathbf{p}_p'|} \approx$$
$$\approx \sqrt{\frac{T_p}{2m_p}} \left[\frac{E_n + m_p}{|\mathbf{p}_n|} \right], \quad \text{if } |\mathbf{p}_p'| \ll m_p. \qquad (2.49)$$

Since $T \ll m_p$ for our case, and given that the factor within square brackets is of order one, the resulting angle turns out to be pretty much $\pi/2$, and conservation of momentum implies that the momentum received by the extra neutron is also is also a vector perpendicular to the beam direction. Since $T_{th} \ll E_n$, the neutron momentum magnitude after the scattering is almost unchanged, and the scattering angle of the neutron is therefore given by:

$$\theta_n \approx \frac{|\mathbf{p}_p'|}{|\mathbf{p}_n|} = \frac{\sqrt{2m_p T_{th}}}{|\mathbf{p}_n|} = \frac{\sqrt{2 \cdot 0.938 \cdot 0.62 \times 10^{-6}}}{10} = 1.1 \times 10^{-4} \text{ rad.} \quad (2.50)$$

Bando n. 1N/R3/SUB/2005

Problem 2.8 A MIP generates, on average, n electron-ion pairs per cm in a gaseous detector at standard pressure. What is the typical value of n, if the gas consists in a argon-isobuthan mixture 60%–40%? Which additional factors acting on the statistics of the produced electrons determine the standard deviation of the signal?

Solution

On average, a MIP releases an amount of energy per unit length described by Eq. (2.3). If the gas is made of argon and isobutane, which have a small ionisation potential $I \approx 12\,\text{eV}$, see e.g. Table 6.1 of Ref. [1] or Ref. [5], the Bethe formula predicts an energy loss per mass surphace of about $2.5\,\text{MeV}\,\text{g}^{-1}\,\text{cm}^2$, see also Fig. 2.1. The density of the gas at STP conditions can be calculated from the law od ideal gases:

$$\rho = \frac{A}{R\,T/P} = \frac{(0.6 \cdot 18 + 0.4 \cdot 58)\,\text{g mol}^{-1}}{8.314\,\text{J mol}^{-1}\,\text{K}^{-1} \cdot 298\,\text{K}/10^5\,\text{Pa}} = \frac{34\,\text{g mol}^{-1}}{2.5 \times 10^4\,\text{cm}^3\,\text{mol}^{-1}} = $$
$$= 1.4 \times 10^{-3}\,\text{g cm}^{-3}. \tag{2.51}$$

The mean excitation energy for the two molecules can be read from Table. 2.2. Taking a weighted average of the two components, we get:

$$n = \frac{|dE/dx|}{(0.6 \cdot 26 + 0.4 \cdot 23)\,\text{eV}} = \frac{2.5\,\text{MeV}\,\text{g}^{-1}\,\text{cm}^2 \cdot 1.4 \times 10^{-3}\,\text{g cm}^{-3}}{24.8\,\text{eV}} = 140\,\text{cm}^{-1}. \tag{2.52}$$

In a gaseous ionisation detector, the primary electrons need to be accelerated by an intense electric fields until they trigger the formation of an avalanche. Indeed, an amount of primary ionisation electrons like in Eq. (2.52) is not sufficient to produce a detectable signal. Since the charge-multiplication process is intrinsically random, it introduces an additional fluctuation in the number of signal carriers. If an electron-ion pair recombines before the formation of the avalanche, or if it gets trapped by the gas molecules to give rise to an ion, it gets lost for later multiplication. Suitable amounts of electronegative gases, like freon, can limit this effect. The gain (see Problem 2.7), and hence the final statistics of signal carriers, depends on the choice of the gas. Noble gases are usually chosen because of their large gain factors. Another typical problem with gaseous detectors is the formation of avalanches in random points of the chamber created by energetic photons emitted by the accelerated electrons. This undesired effect limits the operation rate and resolution of the detector. These effects can be limited by adding appropriate amounts of organic quenchers, like isobutane. Finally, one should remember that the resolution of a gaseous ionisation detector that absorbs all of the particle kinetic energy scales better than $1/\sqrt{n}$ by the so-called *Fano factor*, which for typical gases is in the range $0.05 \div 0.20$, see e.g. Table 6.2 of Ref. [7].

Suggested Readings

An introduction to the physics of electronic avalanches in gas can be found in Refs. [1, 7]. For a more comprehensive review of gaseous detectors, the reader is addressed to Ref. [8].

Problem 2.9 How many electrons does a charged particle produce on average when crossing 100 μm of silicon?

Solution

Let us assume that the charged particle have $z = 1$ and that they behave like a MIP. The energy loss per unit length is given by the Bethe formula of Eq. (2.2). For a MIP, the dependence of dE/dx on the particle energy is mainly through the logarithmic term $\sim \ln \gamma$. Assuming the particle to be in the neighborhood of the global minimum, i.e. $\gamma \approx 4$, we can explicitly compute the right-hand side of Eq. (2.2) for a pure silicon medium, giving:

$$-\frac{dE}{dx} = \left(0.307 \text{ MeV mol}^{-1} \text{ cm}^2\right) \cdot \frac{2.33 \text{ g cm}^3 \cdot 14}{28.1 \text{ g mol}^{-1}} \left[\ln \frac{2 \cdot 0.511 \text{ MeV} \cdot 4^2}{16 \cdot 14^{0.9} \text{ eV}} - 1 \right] =$$
$$= 3.7 \text{ MeV cm}^{-1}, \tag{2.53}$$

which agrees well with the more accurate prediction of 3.9 MeV cm^{-1} [5]. The number of electron-hole pairs produced by the passage of such a particle across a thickness $d = 100$ μm of silicon is therefore given by:

$$n_{eh} = \frac{|dE/dx| \cdot d}{\varepsilon} = \frac{3.7 \text{ MeV cm}^{-1} \cdot 10^{-2} \text{ cm}}{3.6 \text{ eV}} \approx 10^4, \tag{2.54}$$

where we have used the mean excitation energy for silicon as in Table 2.2.

Bando n. 1N/R3/SUB/2005

Problem 2.10 A relativistic electron loses energy by both ionisation and by radiation when moving inside matter.

- How does the energy loss by ionisation and by radiation depend on the material?
- How do they depend on the electron energy?
- The critical energy is defined as the energy at which the two energy losses are equal: which between a muon and an electron has the smallest critical energy?

Solution

The energy loss of relativistic electrons and positrons is discussed in Sect. 2.1. For energies below the critical energy E_c, energy loss by collision with the atomic electrons prevails. The material enters mostly through its electron density $n_e = N_A \rho Z/A$ and the average ionisation potential I. The stopping power is proportional to n_e and depends logarithmically on I. A residual dependence on the atomic number Z comes from the shell and density effects, see e.g. Ref. [2]. For electrons with energy in excess of a few MeV, the rate of energy loss by collision is almost independent of the electron energy, while it goes like T^{-1} at smaller energies.

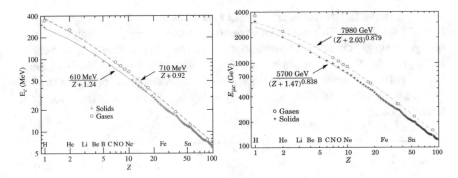

Fig. 2.5 Electron and muon critical energy for the chemical elements. From Ref. [2]

Energy loss by radiation prevails above the critical energy. The material enters through the atomic density $n = N_A \rho A$ and through the atomic number Z. In particular, it is proportional to the combination $Z^2 \rho / A$, as shown by Eq. (2.13). Furthermore, it is proportional to the energy itself, see Eq. (2.12).

Since the energy loss by radiation is inversely proportional to m^2, where m is the mass of the incident particle, see Eq. (2.13), while the energy loss by ionisation is independent of m for sufficiently high energies, it follows that the critical energy must be approximately go as $\sim m^2$, since it is roughly given by the position of the intersection point between two curves in the $(dE/dx, E)$ plane, one of which is roughly constant (energy loss by collision), while the other (energy loss by radiation) is a straight line of slope proportional to m^{-2}. According to this picture, the critical energy for muons, $E_{\mu c}$, is expected to be about 4×10^4 times larger than for electrons. An exact scaling does not hold however, and the critical energy $E_{\mu c}$ is a factor of about 3 smaller than the naive scaling $E_{\mu c} \approx (m_\mu/m_e)^2 E_c$, see e.g. Fig. 2.5 taken from Ref. [2].

Bando n. 13153/2009

Problem 2.11 An electron moving in a material loses energy by a variety of mechanisms. Define the critical energy and explain how it depends on the atomic number Z of the material.

Solution

Energy loss by collision and radiation are discussed in Sect. 2.1. The critical energy E_c is defined as the energy at which the two rates of energy loss become identical. An approximate formula for E_c is given by

$$E_c = \frac{800 \, \text{MeV}}{(Z + 1.2)}, \tag{2.55}$$

see e.g. Ref. [1, 2]. Hence, the critical energy decreases with Z. In particular, it goes like Z^{-1} for $Z \gg 1$. This can be understood by the following argument: the critical energy is roughly given by the position of the intersection point between two curves in the $(dE/dx, E)$ plane, of which one is flat versus energy and goes like $\sim Z$ (energy loss by collision), while the other has a positive slope and goes approximately like $\sim Z^2$ at large values of Z (energy loss by radiation), hence the intersection point should scale as $\sim Z^{-1}$.

Bando n. 5N/R3/TEC/2005

Problem 2.12 Provide an approximate formula for the radiation length X_0 in terms of the atomic and mass numbers of the material.

Solution

An approximate version of X_0 has been derived in Eq. (2.13):

$$X_0 \approx \frac{716\,A}{Z\,(Z+1)\,\ln(287\,\sqrt{Z})}\ \mathrm{g\,cm}^{-2},\qquad (2.56)$$

where A is the mass number in units of $\mathrm{g\,mol}^{-1}$ and Z is the atomic number. Hence, the radiation length scales as $\sim A\,Z^{-2}$, for sufficiently large values of Z.

Bando n. 18211/2016

Problem 2.13 How much energy does an electron with initial energy of 1 GeV lose by crossing a material with thickness equal to one radiation length?

Solution

An energy of 1 GeV is above the critical energy E_c of Eq. (2.11), see Fig. 2.5, therefore the electron loses energy mostly by radiation. The rate of energy loss per unit length is therefore given by

$$\frac{dE}{dx} = -\frac{E}{X_0},\qquad (2.57)$$

where X_0 is the radiation length measured. The electron energy as a function of the traversed length is then obtained by integrating Eq. (2.57) to give:

$$E(x) = E_0\,e^{-x/X_0} \quad \Rightarrow \quad E(X_0) = \frac{E_0}{e} = 0.368\,E_0.\qquad (2.58)$$

The energy lost in the medium is therefore $\Delta E = (1 - 1/e) E_0 = 0.632 E_0$.

Bando n. 13153/2009

Problem 2.14 Determine the law by which a beam of electrons of intensity I_0 gets attenuated while crossing a layer of material of thickness d.

Solution

Electrons lose energy mostly by radiation at high energy, and then by elastic collison with atomic electrons at lower energies. If the beam is monochromatic and the thickness d exceeds the electron range in the material, the beam particles will traverse the full thickness and emerge with an energy distribution centred around a smaller value. Elastic scattering can instead deflect the electron from its original trajectory and remove it from the beam. Let's assume that the reaction which removes electrons from the beam is characterised by a cross section σ and let's denote the density of scattering centres by n. By definition, the probability of interaction per unit length is given by the interaction length of Eq. (1.291), namely $\lambda = 1/(n\sigma)$. If the beam has an intensity $I(x)$ at a depth x, the intensity at a distance $x + dx$ is given by:

$$I(x + dx) = I(x) - I(x)\frac{dx}{\lambda}, \qquad \frac{dI}{I} = -\frac{dx}{\lambda} \quad \Rightarrow \quad I(x) = I_0\, e^{-x/\lambda} \quad (2.59)$$

The intensity varies exponentially with the traversed length.

Bando n. 1N/R3/SUB/2005, Bando n. 13153/2009

Problem 2.15 In which energy interval does Compton scattering dominate in the interaction of photons with matter? What kind of interaction prevails at lower and higher energies? How does it depend on the absorber?

Solution

The interaction of photons with matter is discussed in Sect. 2.1. At low energy, the photoelectric effect (photon absorption with electron emission) is the main interaction mechanism. Compton scattering (incoherent photon-electron scattering) becomes significant for energies above the K-threshold and below a few times $2\,m_e$, after which pair-production dominates. The transition between the photoelectric and Compton-dominated regime depends on the medium (see below). For carbon (lead), the two become of similar size at energies of about 10 (500) KeV, see e.g. Ref. [2].

The absorber type enters mostly through the atomic number Z. The photoelectric cross section for energies in the MeV region is goes as $\sim Z^\beta$ with $\beta = 4 \div 5$. The Compton cross section is instead proportional to the number of electrons per atomi, hence it goes as $\sim Z$. The cross section for pair-production is inversely proportional to the radiation length X_0, hence it is roughly proportional to $\sim Z^2$ for large atomic numbers.

Suggested Readings

Photon interaction in matter is discussed in a large number of textbooks. For a primer, the reader is addressed to Sect. 2.7 of Ref. [1] and to the PDG review [2]. A large amount of tabulated data can be found in Ref. [6, 9].

Bando n. 5N/R3/TEC/2005

Problem 2.16 How does the photoelectric cross section vary as a function of the photon energy? How does it depend on the atomic number Z?

Solution

The interaction of photons with matter is discussed in Sect. 2.1. At low energy, the photoelectric effect (photon absorption with electron emission) prevails. The photoelectric cross section as a function of the photon energy features a number of edges corresponding to the opening of new atomic levels. For energies above the innermost level (K-shell), the cross section steeply falls with energy as $\sim E^{-7/2}$ and it grows with the atomic number as $\sim Z^{\beta}$ with $\beta = 4 \div 5$.

Suggested Readings

See Problem 2.15 and references therein.

Bando n. 18211/2016

Problem 2.17 Determine which process dominates in the photon-matter interaction for the following reactions:

1. 1 MeV photons on Al;
2. 100 keV photons on H_2;
3. 100 keV photons on Fe;
4. 10 MeV photons on C;
5. 10 MeV photons on Pb;

Solution

To solve this exercise, we can refer to Fig. 2.6, taken from Ref. [2], to read the cross section values for carbon and lead, and then use these values, together with the known Z-dependence of the cross sections, in order to extrapolate to other materials. To validate the extrapolation, we can use the values tabulated in Ref. [6].

1. A 1 MeV photon is just below the pair-production threshold. Aluminium has atomic number $Z = 13$. The energy at which Compton and pair-production become similar is about 500 keV in lead and about 10 keV in carbon. Aluminimum must be in-between, therefore Compton scattering has to be by far dominant at such an energy. Indeed, from Ref. [6] we find $\sigma_{\text{Comp}} \approx 3$ barn and $\sigma_{\text{p.e.}} \approx 10^{-3}$ barn.

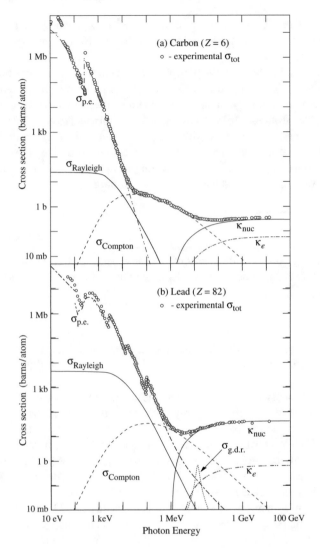

Fig. 2.6 Photon total cross sections as a function of energy in carbon and lead, showing the contributions of different processes. Taken from Ref. [2]

2. A 100 keV photon on hydrogen cannot undergo pair-production. Since Compton scattering dominates the photon-matter interaction at this energy for carbon, it will be *a fortiori* dominant in hydrogen, since the photoelectric cross section decreases as a function of Z much faster compared to the Compton cross section. Indeed, from Ref. [6] we find $\sigma_{Comp} \approx 0.5$ barn and $\sigma_{p.e.} \approx 10^{-6}$ barn.

3. A 100 keV photon on iron cannot undergo pair-production. From Fig. 33.15 of Ref. [2], the photoelectric (Compton) cross section in lead at that energy is around

10^3 barn (10 barn), so that, by assuming a $\sim Z^5$ ($\sim Z$) scaling, we should expect roughly the same cross sections. Indeed, from Ref. [6] we find $\sigma_{\text{Comp}} \approx 12$ barn and $\sigma_{\text{p.e.}} \approx 20$ barn.

4. A 10 MeV photon can undergo pair-production, but Compton scattering is sizable at that energy. Referring to Fig. 33.15 of Ref. [2], one sees that Compton cross section on carbon is larger than pair-production, although the two are still comparable. Indeed from Ref. [6] we find $\sigma_{\text{Comp}} \approx 0.3$ barn and $\sigma_{\text{pair}} \approx 0.8 \times 10^{-1}$ barn.

5. As before, one should expect the pair-production and Compton cross sections to be of the same order. This time, it's the former to be larger because of the $\sim Z^2$ scaling compared to just a $\sim Z$ scaling of Compton scattering. Indeed from Ref. [6] we find $\sigma_{\text{Comp}} \approx 4$ barn and $\sigma_{\text{pair}} \approx 12$ barn.

Bando n. 18211/2016

Problem 2.18 A muon with energy of 400 GeV penetrates vertically into the sea. By which process can it be detected? Estimate the depth at which the muon arrives before decaying.

Solution

A muon of energy $E = 400$ GeV moving in water ($n = 1.33$) emits Cherenkov radiation at a rate of about 200 γ/cm in the wavelength range $[300, 500]$ nm, see Eq. (2.16).

The critical energy for electrons in water is about 80 MeV, see e.g. Ref. [5]. From the $\sim m^2$ scaling of the critical energy with the particle mass, the critical energy for muons is expected to be in excess of 3 TeV, hence far above the initial muon energy of 400 GeV. However, as discussed in Problem 2.10, the naive scaling is only approximate, and the critical energy for muons is about 1 TeV [5], hence still larger than the initial muon energy. From Fig. 33.24 of Ref. [2] we see that the critical energy for oxygen is about 900 GeV, so the same conclusions hold. The dominant energy loss mechanism is therefore by electron collision as described by the Bethe formula of Eq. (2.1). Since $\gamma = E/m = 3.8 \times 10^3 \gg 1$, we can use the approximate formula of Eq. (2.34) to predict the range R in water ($\rho = 1$ g cm^{-3}) to be $R \approx E/C$, where C is a constant that sets the plateau level of the Bethe formula. The stopping power for a MIP muon in water is about 2.0 MeV g^{-1} cm^2 [5]. However, at very large energies, the logarithmic term is non-negligible. Using the value $I = 80$ eV [5], the latter ranges from ≈ 26 at $\gamma = 3.8 \times 10^3$ down to ≈ 12 at $\gamma = 4$ (MIP). Taking an intermediate value of 20, the constant term can be approximated as $2.0 \times 20/12 \approx 3.3$ MeV g^{-1} cm^2. Therefore:

$$R \approx \frac{400 \text{ GeV}}{3.3 \text{ MeV g}^{-1} \text{ cm}^2 \cdot 1 \text{ g cm}^{-3}} = 1.2 \text{ km}. \qquad (2.60)$$

This result is in good agreement with the more accurate estimate of 1.216 km from Table II-28 of Ref [10]. However, the muon is an unstable particle with life-time

$\tau = 2.2 \times 10^{-6}$ s. The range calculation of Eq. (2.60) will hold only if the muon does not decay before coming to a stop. This is indeed the case with high probability. Although the muon momentum progressively changes as the muon penetrates deeper into the sea, time dilatation makes such that the muon decay probability over a fixed length in the Earth frame is significant only for small velocities. At a velocity $\beta = 0.94$, or $\gamma \approx 3$, the muon is at the minimum of the stopping power curve, and the residual energy is dissipated after traversing a length of about

$$\frac{106\,\text{MeV}}{2.0\,\text{MeV g}^{-1}\,\text{cm}^2 \cdot 1\,\text{g cm}^{-3}} \frac{(3-1)^2}{3} = 70\,\text{cm} \ll R. \tag{2.61}$$

Were the muon to conserve $\gamma = 3$, its mean path before decaying would be $\beta c \tau \gamma \approx$ 2 km, so much larger than the residual path before stopping completely.

Discussion

The exploitation of large sea volumes as Cherenkov radiators allows one to study cosmic radiation of very high energy. For example, the IceCube neutrino observatory at the South Pole, is sensitive to the CC interaction of very-high energy neutrinos, which can be detected through their emission of Cherenkov light by an array of PMT's located deep into the ice.

Problem 2.19 An underground experiment located at a depth $d = 1$ km from the top of the mountain measures the momentum of cosmic muons arriving vertically from above. Estimate the muon energy at the top of the mountain if the muon momentum at the detector is $|\mathbf{p}| = 1.0$ TeV.

Solution

Energetic muons lose energy by electron collision and by various forms of electromagnetic radiation, including e^+e^- pair production, bremsstrahlung, and photonuclear interaction. The overall stopping power can be parametrised as

$$-\frac{dE}{dx} = a(E) + b(E)\,E, \tag{2.62}$$

where a and b are slowly varying functions of energy for $E \gtrsim 1$ TeV. Assuming constant values for a and b, Eq. (2.62) can be solved exactly to yield the solution $E_0 = E_0(E, x)$, namely:

$$\begin{cases} -\frac{dE}{dx} = a + b\,E \\ E(0) = E_0 \end{cases}$$

$$-\frac{dE}{a + b\,E} = dx, \quad \ln\left(\frac{1 + E/E_{\mu c}}{1 + E_0/E_{\mu c}}\right) = -b\,x, \quad E_0 = e^{bx}\left(E + E_{\mu c}\right) - E_{\mu c},$$

$$\tag{2.63}$$

where, by definition, $E_{\mu c} \equiv a/b$ is the energy at which energy loss by ionisation equals the energy loss by radiation. Using the values $a = 2.7\,\mathrm{MeV\,g^{-1}\,cm^2}$ and $b = 3.9 \times 10^{-6}\,\mathrm{g^{-1}\,cm^2}$ from Table 29.2 of Ref. [2], and by assuming the standard rock density $\rho = 2.65\,\mathrm{g\,cm^{-3}}$, we get $E_{\mu c} = 0.69\,\mathrm{TeV}$ and:

$$E_0 = \left(\exp\left[3.9 \times 10^{-6}\,\mathrm{g^{-1}\,cm^2} \cdot 2.65\,\mathrm{g\,cm^{-3}} \cdot 10^5\,\mathrm{cm}\right] \cdot 1.69 - 0.69\right)\,\mathrm{TeV} =$$
$$= 4.0\,\mathrm{TeV}. \tag{2.64}$$

Suggested Readings

For more details on cosmic muons and their interaction with matter, the reader is addressed to Sect. 29.4 and Sect. 33.6 of Ref. [2].

Problem 2.20 The vertical flux of cosmic muons with $E_\mu > 1\,\mathrm{GeV}$ at the sea level is about $70\,\mathrm{m^{-2}\,s^{-1}\,sr^{-1}}$, and the muon spectrum goes approximately as $E_\mu^{-2.7}$. Owing to the continuous slowing down and subsequent decay, the muon spectrum underground reduces with depth untill a depth of about 10 km w.e. ($1\,\mathrm{km\,w.e.} = 10^5\,\mathrm{g\,cm^{-2}}$) is attained. At this point, the spectrum settles to a constant value. Explain this behaviour and provide a rought estimate of the muon flux deep underground.

Solution

At a depth d larger than a few km w.e., only muons with energies of order of $E_{\mu c}$ or larger can make their way through the underground soil, see Problem 2.19. In this energy regime, however, the range scales logarithmically with the muon energy at the sea level E_0:

$$R(E_0) \approx b^{-1}\ln\left(1 + \frac{E_0}{E_{\mu c}}\right), \tag{2.65}$$

where a and b are the constants introduced in Problem 2.19. Equation (2.65) implies an exponential suppression of the flux at large depths. At some point, the muon flux becomes so weak that another source of underground muons takes over, namely muon production from charged-current interaction of muon neutrinos with the rock. The latter is almost independent on the depth. For example, let's consider the infinitesimal flux of neutrinos with energy in the range $[E_\nu, E_\nu + dE_\nu]$: they will contribute to the measured muon flux of energy $E_\mu \geq E_{\mathrm{th}}$, where E_{th} is the detector threshold energy, only if the muon interacts with the rock within a distance $r = R - E_{\mathrm{th}}/(dE_\mu/dx)$ from the underground level d (we make the approximation $E_\mu \approx E_\nu$). The probability for such interaction is $r/\lambda \ll 1$, where λ is the interaction length and depends on the neutrino energy, see Eq. (1.291). For $E_{\mathrm{min}} = 1\,\mathrm{GeV}$, the offset $R - r$ is about 200 m. The neutrino spectrum can be assumed to be similar to the muon spectrum, since for every muon, a ν_μ of similar energy is produced, see Problem 1.19. The neutrino-induced flux can be thus estimated to be:

$$\Phi_\mu^{\text{deep}} \approx \int_{E_{\min}}^{E_{\max}} dE_\nu \, \frac{d\Phi_\nu^0}{dE_\nu} \left(\frac{R(E_\nu) - 200\,\text{m}}{\lambda(E_\nu)} \right) \tag{2.66}$$

The maximum energy $E_{\max}$ can be assumed to be of order of $E_{\mu c}$, since for larger energies the range becomes only mildly dependent on the muon energy, see Eq. (2.65), and thus it will contribute by one power less to the muon flux. Although the muon spectrum at $E_\mu \lesssim 10\,\text{GeV}$ decreases slower than $E_\mu^{-2.7}$, for an order-of-magnitude estimate we can assume for simplicity:

$$\frac{d\Phi_\nu^0}{dE_\nu} = (\alpha - 1)(1\,\text{GeV})^{\alpha - 1} \Phi_0 \, E_\nu^{-\alpha}, \tag{2.67}$$

with $\Phi_0 = 70\,\text{m}^{-2}\,\text{s}^{-1}\,\text{sr}^{-1}$ and $\alpha = 2.7$. By using $dE_\mu/dx = 1.9\,\text{MeV}\,\text{g}^{-1}\,\text{cm}^2$ [5] and the cross section of Eq (1.354) for the neutrino-nucleon scattering (with $Q \approx 1$), and neglecting for simplicity the offset of 200 m, we have:

$$\Phi_\mu^{\text{deep}} \sim \int_{1\,\text{GeV}}^{E_{\mu c}} dE_\nu \, \frac{d\Phi_\nu^0}{dE_\nu} \left(\frac{E_\nu}{1.9\,\text{MeV}\,\text{g}^{-1}\,\text{cm}^2 \cdot \rho} \right) \left(\frac{\rho \cdot N_A}{A} \cdot 1.6 \times 10^{-38}\,\text{cm}^2 \frac{E_\nu}{\text{GeV}} \right) =$$

$$= \Phi_0 \cdot 0.23 \times 10^{-12} \left(\frac{\alpha - 1}{3 - \alpha} \right) \frac{(1\,\text{GeV})^{\alpha - 1} \left(E_{\mu c}^{3 - \alpha} - (1\,\text{GeV})^{3 - \alpha} \right)}{\text{GeV}^2} \approx$$

$$\approx 10^{-9}\,\text{m}^{-2}\,\text{s}^{-1}\,\text{sr}^{-1}. \tag{2.68}$$

The result depends only mildly on the choice of $E_{\max}$. This order-of-magnitude estimate is in a decent agreement with the measured spectrum, see e.g. Fig. 2.7 taken from Ref. [2].

Fig. 2.7 Vertical muon intensity *versus* depth. From Ref. [2]

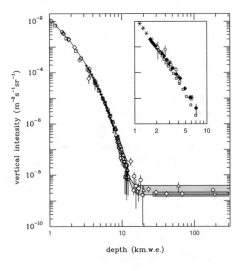

Suggested Readings

More details on cosmic ray fluxes, including a review of theoretical calculation, can be found in Ref. [11]. The reader is addressed to Sect. 29.4 of Ref. [2] for more details on the muon flux underground.

2.2 Particle Identification

Particle identification (PID) is a common problem in particle physics experiments, which are often equipped with a redundance of detectors as to be able to identify the particle type besides measuring their kinematics. As a general rule, the presence of backgrounds and imperfections in the detector makes PID a statistical test rather than a deterministic decision: the probability of correctly identifying a given particle (*efficiency*) has always to be weighted against the probability of wrongly identifying a background event (*fake-rate*). Depending on the particle type and on its energy, a variety of methods can be deployed in experiments. A non-exhaustive list of techniques for PID includes:

- *Measurement of the range.* Each particle loses energy by interaction with matter at a different rate, so that the measured range can be used to differentiate between different particle types. For example, a 10 GeV muon loses energy by collision at a MIP rate of about $11 \, \mathrm{MeV \, cm^{-1}}$, while an electron of the same energy loses energy by radiation at a rate of about $550 \, \mathrm{MeV \, cm^{-1}}$, i.e. about 50 times faster. Hadronic particles interact strongly with the nuclei, with typical interaction lengths of tens of centimetres for condensed materials. Therefore, the capability of muons to penetrate massive detectors exceeds by far larger that of other particles.
- *Measurement of the stopping power.* Even if the particle range is not fully contained within the active volume of a detector, the simultaneous measurement of the stopping power dE/dx and of the particle energy, or momentum, provides a handle to distinguish between different particles. The stopping power can be measured from the energy deposited within the detector. Time Projection Chambers, proportional chambers, nuclear emulsions, solid-state detectors are examples of detectors which can measure the energy loss across the particle trajectory.
- *Cherenkov-light detection.* Relativistic particles can emit Cherenkov lights when moving inside a refractive medium. The angle of emission and the number of emitted photons depend on the particle velocity β as for Eq. (2.15) and Eq. (2.21). A simultaneous measurement of the particle momentum and of the Cherenkov light can be thus used to determine the particle mass.
- *Transition-light detection.* For high-energy particles, Cherenkov detectors as particle identifiers become inefficient, see Eq. (2.21). An alternative to using the β-dependence of Cherenkov detectors is provided by the use of transition radiation detectors, which are sensitive to the light emitted by charged particles while

crossing the separation surface between vacuum and a dielectric material. Since the intensity of the emitted radiation is proportional to the γ-factor of the particle as for Eq. (2.22), particles of a given momentum, but very different mass, like pions and electrons, can be efficiently separated by measuring their transition light.

- *Measurement of the time-of-flight.* A simultaneous measurement of the particle momentum and of the TOF over known distances, allows to determine the particle mass. For unstable particles that decay in reconstructable vertices, the TOF can be measured from the distance traveled by the particle before decaying. Once combined with momentum information, this allows to infere the particle life-time (see Problem 1.32), and hence the particle type.
- *Kinematics.* In scattering experiments where the kinematics of the initial and final state can be measured, four-momentum conservation can be used to infere the mass of the particles involved in the scattering, see e.g. Problem 1.27, 1.28, and 1.62. For unstable particles, the kinematics of the decay products can be used to reconstruct the decay process, from which the mass of the mother particle can be inferred, see e.g. Problems 1.16, 1.20, 1.23, and 1.37.

Problems

Bando n. 13153/2009

Problem 2.21 Mention two methods of identification for charged particles, indicating the range of applicability and their complementarity.

Solution

At small velocities, the simultaneous measurement of the particle momentum $|\mathbf{p}|$ and of its time-of-flight over a known distance, or of the stopping power dE/dx, or of the Cherenkov light emission, represent canonical techniques for PID. However, at higher energies, all these methods become inefficient due to the saturation of the particle velocity to $\beta \to 1$, so that the TOF over a baseline distance L saturates to L/c for all particles, the stopping power (by collision) becomes only logarithmically sensitive to the particle velocity, while for Cherenkov detectors this is due to the fact that the sensitivity to mass differences is suppressed by $|\mathbf{p}|^{-2}$, see Problem 2.25.

At larger energies, one can instead exploit the emission of transition radiation, whose intensity is proportional to the γ-factor of the particle. High-energy electrons can be discriminated from other charged particles thanks to their larger emission of bremstrahlung radiation. In high-energy experiments, a combination of tracking and energy measurements in segmented calorimeters is sometimes used for PID: a calorimeter consisting of an electromagnetic (ECAL) and an hadronic (HCAL) section with independent read-out offers the possibility to separate electrons, which are stopped in ECAL, from hadrons, which interact in both. The attempt to reconstruct and identify each and every particle in a HEP event is called *particle flow* and was pioneered at LEP [12].

Suggested Readings

The PDG review of particle detectors at colliders provides a comprehensive and up-to-date overview of detectors for PID. Introductory textbooks like Ref. [13] are also indicated for a first overview on the subject. Besides the already quoted ALEPH publication [12], the reader is encouraged to read about PID within the particle flow algorithm as implemented in the CMS event reconstruction [14].

Bando n. 13153/2009

Problem 2.22 Discuss a few techniques for neutron detection as a function of the neutron energy.

Solution

Neutrons with energies in excess of a few GeV are best measured by hadronic calorimeters, i.e. devices that degrade the initial hadron energy by initiating a hadronic cascade and measure the visible energy deposited by the cascade particles, which is usually proportional to the incoming neutron energy, see Problem 2.35 for more details.

The detection of fast neutrons relies on the detection of the recoil proton in (n, p) scatterings. This is best achieved by using plastic or liquid organic scintillators, whose molecules contain hydrogen. Given the different fluorescent response of organic compounds to particles of different ionisation power, these materials can also offer n/γ discrimination by pulse-shape analysis.

For thermal neutrons, one usually relies on the nuclear reactions (n, γ) and (n, α), which can be e.g. detected by using liquid, glass, or inorganic scintillators, like Li I (Eu), or gaseous ionisation detectors, like ^{3}He, B F$_3$. The active material is conveniently loaded with suitable nuclei like ^{3}He, ^{6}Li, and ^{10}B, which have large cross sections for the reactions:

$$^3\text{He}\,(n, p)\,\text{t}, \qquad ^6\text{Li}\,(n, t)\,^4\text{He}, \qquad ^{10}\text{B}\,(n, \alpha)\,^7\text{Li}^{(*)}, \qquad (2.69)$$

respectively. The kinetic energy of the emitted particles (protons, tritium, α-particles, Li ions) peakes at values determined by the Q-value of the reactions, thus allowing to separate the neutron signals from other backgrounds, most notably by photon interactions.

Suggested Readings

Chapter 7.7 of Ref. [1] describes the pulse-shape technique with scintillators and provides an introduction to various experimental techniques for neutron detection.

Bando n. 1N/R3/SUB/2005

Problem 2.23 In order to separate K^+ and π^+ in a momentum window between 700 MeV and 4 GeV, one can use two threshold Cherenkov detectors operated in series. Neglecting possible inefficiencies of the detectors near the threshold, determine which values of the refraction index can be chosen, and propose a suitable radiator.

Discussion

Although not mentioned explicitly, Cherenkov detectors are often integrated with spectrometers or other detectors that can measure the momentum of the particle. For example, Cherenkov detectors can be employed to select particles of a given type from a composite beam of given momentum.

Solution

The momentum acceptance of the experiment provides four threshold velocities and as many refraction indexes, namely:

$$\begin{cases} n < 1.0006 & \text{no } \pi \text{ emit} \\ n > 1.0195 & \text{all } \pi \text{ emit} \\ n < 1.0076 & \text{no } K \text{ emit} \\ n > 1.22 & \text{all } K \text{ emit} \end{cases} \tag{2.70}$$

With two counters at hand, one could set counter A at a value of $n_A = 1.0195$, so that no signal there would imply that the particle is a kaon (K-tag), and counter B at a value $n_B = 1.0076$, so that a signal in that counter would imply that the particle is not a kaon (π-tag). With this scheme one has three possibilities, summarised in Table 2.5. The third row (all counters with no-signal) represents a useful event only if the experiment is equipped with an independent trigger (e.g. a scintillator located along the beam direction). However, there remains an ambiguity for the case where only counter A records a signal. If one further assumes that the particle momentum can be measured, then the ambiguity is lifted. Indeed, if one considers pions and kaons with velocities in the range $[1/n_A, 1/n_B]$, the corresponding momenta span two non-intersecting ranges:

$$\frac{1}{n_A} < \beta < \frac{1}{n_B} \quad \Rightarrow \quad |\mathbf{p}| \in \begin{cases} [0.70, 1.12] \text{ GeV} & \pi \\ [2.49, 4.0] \text{ GeV} & K \end{cases} \tag{2.71}$$

so that a simultaneous measurement of the particle momentum and of the Cherenkov counters can discriminate between the two particles. Figure 2.8 shows the critical index $1/\beta$ for the two particle types as a function of $|\mathbf{p}|$. The dashed lines indicate the indexes chosen for counters A and B, while the vertical arrows mark the upper and lower momenta at which pions fail to generate a signal in B and kaons generate a signal in A, respectively. Concerning the choice of radiator medium, we can refer to Table 2.4 to identify possible candidates. In particular, we see that a value of $n - 1 \approx 2 \times 10^{-2}$ can be obtained for example by using aerogels, while $n - 1 \approx 7 \times 10^{-3}$ can

Table 2.5 Possible outcomes of a single-particle event using two threshold Cherenkov detectors in series with $N_A > n_B$

A	B	Particle
1	1	π
1	0	π or K
0	0	K
0	1	Not possible

Fig. 2.8 The critical index $1/\beta$ for the two particle types as a function of $|\mathbf{p}|$. The dashed lines indicate the indexes chosen for counters A and B, while the vertical arrows mark the upper and lower momenta at which pions fail to generate a signal in B and kaons generate a signal in A, respectively

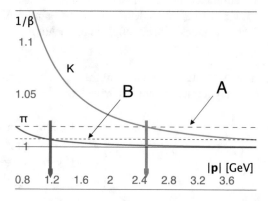

be obtained by using e.g. pentane (C_5H_{12}) or perfluoropentane (C_5F_{12}) of appropriate temperature and pressure.

Bando n. 13153/2009

Problem 2.24 Explain how the Cherenkov threshold depends on the refraction index of the medium. Three particles of different mass but same momentum $|\mathbf{p}|$ cross a system of two Cherenkov detectors arranged in series. How can the three particles be identified?

Solution

The Cherenkov threshold is the velocity β that equals the group velocity of light in the medium, i.e. $\beta = 1/n$, where n is the refraction index. By definition, vacuum has $n = 1$, and $n > 1$ for any other medium, see Table 2.4 for a few representative materials.

Given two threshold Cherenkov detectors A and B operated in series, the identification of three particles of different mass but same momentum $|\mathbf{p}|$, such that the three particles have velocities $\beta_1 < \beta_2 < \beta_3$, can be achieved by setting the refraction index of the two counters at $n_A = 1/\beta_1$ and $n_B = 1/\beta_2$, so that:

- particle (1) is below threshold in both counters ($\beta_1 \leq 1/n_A,\ 1/n_B$), thus producing no signal in any of the two counters.

- particle (2) is above threshold in counter A ($\beta_2 > 1/n_A$), but below threshold in counter B ($\beta_2 \leq 1/n_B$), thus producing a signal in only one counter;
- particle (3) is above threshold ($\beta_3 > 1/n_A$, $1/n_B$) in both detectors, thus producing a signal in both counters;

An analysis of the signal output in the two counters can thus reveal which of the three particles has crossed the detector. This configuration also maximises the light yield when the particle is above threshold.

Problem 2.25 A Cherenkov imaging detector measures the angle θ of Cherenkov photons with a resolution $\sigma_\theta = 2$ mrad. What is the largest beam momentum $|\mathbf{p}|$ such that kaons and pions can be discriminated to better than 3σ by the angular measurement only, if the Cherenkov radiator consists of fused silica ($n = 1.474$) or fluorocarbon gas ($n = 1.0017$)?

Solution

Let the Cherenkov angle be denoted by θ. A separation to better than 3σ amounts to require $\Delta\theta/\sigma_\theta \geq 3$. By approximating finite differences by their differentials, we get:

$$\frac{\Delta\theta}{\sigma_\theta} \approx \frac{d\theta}{\sigma_\theta} = \frac{1}{\sigma_\theta \sin\theta} d\cos\theta = \frac{\beta}{\sigma_\theta\sqrt{\beta^2 n^2 - 1}} d\left(\frac{1}{\beta}\right) = \frac{\beta^2 \, dm^2}{2\,\sigma_\theta\sqrt{\beta^2 n^2 - 1}\,|\mathbf{p}|^2}$$

$$\approx \frac{|m_K^2 - m_\pi^2|}{2\sigma_\theta\sqrt{n^2 - 1}\,|\mathbf{p}|^2}. \tag{2.72}$$

Hence, the largest momentum for which the statistical separation is in excess of $N_\sigma = 3\sigma$ is provided by:

$$|\mathbf{p}| < \frac{|m_K^2 - m_\pi^2|}{\left[N_\sigma \cdot 2\sigma_\theta\sqrt{n^2 - 1}\right]^{\frac{1}{2}}} = \begin{cases} \dfrac{0.474\ \text{GeV}}{\sqrt{3\cdot2\cdot2\times10^{-3}\cdot\sqrt{1.474^2-1}}} = 4.2\ \text{GeV} \quad \text{silica} \\[2ex] \dfrac{0.474\ \text{GeV}}{\sqrt{3\cdot2\cdot2\times10^{-3}\cdot\sqrt{1.0017^2-1}}} = 18\ \text{GeV} \quad \text{fluorocarbon} \end{cases}$$

$$\tag{2.73}$$

Suggested Readings

This problem is inspired by Sect. 34.5 of Ref. [2]. The reader is addressed to this reference for more information on the subject.

Problem 2.26 Tellurium dioxide (TeO_2) crystals ($n = 2.4$, $\rho = 6\,g\,cm^{-3}$) have been used to search for the putative neutrinoless double-beta decay $^{130}_{52}Te \to ^{130}_{54}Xe$ in bolometric calorimeters. The experimental signature is provided by an energy deposit around 2.53 MeV. A major background to this process is represented by α-decays of radioactive contaminants. Show that the simultaneous measurement of Cherenkov photons and calorimetric energy would allow to separate α particles from signal events. Estimate the mean number of Cherenkov photons with wavelengths in the range [350, 600] nm produced by a signal event in a few centimetres long crystals.

Discussion

Differently from an ordinary double-β decay ($2\nu\beta\beta$), where a nucleus $^A_Z X$ decays to $^A_{Z+2}Y + 2\nu + 2e^-$, a neutrinoless double-β decay ($0\nu\beta\beta$) does not produce neutrinos in the final state. The Q-value of the reaction, see Problem 1.39, is entirely taken by the two electrons: their energy sum is therefore a line around Q smeared by the detector resolution. This also implies that the electron energies are fully anticorrelated. The theoretical energy distributions for this decay can be found in Ref. [15]. Alpha particles of a few MeV energy, typical of radioactive decays, behave like background events by releasing their energy in the calorimeter.

Solution

In order to prove that the electrons radiate Cherenkov light while the α particles do not, it suffices to verify that the threshold velocity $\beta = 1/n = 0.717$ in TeO_2 is above the velocity of α's, but below the velocity of at least one of the electrons. Assuming $T_\alpha = 2.53$ MeV, one has

$$\beta_\alpha \approx \sqrt{\frac{2T_\alpha}{m_\alpha}} = \sqrt{\frac{2 \cdot 2.53\,\text{MeV}}{3.73\,\text{GeV}}} = 0.037 < \beta, \qquad (2.74)$$

while for a $0\nu\beta\beta$ decay:

$$\max \beta_e > \sqrt{1 - \left(\frac{m_e}{Q/2 + m_e}\right)^2} = \sqrt{1 - \left(\frac{0.511\,\text{MeV}}{1.77\,\text{MeV}}\right)^2} = 0.958 > \beta. \quad (2.75)$$

To good approximation, the total range and the number of Cherenkov photons are independent of the energy sharing between the two electrons, thanks to the anti-correlation between the two energies. Indeed, for $\gamma \gg 1$, the range is a linear function of energy as for Eq. (2.32). In the case of interest, though, the average kinetic energy is comparable to m_e, so the linearity is lost. However, a numerical investigation shows that the total range is constant to within 15% over the allowed electron spectrum, and is larger when the energy sharing is more asymmetric. Furthermore, Eq. (2.32) is expected to underestimates the true range for small values of γ, and one should rather use the full calculation. To circumvent the lack of tabulated data and the mild dependence on the kinematics, we consider a particular decay configuration, namely $T_1 = 1.0\,\text{MeV}$ and $T_2 = Q - T_1 \approx 1.5\,\text{MeV}$. We then approximate the stopping power by averaging the tabulated values for two similar materials: Na I, which contains Iodine, a Tellurium neighbour in the periodic table, and Ti O_2, which is also a metal dioxide. At $T = 1\,\text{MeV}$, Ref. [16] gives:

$$R_1(\text{Na I}) = 0.69\,\text{g cm}^{-2}, \qquad R_1(\text{Ti O}_2) = 0.55\,\text{g cm}^{-2}, \qquad (2.76)$$

Taking the mean, we get $R_1 \approx 0.64\,\text{g cm}^{-2}$, or $0.10\,\text{cm}$. There are no values tabulated for $T = 1.5\,\text{MeV}$, but we can use the scaling predicted by Eq. (2.32), giving a ratio $R_2/R_1 = 1.69$. Hence, $R_2 \approx 0.175\,\text{cm}$. The light output in the wavelength window $[350, 600]\,\text{nm}$ can be estimated by using Eq. (2.16) with $\langle \sin^2 \theta \rangle \approx 1 - 1/n^2$, giving:

$$N_\gamma \approx (0.10 + 0.175)\,\text{cm} \frac{1.15 \times 10^3\,\text{cm}^{-1}}{\sqrt{600 \cdot 350/400}} \left(1 - \frac{1}{2.4^2}\right) \frac{600 - 350}{\sqrt{600 \cdot 350}}$$
$$= 46 + 79 = 125, \qquad (2.77)$$

which agrees with the more accurate expectation of Ref. [17], which averages the range over the proper energy spectrum.

Suggested Readings

The idea of exploiting Cherenkov radiation in bolometric detectors has been first proposed in Ref. [17], from which the problem is largely inspired.

Problem 2.27 A threshold Cherenkov detector is used to separate muons from pions in a beam with momentum $|\mathbf{p}| = 150\,\text{MeV}$. What values of the refraction index n can be used?

Solution

The condition for which muons emit Cherenkov light, while pions do not, is given by:

$$1/\beta_\mu < n < 1/\beta_\pi \quad \Leftrightarrow \quad \sqrt{\left(\frac{m_\mu}{|\mathbf{p}|}\right)^2 + 1} < n < \sqrt{\left(\frac{m_\pi}{|\mathbf{p}|}\right)^2 + 1},$$

giving the result: $1.22 < n < 1.37$.

Bando n. 1N/R3/SUB/2005

Problem 2.28 An experiment needs to distinguish pions from kaons of momentum $|\mathbf{p}| = 2\,\text{GeV}$ by measuring the time flight on a $L = 2\,\text{m}$ baseline. The instrumentation has a time resolution $\sigma_t = 0.2\,\text{ns}$. Can each particle be identified? With which precision can the pion fraction be determined?

Solution

The time-of-flight (TOF) for pions and kaons in the beam is given by:

$$t = \frac{L}{\beta c} = \frac{L}{c}\sqrt{1 + \frac{m^2}{|\mathbf{p}|^2}} = \frac{2\,m}{3 \times 10^8\,\mathrm{m\,s^{-2}}} \begin{cases} \sqrt{1 - \left(\frac{0.139}{2}\right)^2} = 6.68\,\mathrm{ns} \quad \pi \\ \sqrt{1 - \left(\frac{0.494}{2}\right)^2} = 6.87\,\mathrm{ns} \quad K \end{cases} \quad (2.78)$$

Since $\Delta t = 0.19\,\mathrm{ns} \approx \sigma_t$, particle-by-particle identification is affected by a large statistical uncertainty, i.e. the Type-II error is large for any given efficiency to identify the correct particle type. For example, if we decided to tag a particle as a K if the TOF is in excess of $6.87 - 1\sigma_t = 6.67\,\mathrm{ns}$, the selection efficiency would be 84%, for a fake-rate of about 50%. Even though an event-by-event classification is not very accurate, the pion (or kaon) fraction of the beam can be estimated with large accuracy for a sufficiently large number of measurements. Assuming N independent and gaussian distributed measurements $\mathbf{X} = \{X_i\}$, the maximum-likelihood (ML) estimator of the pion fraction $\bar{\varepsilon}_\pi$ is given by the solution of the equation:

$$0 = \left. \frac{\partial L(\mathbf{X}, \varepsilon_\pi)}{\partial \varepsilon_\pi} \right|_{\hat{\varepsilon}_\pi},$$

$$\text{with} \quad L = \prod_{i=1}^{N} f(X_i, \varepsilon_\pi) = \prod_{i=1}^{N} [\varepsilon_\pi \, \mathcal{N}(X_i \mid t_\pi, \sigma_t) + (1 - \varepsilon_\pi)\, \mathcal{N}(X_i \mid t_K, \sigma_t)]$$

$$(2.79)$$

The classical theory of estimators predicts that the asymptotic variance of the ML estimator is given by

$$\mathrm{Var}\left[\hat{\varepsilon}_\pi\right] = \frac{1}{N\,I(\hat{\varepsilon}_\pi)}, \quad \text{with} \quad I(\hat{\varepsilon}_\pi) = \mathrm{E}\left[-\frac{\partial^2 \ln f(x, \varepsilon_\pi)}{\partial^2 \varepsilon_\pi}\right], \quad (2.80)$$

see Sect. 4.1. The information can be computed numerically using a simple program for different values of ε_π, see Appendix 2.3. The result is a number of $\mathcal{O}(1)$: for example, for $\varepsilon_\pi = 0.1\ (0.3)$ one gets $I = 1.03\ (0.80)$. Hence, the standard deviation on the pion fraction will be given by:

$$\sigma_{\hat{\varepsilon}_\pi} \approx \frac{1}{\sqrt{N}}. \quad (2.81)$$

Problem 2.29 In 1987, the water Cherenkov detector Kamiokande-II in the Kamioka mine (Japan), detected a neutrino burst that was attributed to a supernova event occurred at a distance $d = 5.5 \times 10^4\,\mathrm{kpc}$ from the Earth. The energy and arrival time at the detector could be measured for those (anti)neutrinos that interacted via the charged-current (CC) scattering $\bar{\nu}\, p \rightarrow n\, e^+$, or by the electron-scattering (ES)

Fig. 2.9 Scatter plot of energy and time for the twelve supernova candidate events recorded by Kamionkande in 1987 (from Ref. [18])

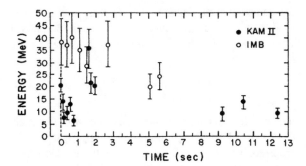

reaction $\nu_e\, e^- \to \nu_e\, e^-$, within the fiducial volume of the detector. During a time interval $\Delta t = 12$ s, a total of 12 events were registered. The time vs energy diagram of the signal events is reported in the Fig. 2.9.

- The Kamiokande experiment could not distinguish electrons from positrons by using the sole Cherenkov light. How was it then possible to separate ν_e from $\bar{\nu}_e$?
- Explain how the antineutrino energy $E_{\bar{\nu}}$ could be measured from the positron energy E_{e^+}.
- Determine a lower bound to the ν_e lifetime.
- Using the data reported in the plot, estimate an upper bound to the electron neutrino mass m_{ν_e}.

Discussion

As of 1987, the Kamiokande-II experiment consisted of a cylindric water tank containing over 2000 t of water instrumented with uniformly distributed PMT's covering about 20% of the total surface. The PMT's were sensitive to the Cherenkov light in the range $300 \div 500$ nm. At these wavelengths, the light attenuation length exceeds 50 m, thus allowing an efficient light collection all across the fiducial volume. The event trigger, production vertex, direction, and energy of the particles were reconstructed by using the charge and time stamp of all PMT with a signal above the noise. The single-PMT time resolution was 13 ns, while the relative energy resolution was estimated from simulation to be about 20%. An electron neutrino with energy of about 10 MeV interacts mostly through ES on the atomic electrons. The CC interaction with the transmutation $^{16}_6\text{O} \to \,^{16}_7\text{F}$ is instead suppressed by the large mass difference $\mathcal{B}(^{16}_6\text{O}) - \mathcal{B}(^{16}_7\text{F}) \approx 16$ MeV. Conversely, an electron antinutrino interacts mostly through the CC reaction $\bar{\nu}\, p \to n\, e^+$, provided $E_{\bar{\nu}} \gtrsim 2$ MeV. The main background to ~ 10 MeV electrons and positrons is represented by cosmic muons, β-decays of unstable isotopes polluting the water, and by γ/n radiation from the cavern walls.

Solution

The separation between electrons and positrons is possible on a statistical basis. Indeed, the CC scattering for antineutrino energies $E_{\bar{\nu}} \approx 10$ MeV is isotropic in the laboratory frame. This can be proved as follows. First, one notices that the velocity of the centre-of-mass frame is $\beta = E_{\bar{v}}/(E_{\bar{v}} + m_p) \approx 10^{-2}$, so that the centre-of-mass is almost at rest in the laboratory frame. In the latter, the dynamics is governed by the exchange of a virtual W boson, as described by the Fermi Lagrangian:

$$\mathscr{L}_F = \frac{G_F}{\sqrt{2}} \cos\theta_C \left[\bar{n}\gamma_\mu(1 - \alpha\gamma_5)p\right]\left[\bar{v}\gamma^\mu(1 - \gamma_5)e\right]. \tag{2.82}$$

The amplitude squared can be obtained with the usual Casimir's tricks. By taking $\alpha = -1$, it becomes proportional to $(p_{e^+} p_p)(p_{\bar{v}} p_n) \approx E_{e^+} E_{\bar{v}} m_p m_n$, if the neutron recoil is neglected compared to the nucleon mass. In this case, E_{e^+} is also a constant, hence the amplitude squared itself is constant. From Problem 1.53 and the considerations above, we can see that the cross section is roughly isotropic in the laboratory frame. This is not the case for the ES, since the centre-of-mass velocity is now $\beta = E_{\bar{v}}/(E_{\bar{v}} + m_e) \approx 1$. which gives rise to a very forward-peaked differential cross section in the laboratory frame, see Problem 1.15.

For the antineutrino scattering, energy conservation implies

$$E_{\bar{v}} + m_p = E_{e^+} + m_n \quad \Rightarrow \quad E_{\bar{v}} \approx E_{e^+} + \underbrace{(m_n - m_p)}_{1.3 \text{ MeV}}. \tag{2.83}$$

The neutrino lifetime, τ_v, has to be large enough so that the neutrinos can make it to Earth, i.e.:

$$\tau_v \gtrsim \frac{d}{c\,\gamma_v} = \frac{5.5 \times 10^4 \text{ pc}}{c\,(E_v/m_v)} = \frac{5.5 \times 10^4 \cdot 3.3\,c \cdot y}{c\,(E_v/m_v)} = 1.8 \times 10^5 \left(\frac{m_v}{E_v}\right) y, \tag{2.84}$$

where we have used the relation $1 \text{ pc} \approx 3.3\,c \cdot y$.

If the neutrino burst starts at the time $t = 0$, the arrival time at the detector is:

$$t = \frac{d}{\beta_v c} = \frac{d}{c} \frac{1}{\sqrt{1 - (m_v/E_v)^2}} \approx \frac{d}{c}\left[1 + \frac{1}{2}\left(\frac{m_v}{E_v}\right)^2\right]. \tag{2.85}$$

Two neutrinos of energies E_1 and E_2, emitted at the same time $t = 0$, will arrive at destination with a time separation:

$$t_1 - t_2 = \frac{d}{2c} m_v^2 \left(\frac{1}{E_1^2} - \frac{1}{E_2^2}\right). \tag{2.86}$$

From the recorded data, we observe the presence of a few neutrino events separated by about 10 s from the the first burst events, which is larger than the expected duration of a supernova burst (a few seconds), is an indication that neutrinos have a mass, since otherwise they would have arrived all in one shot. The presence of two populations of events, one located within the first second, and the other around $t = 2$ s, which are not distributed according to Eq. (2.85), indicates, though, that the pattern of neutrino emission from the supernova has some non-trivial time dependence, i.e. one cannot assume a perfectly synchronous burst. Yet, some of the neutrinos must have been created simultaneously, and with some broad spectrum of energies, so that any difference in arrival time has to be attributed to the non-zero neutrino mass. A conservative upper limit on m_ν can thus be obtained by considering, those events that feature the largest energy difference $|\Delta E|$ among the first and last arrived events, respectively. From the plot, we take e.g.: $(E_1, t_1) = (35\,\text{MeV}, 1.5\,\text{s})$ and $(E_2, t_2) = (10\,\text{MeV}, 12.5\,\text{s})$. Inverting Eq. (2.86), we have:

$$m_\nu \lesssim \sqrt{\frac{2c\,(t_1 - t_2)}{d}}\,\frac{E_1 E_2}{\sqrt{E_2^2 - E_1^2}} \approx 20\,\text{eV}. \tag{2.87}$$

Suggested Readings

This problem is inspired by the Kamiokande publication of Ref. [18].

Bando n. 18211/2016

Problem 2.30 A ν_μ beam with an energy of $30\,\text{GeV}$ enters a detector containing liquid Ar. A fraction of the events features a few metres long track starting from the interaction point, while, for a smaller fraction of the events, all tracks are contained within a small volume. Explain this behaviour.

Solution

As already discussed in Problem 1.64, neutrinos can undergo interactions with both the nuclei and and the atomic electrons, the latter having a cross section suppressed by a factor of m_e/m_N. In both cases, the neutrinos can interact via either the charged current, $\nu_\mu\,X \to \mu^-\,Y$, or the neutral current interaction, $\nu_\mu\,X \to \nu_\mu\,X'$. The EWK theory predicts the ratio between neutral and charged current cross section in terms of the Weinberg angle θ_W to be:

$$\left(\frac{\sigma_{NC}}{\sigma_{CC}}\right)_\nu = \frac{1}{2} - \sin^2\theta_W + \frac{20}{27}\sin^4\theta_W \approx 0.31, \tag{2.88}$$

see e.g. Ref. [19]. When a neutrino of energy $E_\nu = 30\,\text{GeV}$ interacts via CC, it produces a muon of similar energy, which being a MIP, is highly penetrating in the Ar medium and can be therefore identified as a long track. Conversely, in the occurrence of a NC interaction, the only detectable signal is provided by the recoil of the struck nucleus. Since DIS prevails in this energy regime, the interaction is inelastic and results in a number of hadronic particles which, being much heavier than the muon and less energetic, have smaller range, thus appearing as a set of short tracks emerging from the interaction point.

Suggested Readings

The reader is addressed to Chap. 12 of Ref. [19] for more information on neutrino interactions in matter.

Problem 2.31 A charged particle is moving inside a uniform magnetic field of intensity $B = 1.0\,\mathrm{T}$. The radius of curvature of the track is $R = 7.25\,\mathrm{m}$ with negligible error. The kinetic energy of the particle is measured to be $T = (2.00 \pm 0.03)\,\mathrm{GeV}$. Determine which type of particle is most probably being measured.

Solution

The charge sign is fixed by the direction of curvature. The particle momentum $|\mathbf{p}|$ is instead given by the formula:

$$|\mathbf{p}| = 0.3\,|z|\,(B/\mathrm{T})\,(R/\mathrm{m})\ \mathrm{GeV} = 2.20\,|z|\,\mathrm{GeV}, \qquad (2.89)$$

where z is the particle charge in units of the proton charge e, see Problem 3.3. The particle mass m is therefore given by:

$$m^2 = (T + m)^2 - |\mathbf{p}|^2, \qquad m = \frac{|\mathbf{p}|^2 - T^2}{2T} = \frac{(2.20 \cdot z)^2 - (2.00)^2}{2 \cdot 2.00}\,\mathrm{GeV}. \qquad (2.90)$$

The uncertainty on m can be obtained by propagating the uncertainty on T:

$$\Delta m = \left|\frac{\partial m}{\partial T}\right| \Delta T = \frac{1 + |\mathbf{p}|^2/T^2}{2}\,\Delta T = \frac{1 + (2.20 \cdot z/2.00)^2}{2} \cdot 0.03\,\mathrm{GeV}. \quad (2.91)$$

Stable, non-exotic particles have integer charges. We can therefore try different *ansatz* values of $|z|$ and compare the result with the known spectrum of particles. For $|z| = 1$, Eq. (2.90) gives $m = (210 \pm 30)\,\mathrm{MeV}$, which does not match any known particle within the experimental uncertainty. For $|z| = 2$, one has $m = (3.84 \pm 0.09)\,\mathrm{GeV}$, which is compatible with the mass of the α particle $m_\alpha = 3.73\,\mathrm{GeV}$ at the 1σ level.

Bando n. 13153/2009

Problem 2.32 Describe which methods could be used to measure lifetimes of order 10^9 years, 10^{-12} s, and 10^{-22} s.

Solution

Lifetimes of order 10^9 years are typical of radioactive decays. Such lifetimes can be measured by counting the number of decays in a sample and in a given time interval Δt. Let N_C be the number of countings after background-subtraction. Under the assumption $\tau \gg \Delta t$, the lifetime can be measured from the relation:

$$\tau = \frac{V \rho N_A}{A} \frac{\Delta t}{N_C}, \tag{2.92}$$

where V is the volume of the sample being observed.

Lifetimes of order 10^{-12} s are characteristics of weakly decaying particles, like D and B mesons, or τ leptons. Since $c = 3 \times 10^2$ μm/ps, the decay vertexes of such particles are of order 300 μm, when the particles are produced at relativistic energies. Silicon detectors, with intrinsic spatial resolutions of a few tens of microns or better, see Problem 2.43, are ideal candidates to build vertex detectors with sufficient resolution to resolve such decays.

Lifetimes of 10^{-22} s are characteristics of strongly decaying particles, like the ρ and ω mesons, or the Δ baryon. The distance of flight is far too small to be measurable by any position-measuring device. Such lifetimes are therefore indirectly estimated from the decay width Γ of Eqs. (1.186), as measured from the invariant mass distributions of the decay products, or from the production cross section.

2.3 Functioning of Particle Detectors

Particle detectors record the passage of particles. Depending on the detector type and on the form of radiation it is sensitive to, detectors can be used to measure the position and time of arrival of a given particle at the detector location, the energy and direction of the incoming particle, and sometimes even identify the type of particle. Detectors are usually composed of an active volume, which interacts with the particle, and a readout component, hosting the electronics required to generate an electric signal, provide signal amplification to improve the signal-over-noise ratio, and finally shape the signal according to some logic suitable for later processing in the experiment or for persistent data storage. In modern experiments, detectors are commonly operated by computers, which supervise their correct functioning and take care of data acquisition. The field of particle detection is vast and finds application that range from pure research to industry. No attemp is made here to give a comprehensive overview on this subject. The selected problems want to discuss the main technologies and introduce general concepts, like resolution, efficiency, dead time.

Problems

Bando n. 1N/R3/SUB/2005

Problem 2.33 In an electromagnetic calorimeter, the stochastic contribution to the resolution is $0.07/\sqrt{E}$. Can we conclude that the energy resolution for an electron of energy $E = 50\,\text{GeV}$ is 1%?

Discussion

Electromagnetic calorimeters are detectors that measure the kinetic energy of charged particles by exploiting one or more interaction mechanisms between charged particles and matter, including fluorescence, Cherenkov light emission, and ionisation. In general, only a fraction of the total initial energy is converted into a visible signal: the proportionality between the measured signal and the total energy allows to measure the latter, after a proper calibration is performed. Electromagnetic calorimeters can be broadly classified into two categories: *homogeneous* and *sampling*, depending on whether the active medium is composed of the same material, or interleaved with layers of inactive absorbers which degrade the energy of the incoming particle. The total energy resolution depends on the choice of active material, which determines the statistics of signal carriers per unit of deposited energy (e.g. the statistics of scintillation photons), on the signal generation and electronics (efficiency of the photodetector, electronic noise), and on other geometrical properties of the detector (e.g. uniformity, dependence of the response with the particle impact point, etc.). In most applications, the relative energy resolution can be parametrised in terms of these three contributions as:

$$\frac{\sigma(E)}{E} = \frac{a}{\sqrt{E}} \oplus \frac{b}{E} \oplus c, \tag{2.93}$$

where the symbol $\oplus$ indicates sum in quadrature. The three contributions are called *stochastic*, *noise*, and *constant term*, respectively. As a general rule, homogeneous calorimeters shine for their small stochastic term of order 1% in units of $1/\sqrt{E/\text{GeV}}$, while for sampling calorimeters the stochastic term is in the range $5 \div 20\%$, in the same units. The importance of the noise term a depends on the signal collection type: scintillation and Cherenkov calorimeters coupled to high-gain PMT suffers the least from the electronic noise, while the noise is usually larger for calorimeters that collect the signal in the form of charge (e.g. semiconductive, gas sampling, and noble-gas calorimeters), since a preamplifier is the first element in the readout chain. For this contribution to be subleading in the GeV range, the parameter b needs to be kept at the 100 MeV level per channel. For use in high-energy experiments, where particles with energies of hundreds of GeV need to be measured, the constant term ends up to be the limiting factor to the ultimate energy resolution. As an example, the electromagnetic calorimeters employed by the CMS and ATLAS experiments at the LHC are built with different technologies, but achieve similar physics performances, overall. The CMS detector makes use of a homogeneous scintillation calorimeter based on $PbWO_4$ crystals. A test beam on a small prototype yielded a stochastic term of $3.3\%/\sqrt{E/\text{GeV}}$, a noise term of $0.19/(E/\text{GeV})$, and a local constant term of 0.27%. When averaged over the full detector acceptance, the goal constant term needs to be kept below 0.5%, which is challenging since the whole detector is composed of about hundred thousand crystals that need to be inter-calibrated. This problem is somehow relieved by the ATLAS setup, which uses instead a sampling liquid-Ar calorimeter, at the price of increasing the stochastic term. A test beam on a prototype yielded a stochastic term of $10\%/\sqrt{E/\text{GeV}}$, a noise term of $0.25/(E/\text{GeV})$, and a local constant term of 0.3%.

Solution

As discussed above, the energy resolution of an electromagnetic calorimeter depends on the energy as in Eq. (2.93). For an electron with $E = 50\,\text{GeV}$ and a calorimeter with $a = 7\%$, the stochastic term is $7\%/\sqrt{50} \approx 1\%$. The latter has to be added in quadrature to the constant and noise term to obtain the total relative energy resolution. We can estimate an upper limit to the noise and constant terms such that they do not contribute individually to the total relative resolution by more than a certain fraction f, that we can conventionally set to e.g. $f = 0.1$. With this choice:

$$\frac{\sigma(E)/E - 1\%}{1\%} < 0.1 \quad \Rightarrow \quad \begin{cases} \frac{1}{2}\left(\frac{b/50\,\text{GeV}}{1\%}\right)^2 \lesssim 0.1, & b \lesssim 220\,\text{MeV} \\ \frac{1}{2}\left(\frac{c}{1\%}\right)^2 \lesssim 0.1, & c \lesssim 0.5\% \end{cases} \tag{2.94}$$

We can therefore conclude that the energy resolution for an electron of energy $E = 50\,\text{GeV}$ is about 1% provided that the noise and constant term are below about $200\,\text{MeV}$ and 0.5%, respectively.

Suggested Readings

A succint but complete review of calorimetry in particle physics can be found in Ref. [20]. More informations on the state-of-the-art in calorimetry can be found in the PDG review [2] and references therein.

Bando n. 1N/R3/SUB/2005

Problem 2.34 A relativistic electron releases energy in a block of BGO, generating a signal of about 10^6 p.e./GeV, while the signal generated in a block of lead glass of the same size is only 10^3 p.e./GeV. How can such a difference be explained?

Discussion

Both BGO and lead glass feature a radiation length X_0 of about $1\,\text{cm}$ and a critical energy of about $10\,\text{MeV}$ [5]. An electron of few GeV energy loses energy mostly by radiation. The emitted bremsstrahlung photons undergo pair-production, with subsequent photon emission. The resulting electromagnetic shower is characterised by an energy profile

$$\frac{dE}{dt} = E_0 b \frac{(bt)^{a-1} e^{-bt}}{\Gamma(a)}, \tag{2.95}$$

where $t = x/X_0$ and a and b are constants that depend on the material. Simplifying the shower development as a series of $1 \to 2$ branches ($e^\pm \to e^\pm \gamma$ and $\gamma \to e^+ e^-$)

with equal energy sharing and separated by a distance X_0, so that the energy per constituent at a depth t is $E/2^t$, it follows that the total track length $L(t)$ from electrons, positrons, and photons, after traversing t radiation lengths is given by

$$L(t) = 2^t X_0. \tag{2.96}$$

The maximum number of radiation lengths t_{max} is determined by the condition that the electron/positron energy falls below the critical energy E_c, i.e. $t_{max} = \ln(E/E_c)/\ln 2$, and

$$L = 2^{\frac{\ln E/E_c}{\ln 2}} X_0 = \left(\frac{X_0}{E_c}\right) E. \tag{2.97}$$

A more refined treatment of shower development, will still predict the total track length L to be *proportional* to the initial energy. Along their path, electrons and positrons excite the fluorescent levels of the crystal, characterised by an average excitation energy ε, so that the total photon output N_γ is still proportional to the initial energy E.

Solution

BGO, an acronym for $(Bi_2 O_3)_2 (Ge O_2)_3$, is a scintillating crystal. The mean excitation energy per photon is reported in Table 2.2 and is about $300\,eV/\gamma$, or $3 \times 10^6\ \gamma/GeV$, which is in the ballpark of the value reported by the problem (the ultimate p.e. statistics depends on the PMT collection and quantum efficiency). Lead glass (Pb O) is an amorphous material and does not scintillate. It has a large refraction index ($n \approx 1.8$) and is transparent to visible wavelengths, which makes it a good Cherenkov radiator. Assuming a quality factor N_0 of about $90\,cm^{-1}$, see Eq. (2.21), and a total charged track length as in Eq. (2.97), an upper limit to the number of p.e. per GeV can be estimated as:

$$\frac{N_{p.e.}}{E} = \frac{N_{p.e.}}{L}\frac{L}{E} \approx 90\,cm^{-1} \langle \sin^2 \theta_c \rangle \cdot (2/3)\frac{X_0}{E_c} =$$
$$= 90\,cm^{-1} \cdot 0.69 \cdot (2/3)\frac{1.3\,cm}{10\,MeV} = 5 \times 10^3/GeV, \tag{2.98}$$

where the factor of $2/3$ accounts for the fact that only electrons and positrons produce Cherenkov light. This estimate does not account for the fact that the simple shower model is not well representative of the energy distribution within the shower: the bremsstrahlung cross section $d\sigma/d\nu$ for emitting one photon with frequency ν is approximately proportional to ν^{-1}, see e.g. Eq. (2.68) of Ref. [1], so that the secondary $e^+ e^-$ pairs from γ conversion are preferably soft, with implications on the total Cherenkov light yield. A more accurate estimation would yield a smaller value $N_{p.e.}/E \approx 10^3/GeV$ [20].

The difference between the two materials can be therefore ascribed to the different mechanism by which photoelectrons are produced in the two materials.

Suggested Readings

The reader is addressed to Ref. [20] for a primer on calorimetry for particle physics.

Bando n. 13153/2009

Problem 2.35 Measuring the energy of hadronic particles through calorimetric methods is a fundamental ingredient in HEP experiments. When a hadron produces a shower, on average 30% of the initial energy is transformed into "invisible" energy. Indicate which mechanisms are responsible for the production of invisible energy and discuss at least one method to recover it.

Discussion

The physics of hadronic cascades is by far more involved compared to the development of electromagnetic showers due to the richness of interactions that hadronic particles undergo when crossing matter. The interaction of a high-energy hadron with a typical calorimetric material, like iron, lead, or copper, involves the production of energetic secondary hadrons through strong interactions with typical interaction lengths of about $35\, A^{1/3}\, \mathrm{g\, cm^{-2}}$, followed by the degradation of their energy by nuclear reactions that produce nuclear excitation, evaporation, spallation, fission, etc., resulting in particles with characteristic nuclear energy ($100\,\mathrm{keV} \div$ a few MeV). The low energy spectrum of the hadronic cascade is dominated by neutrons, photons, electrons and positrons, the latter produced by the interaction of photons with matter. Photons are produced by two main mechanisms: from $\pi^0 \rightarrow \gamma\,\gamma$ and from nuclear de-excitations and (n, γ) reactions. The latter can come delayed up to 1 μs with respect to the primary interaction, and overall account for about 30% of the total cascade energy. Since the number of high-energy interactions that produce pions increases with energy, the fraction of energy drained away in the form of $\pi^0 \rightarrow \gamma\,\gamma$ photons increases with energy. The hadronic shower in usually initiated inside the so-called *radiator*, whereas the energy measurement is performed in the active material that samples the cascade. Both the hadronic and electromagnetic component of the cascade contribute to the energy measurement in the active material, although with different efficiencies. Let η_e (η_h) be the efficiency of detecting the energy contained in the electromagnetic (hadronic) component. The total energy measured by the interaction of a high-energy hadron with initial energy E is therefore given by:

$$E_{\mathrm{vis}}^h = [\eta_e\, F_{\pi^0}(E) + \eta_h\, F_h(E)]\, E = \eta_e \left[1 + \left(1 - \frac{\eta_h}{\eta_e}\right) F_h(E)\right] E, \qquad (2.99)$$

where $F_h = 1 - F_e$ is the hadronic energy fraction, which depends on the initial hadron energy [20]. The ratio between the response to an hadron h and to an electromagnetic particle, like an electron, is therefore:

$$\frac{E_{\mathrm{vis}}^h}{E_{\mathrm{vis}}^e} \equiv \left(\frac{e}{\pi}\right)^{-1} = 1 + \left(1 - \frac{\eta_h}{\eta_e}\right) F_h(E). \qquad (2.100)$$

Since $\eta_h \neq \eta_e$ in general, and because of the dependence of F_h with energy, Eq. (2.100) implies that

- the energy response of a hadronic calorimeter is in general non-linear;
- the energy resolution is worse than for an electromagnetic calorimeter due to the stochastic fluctuations on F_h;
- the energy response is not gaussian.

For example, in a homogeneous calorimeter, $e/\pi \approx 1.4$ as a result of the lower efficiency of detecting the hadronic component. This problem can be greatly mitigated by tuning the ratio η_h/η_e to unity, i.e. by *compensating* the calorimeter for the intrinsically different response to the hadronic component.

Solution

The origin of invisible energy in hadronic cascades can be tracked down to the production of delayed photons, soft neutrons that undergo nuclear reactions giving low-range particles, and to the production of nuclear binding energy, which is again drained away in the form of low-range nuclear decays. Although such energy is not measurable, it is possible to compensate for it in a statistical sense by decreasing the sensitivity of the detector to the electromagnetic component. For example, in a sampling calorimeter made of high-Z material like brass, uranium, or lead, interleaved with a plastic organic scintillator, the response to the electromagnetic cascade gets reduced proportionally to the sampling fraction, i.e. the fraction of active material. The latter can be tuned by varying the thickness of the scintillator layers. On the contrary, the response of the scintillator to fast neutrons is only marginally affected, since a recoil proton with $T \sim 1$ MeV has a range of a few tens of microns, see Eq. (2.35), hence it will always interact in the active material regardless of its thickness. By tuning the e/π ratio to unity, the energy resolution can be grearly enhanced.

Suggested Readings

The review article [20] gives a concise but clear discussion of the phenomenology of hadron cascades, with quantitative description of compensation in real detectors.

Bando n. 18211/2016

Problem 2.36 Which processes among pair-production, Compton scattering, and photoelectric effect, are non-negligible in the interaction of γ emitted by a ^{60}Co source with a Ge detector? Which process has necessarily to happen in order to measure the total photon energy?

Discussion

Thanks to the large Z value and the small excitation energy, see Table 2.2, Ge detectors place among the most precise detectors for γ spectroscopy below a few MeV. When dealing with γ radiation, an important property of the detector is the photopeak efficiency, i.e. the efficiency of detecting a photon which is entirely absorbed by photoelectric effect. For Ge detectors and photons of order 1 MeV energy, the photo-peak efficiency is $\lesssim 1\%$, see e.g. Fig. 10.20 of Ref. [1].

Solution

In its β-decay chain, the ^{60}Co isotope produces two monochromatic photon lines of energy 1.17 and 1.33 MeV, hence just above the pair-production threshold $E_{\text{th}} = 2m_e \approx 1.02$ MeV. The K-shell for Ge is located at 11 keV [9], hence the photoelectric effect is expected to be small for the ^{60}Co photons, while Compton scattering should be the dominant interaction mechanism. Indeed, from Ref. [6], we find $\sigma_{\text{p.e.}} \approx 5 \times 10^{-2}$ barn, $\sigma_{\text{Comp}} \approx 6$ barn, and $\sigma_{\text{pair}} \approx 10^{-2}$ barn for $E_\gamma = 1.25$ MeV. If the photon undergoes Compton scattering, only the energy deposited by the recoil electron can be measured by the detector. The interaction length for photons in Ge is given by:

$$\lambda_{\text{Comp}} = (n\,\sigma_{\text{Comp}})^{-1} = \left(\frac{5.3\,\text{g cm}^{-3} \cdot 6 \times 10^{23}\,\text{mol}^{-1}}{72\,\text{g mol}^{-1}} \cdot 6\,\text{barn} \right)^{-1} \approx 4\,\text{cm},$$

$$(2.101)$$

so there is a finite probability that the photon undergoes one Compton scattering only before leaving the active volume, if the latter is a a few mm thick, like in practical Ge detectors. The maximum electron recoil energy is given by Eq. (1.139), namely:

$$T_{\text{max}} = E_\gamma \frac{2k}{1+2k} = 0.96,\ 1.1\,\text{MeV}, \qquad (2.102)$$

where $k = E_\gamma/m_e$. For example, the range in Ge for an electron of kinetic energy 1.1 MeV is about 1.2 mm [5], hence there is a non-negligible chance that the recoil electron escapes the active volume. The same holds for the photoelectrons, which have energies $E_\gamma - \mathcal{B} \approx 1.16$ and 1.32 MeV, and ranges below 2 mm.

The only reactions that guarantee a full energy measurement are therefore the photoelectric effect (probability $\approx 1\%$), with full electron confinement, and pair-production (probability $\approx 0.2\%$). In the latter case, the emitted $e^\pm$ have an energy of about $(E_\gamma - 2m_e)/2 \approx 75$ and 150 keV and ranges of about 25 and 85 μm, respectively, and are therefore very likely to be fully contained in the active volume. After annihilation with an atomic electron, the $2m_e$ rest energy of the e^+e^- pair restores the full energy measurement if the two photons from positronium annihilation interact with the active material (the interaction length for 0.5 MeV photons is about 2.4 cm).

Bando n. 18211/2016

Problem 2.37 Estimate the contribution to the energy resolution (FWHM) due to the stochastic fluctuations in silicon calorimeters generated by photons of energy 2 keV, 6 keV, and 15 keV.

Discussion

If the measured energy E is distributed according to a Gaussian law with mean $\mu = 0$ and standard deviation σ, the FWHM resolution is defined as the interval such that the p.d.f equals half of its value at the mean position μ, i.e.:

$$\mathcal{N}(x_\pm; \; \mu, \sigma) = \frac{1}{2}\mathcal{N}(0; \; \mu, \sigma) \quad \Rightarrow \quad x_\pm = \pm\sqrt{2\ln 2}\,\sigma \approx \pm 1.177\,\sigma$$

$$\sigma_{\text{FWHM}} = (x_+ - x_-) = 2.35\,\sigma \qquad (2.103)$$

When dealing with energy resolution with particle detectors, an important concept is the so-called *Fano factor* (F). If a particle produces on average $N = E/\varepsilon$ signal carriers through independent random interactions characterised by probability p, the stochastic fluctuation in this number is $\sqrt{N}$ from Poisson statistics, and the relative energy resolution is $1/\sqrt{N}$. However, if the detector cannot but absorb all of the particle energy by converting it into detectable signal carriers, the multiplicity of the latter is ideally fixed to N and there would be no stochastic fluctuations at all. This is seldom the case, since there is in general a partioning of the energy transferred by the particle to the active material into more channels, some of which may not produce signal carriers. Indeed, in some circumstances it is observed that the relative energy variance is smaller than the Poisson expectation by an empirical factor F, with $F < 1$, i.e.:

$$\frac{\sigma}{E} = \sqrt{\frac{F\,\varepsilon}{E}} \qquad (2.104)$$

Semiconductors that absorb the full particle energy into eh-pairs, feature a Fano factor of about 0.12. The Fano factor for ionisation detectors has been discussed in Problem 2.8. More informations can be found in Chap. 4 of Ref. [7].

Solution

At energies below 15 keV, the photoelectric effect dominates the interaction of photons with silicon, see e.g. Ref. [6]. We can therefore assume that the photon interacts with one atom by emitting an electron of a few keV energy. The photo-produced electron loses energy by collision loss and creates additional electon-hole pairs along its track. At $E = 2$ keV, the photoelectron will most likely originate from a K-shell emission. Since the K-edge in silicon is at 1839 eV [9], the resulting photo-electron will be rather soft as for Eq. (2.23). However, the ionised atom is in an excited state, which will bring to the emission of either K-α and K-β photons, which undergo photoelectric effect from L-shells with the emission of secondary photoelectrons, or to the emission of short-range Auger electrons [5]. In any case, the secondary particles will release energy in the active medium, so that one can still assume that the whole photon energy is absorbed with little energy partitioning. This reduces the standard deviation of the number of electron-hole pairs N_{eh} from the Poisson expectation of $1/\sqrt{N_{eh}}$ to $\sqrt{F/N_{eh}}$, with $F \approx 0.12$ for silicon. The mean excitation energy is $\varepsilon = 3.6\,\text{eV}$, see Table 2.2. We can therefore estimate the FWHM of the measured signal to be:

$$\sigma_{\text{FWHM}} = 2.35\sqrt{\frac{F\varepsilon}{E}} = 2.35\sqrt{\frac{0.12 \cdot 3.6\,\text{eV}}{E}} = \begin{cases} 3.4\% & E = 2\,\text{keV} \\ 2.0\% & E = 6\,\text{keV} \\ 1.3\% & E = 15\,\text{keV} \end{cases} \quad (2.105)$$

Suggested Readings

Reference [21] discusses in more detail the use of silicon detectors for γ spectroscopy, with examples of measured spectra from nuclear candles. A broader discussion on the phenomenology of photoelectric absorption in matter can be found in Ref. [7].

Bando n. 1N/R3/SUB/2005

Problem 2.38 A piece of Na I (Tl) scintillator, read-out by a phototube, is used to measure the ^{137}Cs line: estimate the energy resolution by listing the contributing factors.

Solution

The energy resolution for a coupled scintillator-phototube detector is described by Eq. (2.45). The ^{137}Cs isotope produces a monochromatic X-ray emission with energy $E = 661\,\text{keV}$. The main contribution to the energy resolution comes from the statistics of photoelectrons, which depends on the mean number of photons $n_\gamma = E/\varepsilon$, where ε is the mean excitation energy, see Table 2.2, and on the overall efficiency of the photocathode. The electronic noise plays also an important role. An other contribution may come from the dependence of the response with the photon impact point and from an imperfect shower containment. Assuming $\varepsilon_Q\,\varepsilon_C = 0.2$ for a typical PMT, see e.g. Table 34.2 of Ref. [2], and negligible noise from the electronics and amplification statistics ($f_N = 1$, $G \gg 1$), the relative energy resolution (FWHM) can be estimated to be:

$$\frac{\sigma_{\text{FWHM}}}{E} = 2.35\sqrt{\frac{\varepsilon}{E \cdot \varepsilon_Q\varepsilon_C}} = 2.35\sqrt{\frac{22\,\text{eV}}{661\,\text{keV} \cdot 0.2}} = 3.0\%, \quad (2.106)$$

see Problem 2.37. No Fano factor has been accounted for in Eq. (2.106), since there is no evidence for its presence in scintillators.

Bando n. 5N/R3/TEC/2005

Problem 2.39 Estimate the energy resolution at $140\,\text{keV}$ of a photo-detector equipped with Na I (Tl) crystals.

Solution

We can refer to Problem 2.38 for determining the energy resolution of a similar setup. Assuming $\varepsilon_Q\,\varepsilon_C = 0.2$ for a typical PMT and negligible noise from the electronics and amplification statistics, the relative energy resolution (FWHM) can be estimated to be:

$$\frac{\sigma_{\text{FWHM}}}{E} = 2.35\sqrt{\frac{\varepsilon}{E \cdot \varepsilon_Q\,\varepsilon_C}} = 2.35\sqrt{\frac{22\,\text{eV}}{140\,\text{keV} \cdot 0.2}} = 6.6\%, \qquad (2.107)$$

where $\varepsilon = 22\,\text{eV}$ is the mean excitation energy for Na I (Tl), see Table 2.2. See Problem 2.37 for the definition of FWHM.

Bando n. 18211/2016

Problem 2.40 A scintillator emits 10^4 γ/MeV. Calculate the resolution (FWHM) for a 4 MeV particle assuming a total light collection efficiency $\varepsilon_C\,\varepsilon_Q{=}1$.

Solution

The energy resolution of the detector is described by Eq. (2.45). Assuming $\varepsilon_Q\,\varepsilon_C = 1$ and negligible noise from the electronics and amplification statistics, the relative energy resolution (FWHM) can be estimated to be:

$$\frac{\sigma_{\text{FWHM}}}{E} = 2.35\sqrt{\frac{\varepsilon}{E}} = 2.35\sqrt{\frac{10^{-4}\,\text{MeV}}{4\,\text{MeV}}} = 1.2\%. \qquad (2.108)$$

See Problem 2.37 for the definition of FWHM.

Bando n. 13153/2009

Problem 2.41 Calculate the energy resolution for photons of energy E measured by a solid state detector with ionisation energy ε, leakage current I_d, and integration time of the associated electronics equal to T_S.

Solution

If the photon energy is intirely absorbed by the detector, the mean signal charge Q collected at the electrodes of the p-n junction and its standard deviation are given respectively by:

$$Q = \frac{E}{\varepsilon}\,e, \qquad \sigma_Q = \sqrt{\frac{F\,E}{\varepsilon}}\,e, \qquad (2.109)$$

where $F \approx 0.12$ is the Fano factor in silicon. In the integration time T_S taken by the electronics to shape the signal, the leakage current contributes to the noise via an equivalent squared-charge:

$$Q_n^2 = 2\,e\,I_d\,T_S, \qquad (2.110)$$

see e.g. Sect. 34.8 of Ref. [2]. Since the noise from the leakage current and the statistical fluctuation in the number of signal carriers are uncorrelated, the relative energy resolution is given by the sum in quadrature:

$$\frac{\sigma_E}{E} = \frac{\sqrt{\sigma_Q^2 + Q_n^2}}{Q} = \sqrt{\frac{\varepsilon}{E}}\sqrt{F + \left(\frac{2\,I_d\,T_S}{e}\right)\frac{\varepsilon}{E}} \qquad (2.111)$$

Suggested Readings

For a concise overview of low-noise front-end electronics for particle detectors, the reader is addressed to Sect. 34.8 of Ref. [2].

Bando n. 1N/R3/SUB/2005

Problem 2.42 The drift velocity of electrons in some gas mixture is $v = 5\,\text{cm}/\mu\text{s}$. What does it imply for a multiwire chamber with wire spacing $s = 2\,\text{mm}$, and what for a drfit chamber read-out by a TDC with 500 MHz clock?

Discussion

Multiwire chambers have been briefly discussed in Problem 2.52. Drift tubes (DT) are gaseous ionisation detectors that measure the time taken by the primary ionisation electrons to drift from their point of formation up to the anode. For ions moving in a gas, the drift velocity v is roughly proportional to the electric field intensity:

$$v = \mu\,E, \qquad (2.112)$$

where μ is called mobility and depends on the pressure P and temperature T of the gas, while it is almost independent of the electric field. Electrons can instead reach much higher velocities compared to ions, and the mobility μ depends on E in such a way that a saturation of the velocity at values of order 50 μm/ns is reached for $E \sim 1\,\text{kV/cm}$ at STP. By making the electric field as uniform as possible in the drift region, Eq. 2.112 implies a proportionality between the distance from the anode of the primary ionisation position and the drift time. The latter is defined as the time interval between a fast trigger, that provides the start time to the clock, and the time of formation of the electric signal at the anode. Drift tubes are built according to this concept. Typical position resolutions achievable with DT are 100 μm over few drift lengths d of a few cm. The position resolution is determined by the sum in quadrature

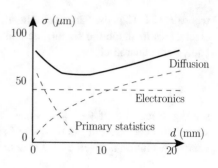

Fig. 2.10 Typical position resolutions in a drift chamber as a function of the drift length d. The total resolution is broken up into three main contributions: statistics of the primary ionisation, noise from the electronics, and charge diffusion

of three dominant contributions: the statistics of primary ionisation (relevant at small d), the electronic noise (independent from d), and electron diffusion (proportional to $\sqrt{d}$). Figure 2.10 provides a qualitative description of the position resolution as a function of the drift length d.

Solution

A MWPC with wire spacing $s = 2\,\mathrm{mm}$, has a spatial resolution along the coordinate y orthogonal to the wires:

$$\sigma_y^{\mathrm{MW}} = \frac{s}{\sqrt{12}} = \frac{2\,\mathrm{mm}}{\sqrt{12}} \approx 580\,\mu\mathrm{m}. \tag{2.113}$$

The factor $\sqrt{12}$ accounts for the fact that the particles arrive at the detector uniformly distributed across y. The time resolution is therefore given by

$$\sigma_t^{\mathrm{MW}} = \frac{\sigma_y^{\mathrm{MW}}/2}{v} = \frac{580\,\mu\mathrm{m}/2}{5\,\mathrm{cm}/\mu\mathrm{s}} \approx 5.8\,\mathrm{ns} \tag{2.114}$$

In Eq. (2.114), the factor of $1/2$ at the numerators comes from the fact that a primary ionisation generated outside of the $\pm s/2$ range from a given wire will be detected by one of the two neighbouring wires. For a DT readout by a time-to-digital converter (TDC), the TDC clock period f^{-1} sets a minimum time resolution

$$\sigma_t^{\mathrm{DT}} = \frac{f^{-1}}{\sqrt{12}} = \frac{2\,\mathrm{ns}}{\sqrt{12}} = 0.58\,\mathrm{ns}. \tag{2.115}$$

Again, one has to divide by $\sqrt{12}$ since the actual arrival time at the anode is uniformly distributed across the time interval f^{-1} between subsequent clocks. The position uncertainty induced by the TDC clock is therefore given by:

$$\sigma_{y,\,\mathrm{clock}}^{\mathrm{DT}} = \sigma_t^{\mathrm{DT}} \cdot v = 0.58\,\mathrm{ns} \cdot 5\,\mathrm{cm}/\mu\mathrm{s} \approx 29\,\mu\mathrm{m}. \tag{2.116}$$

This term contributes to the electronic noise shown in Fig. 2.10. The overall position resolution depends however on other factors, as discussed above. Typical position resolutions of a conventional DT is about 100 μm, which is anyway smaller than the one from a typical MWPC.

Suggested Readings

For a comprehensive review of DT, the reader is addressed to Ref. [22].

Bando n. 1N/R3/SUB/2005

Problem 2.43 A depleted microstrip silicon detector has a strip pitch of 50 μm and operates without charge division. What is its spatial resolution?

Discussion

A *silicon microstrip* is a solid-state detector consisting of a wafer of doped silicon, for example, of a high-resistivity n-type with typical thickness of about 300 μm, with p-n junctions shaped in the form of long and thin parallel strips separated by a distance (pitch) ranging between 20 and 200 μm. In a possible setup, one surface of the wafer is grounded and the strips are implanted on the opposite side and connected to the bias voltage via DC or AC coupling. The junction may be realised by p^+-type silicon and, for a typical wafer thickness, it gets completely depleted by a bias voltage of order 100 V. A MIP loses $1.66 \cdot 2.33$ MeV/cm ≈ 3.87 MeV/cm in silicon [5]. Given that the average excitation energy is $\varepsilon = 3.6$ eV, a total of 3×10^4 eh-pairs are produced on average across a 300 μm-thick junction. The signal carriers drift under the effect of the bias voltage and the induced charge is measured by the front-end electronics.

The *charge division* method consists in an analog measurement of the signal from the strips close to the one which recorded the hit, i.e. the one with the largest signal yield. The centre-of-mass of the strip charges $\bar{x} = \sum_i Q_i x_i / \sum_i Q_i$, where i runs over the strips and x_i (Q_i) are the strip positions (measured signal), provides an estimator of the impact position with typical resolution of about

$$\sigma_x^{\text{ana}} \sim \frac{d}{\text{SNR}}, \tag{2.117}$$

where d is the strip pitch and SNR is the signal-over-noise ratio. This can be easily proved by using the standard propagation of error for uncorrelated measurements, see Eq. (4.73):

$$\bar{x} = \frac{\sum_i Q_i x_i}{\sum_i Q_i} \quad \Rightarrow \quad \sigma_{\bar{x}}^2 = \sum_j \left| \frac{\partial \bar{x}}{\partial Q_j} \right|^2 \delta Q_j^2 = \sum_j \frac{(x_j - \bar{x})^2}{(\sum_i Q_i)^2} \delta Q_j^2 = d^2 \frac{\sum_j \delta Q_j^2}{\left(\sum_j Q_j \right)^2},$$

$$\sigma_{\bar{x}} = \frac{d}{\text{SNR}}, \quad \text{with} \quad \text{SNR} = \frac{\sum_j Q_j}{\sqrt{\sum_j \delta Q_j^2}} \equiv \frac{S}{N}. \tag{2.118}$$

Conversely, if the strips can be read in digital mode only, the position resolution is given by the strip pitch:

$$\sigma_x^{\text{dig}} = \frac{d}{\sqrt{12}}, \quad (2.119)$$

Additional sources of uncertainty affecting the collection of charge carriers, like thermal diffusion, multiple-scattering, δ-rays, should be also considered for realistic detectors.

Solution

In the absence of charge division, the spatial resolution of a microstrip detector is primarily determined by the pitch size d. Since the particle flux can be asumed to be uniformly distributed across the microstrip detectors, we can estimate the spatial resolution (FWHM) as:

$$\sigma_{\text{FWHM}}^x = 2.35 \frac{d}{\sqrt{12}} = 34 \ \mu\text{m}, \quad (2.120)$$

where the factor of $1/\sqrt{12}$ comes from the assumed flux uniformity, see Problem 2.42.

Suggested Readings

For a first introduction to microstrip detectors, the reader is addressed to Sect. 10.6 of Ref. [1].

Bando n. 18211/2016

Problem 2.44 A silicon detector is made of a pixels with dimension 100 μm × 200 μm. What is the smallest spatial resolution in the two dimensions, if the detector has digital readout?

Discussion

Pixel detectors are semiconductive detectors where the active volume is segmented in small picture elements (pixels), which are indépendently read-out. Planar pixel detectors are commonly employed in HEP experiments as vertex detectors, thanks to their superior spatial resolutions in two dimensions, which allows for a small occupancy even at the closest distance to the interaction point, and their close-to-ideal efficiency to detect the passage of ionising particles.

Solution

If the detector is operated in digital readout, see Problem 2.43, a lower bound to the spatial resolution (FWHM) in the two directions is given by:

$$\begin{cases} \sigma_{\text{FWHM}}^x = 2.35 \frac{d_x}{\sqrt{12}} = 68 \,\mu\text{m} \\ \sigma_{\text{FWHM}}^y = 2.35 \frac{d_y}{\sqrt{12}} = 136 \,\mu\text{m} \end{cases} \tag{2.121}$$

where the factor of $1/\sqrt{12}$ comes from the assumed flux uniformity, see Problem 2.42.

Suggested Readings

For a comprehesive introduction to pixel detectors in HEP experiments, the reader is addressed to Ref. [21].

Bando n. 18211/2016

Problem 2.45 Why is a diode used as radiation detector usually operated with an inverse bias?

Solution

A p-n junction operated at inverse bias give rise to an active region depleted from mobile charge where an intense electric field can sweep out free charges liberated by a ionising particle. The thickness of the depletion zone for the case of a silicon *p-n* junction realised by a p^+-doped material put into contact with a lightly doped n region, is approximately given by:

$$W = 0.5\sqrt{\left(\frac{\rho_n}{\Omega\,\text{cm}}\right)\left(\frac{V_0 + V_{\text{bias}}}{\text{V}}\right)} \,\mu\text{m}, \tag{2.122}$$

where ρ_n is the resistivity of the *n*-type region, $V_0 \sim 1\,\text{V}$ is the barrier voltage, and V_{bias} is the bias voltage, see e.g. Ref. [2]. The importance of applying an inverse bias to the junction as to enlarge the active volume is made clear by Eq. (2.122). For example, for typical values $\rho_n = 2 \times 10^4\,\Omega$ cm, the thickness of the depletion region would change from 70 to 700 μm, if a reverse bias $V_{\text{bias}} = 100\,\text{V}$ is applied.

Suggested Readings

An introduction to the physics of semiconductors for particle detectors can be found in Chap. 20 of Ref. [1].

Bando n. 13705/2010

Problem 2.46 Consider a D^0 meson produced with an energy of 20 GeV. Determine the spatial resolution necessary to measure the production and decay vertex position, and indicate which detectors are best suited for an efficiency exceeding 90%.

Solution

The D^0 meson decays via the electroweak interaction with a lifetime of about 0.41 ps, corresponding to $c\tau \approx 120 \, \mu m$ [2]. A good channel to reconstruct its decay is $D^0 \rightarrow K^{\pm}\pi^{\mp}$. The probability of surviving up to a distance d ot more from its production vertex is given by Eq. (1.174):

$$P[x \geq d] = \exp\left[-\frac{mc}{|\mathbf{p}|}\frac{d}{c\tau}\right] = \exp\left[-\frac{1}{\sqrt{\gamma^2 - 1}}\frac{d}{c\tau}\right] \quad (2.123)$$

Requiring this probability to be at least 90% is equivalent to impose that the flight distance should be in excess of:

$$d_{90\%} = (-\ln 0.9)\, c\,\tau\,\gamma = (0.105 \cdot 123 \cdot 11) \, \mu m \approx 140 \, \mu m, \quad (2.124)$$

where we have used the fact that $\gamma \approx 11$ is large. Therefore, if we want to reconstruct at least 90% of the D^0 decays from their decay vertex, the vertex resolution must be smaller than about 140 μm. This can be easily achieved by silicon-based vertex detectors, either pixel- or mictrostrip-based.

Discussion

For $E = 20 \, \text{GeV}$, the decay products have energy of about 10 GeV each. In this regime, multiple scattering usually dominates the tracking resolution when using silicon detectors with pixel/pitch size $\lesssim 100 \, \mu m$, see Problem 3.9. The *impact point resolution* (σ_{ip}) is the uncertainty on the position of closest approach of the track extrapolation to the primary vertex point (PV), and is related to the resolution on the position of the secondary vertex (SV), see Fig. 2.11. Modulo resolution effects, the quantity

$$s_{ip}^j = \text{sign}\left[\mathbf{ip}_j \cdot (\mathbf{SV} - \mathbf{PV})\right] \quad (2.125)$$

should be positive for tracks emerging from the same secondary vertex. Conversely, the detector rsolution smears the impact point of tracks emerging from the PV around zero, with equally likely values of s_{ip}. This property can be exploited to define tagging algorithms for displaced vertexes and experimental methods to measure their efficiency in data [24, 25]. Assuming a MS-dominated regime, the impact point resolution is given by:

$$\sigma_{ip} \approx r_1\sqrt{\langle\theta_1^2\rangle}, \quad (2.126)$$

where r_1 is the distance of the innermost silicon layer from the interaction point and $\langle\theta_1^2\rangle$ is given by Eq. (2.8). For example, assuming the design of the CMS pixel detector, one has $r_1 = 4.4 \, \text{cm}$ and a MS mean angle of about 2×10^{-4} rad at $|\mathbf{p}| = 10 \, \text{GeV}$, giving $\sigma_{ip} \approx 10 \, \mu m$, see Ref. [26]. A more realistic simulation, which

Fig. 2.11 Cartoon illustrating the two-body decay of a D^0 meson. The distance of closest approach of the extrapolation of the daughter particles trajectories to the primary vertex (PV) is called *impact parameter*

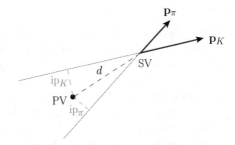

includes measurement uncertainty and MS in the beam pipe, gives about a factor of 2 larger resolution, which still satisfies the constraint of Eq. (2.124).

Suggested Readings

For an overview of tracking and vertexing performances at the LHC, the reader is recommended to read the review article [26].

Bando n. 18211/2016

Problem 2.47 Explain why Ge sensors need to be cooled, while Si sensors do not.

Solution

Germanium detectors are commonly operated at liquid nitrogen temperature ($T = 77$ K) to reduce the leakage current I_d due to thermal excitation, and hence the electronic noise and power consumption, see Problem 2.41). The bias current depends exponentially on the temperature T:

$$I_d(T) \propto T^2 \exp\left[-\frac{E_{gap}}{2\,k_B\,T}\right] \quad \Rightarrow \quad \frac{I_d(T_2)}{I_d(T_1)} = \left(\frac{T_2}{T_1}\right)^2 \exp\left[-\frac{E_{gap}}{2\,k_B}\left(\frac{1}{T_2} - \frac{1}{T_1}\right)\right],$$
$$\tag{2.127}$$

where E_{gap} is the energy gap, see e.g. Sect. 34.7 of Ref. [2]. Although the same effect exists in silicon, the energy gap in the latter is larger than in germanium. For example, at room temperature, one has $E_{gap} = 1.1$ eV for silicon and 0.7 eV for germanium, corresponding to a factor of 3×10^3 larger leakage current for the latter.

Bando n. 18211/2016

Problem 2.48 Which property of a SiPM makes it a preferable solution cimpared to a conventional PMT for an integrated imaging PET-MRI system?

Discussion

Silicon photomultipliers (SiPM), also known as *Pixelized Photon Detectors* (PPD) are photodetectors composed by an array of pixel-size photodiodes with typical size ranging from $25 \times 25 \ \mu m^2$ to $100 \times 100 \ \mu m^2$, packed over a small area, typically from $0.5 \times 0.5 \ mm^2$ to $5 \times 5 \ mm^2$, and operated in Geiger mode, i.e. with a bias voltage in excess of the break-down voltage. When a eh-pair is created in the depleted region, the intense electric field triggers the formation of an avalanche. The high bias voltage provides large gains per incident photon and per pixel, but proportionality between the number of photons impinging on a given cell and the collected charge is lost. The proportionality with the total input photons is restored by summing the binary cell outputs from the full array.

Solution

SiPM's represent a convenient alternative to PMT's for applications in environment with intense magnetic field, like in *positron emission tomography-magnetic resonance imaging* (PET-MRI) applications, since the amplification stage in a SiPM does not require the photoelectrons to be accelerated along the dynode of conventional PMT's, which suffers from the presence of magnetic fields, for example by altering the gain.

Suggested Readings

For an introduction to SiPM's, the reader is addressed to the dedicated PDG review [2] and references therein.

Bando n. 1N/R3/SUB/2005

Problem 2.49 Order the following detectors by decreasing dead time: silicon, plastic scintillator, drift chamber. Which one would you chose for a time measurement with resolution of a few hundred ps?

Discussion

The *dead time* τ is the time required by a detector to process one event and be ready to accept a new event. Depending whether the detector is sensitive or not to a new event while processing the previous one, two types of dead-time exist: *extendable* or *not-extendable*. In the first case, if we assume that the first event occurred at time t_0, the arrival of a new event at a time $t_1 < t_0 + \tau$ shifts the time at which the detector is ready to accept and process a new event to at least $t_2 = t_1 + \tau$. In the second case, the new event does not change the detector state at all, and the subsequent event can be accepted and processed at any time $t_1 \geq t_0 + \tau$, regardless of what happens meanwhile. See Problem 3.38 for more details.

Solution

Plastic scintillators are generally faster than inorganic scintillators, with decay times of a few ns, see e.g. Table 7.1 of Ref. [1]. Fast photodetectors can also have risetimes below 1 ns, see e.g. Ref. [2]. A coupled scintillator-photodetector system is ready to accept and process a new event after the fluorescent excitation from the previous event have decayed to the ground level, which can take about 10 ns for fast scintillators.

A silicon strip or pixel detector has time resolutions of a few ns, but the time needed to collect the full charge released in the depleted zone can take a few tens of nanoseconds (10 ns for electrons and 25 ns for holes in a 300 μm thich detector, see Sect. 34.7 of Ref. [2]). The readout electronics further increases the processing time to at least 50 ns.

In a drift chamber, the dead time is mostly due to the time taken by the primary ionisation electrons (ions) to drift to the anode (cathode), see Problem 2.42. For a typical electron velocity of 5 cm/μs, the time needed to drift over 1 cm is about 200 ns. During this time, a new event would cause pile-up and confusion on the time measurement.

A time measurement with a few hundreds ps time resolution is best accommodated with plastic scintillators coupled to fast photo multipliers, like microchannel plate (MCP) or gas electron multipliers (GEM), with a fast sampling frequency of the readout electronics as to allow for the full pulse shape reconstruction.

Suggested Readings

More details on the dead time of particle detectors, including techniques for measuring it in the laboratory, can be found in Sect. 5.7 of Ref. [1]. Table 34.1 of Ref. [2] summarises the typical resolutions and dead times of common charged particle detectors.

Bando n. 18211/2016

Problem 2.50 The mean counting rate on single electrode for a given detector is 150 kHz. Estimate an upper bound to the processing time of the analog pre-amplifier and shaper, if the pile-up probability has to be maintained below 3%.

Solution

For what concerns the pileup of multiple events, we can use the same line of thought used to relate the true and measured rate in a non-paralyzable system, see Problems. 2.49 and 3.38. Referring to Eq. (3.186) with $\varepsilon = 1$, we can therefore invert the equation and express the true rate ν as a function of the measured rate m and of the dead time τ, i.e.

$$\nu = \frac{m}{1 - m\,\tau}. \tag{2.128}$$

Requiring that the pile-up is less than $\delta = 3\%$ amounts to require that the ratio between the measured rate and the true rate is larger than $1 - \delta$, or equivalently:

$$\frac{m}{v} = 1 - m\tau > 1 - \delta \quad \Rightarrow \quad \tau < \frac{\delta}{m} = \frac{0.03}{1.5 \times 10^5 \,\text{Hz}} = 200 \,\text{ns}. \quad (2.129)$$

Bando n. 18211/2016

Problem 2.51 A proportional cylindrical tube has inner radius R, wire radius r, and anodic tension V_0. What is the value of the electric field at a distance $d \leq R$ from the anode?

Solution

Let's assume that the anode is connected to a potential $V_0 > 0$ and that the cathode is grounded. The wire acquires a charge with uniform linear density. By using the cylindrical symmetry of this configuration, it is easy to prove that the electric field must be radial, i.e. $\mathbf{E} = E(d)\,\mathbf{e}_r$. By virtue of Gauss law, the field intensity $E(r)$ must scale as d^{-1}, i.e.

$$E(d) = \frac{c_0}{d}, \quad (2.130)$$

where c_0 is a constant that depends on the boundary conditions. Since $E = -\nabla V$, the electric potential $V(d)$ must be proportional to $\ln d$. Together with the boundary conditions at the two electrodes, this fully determines the potential to be:

$$V(d) = \frac{V_0}{\ln(r/R)} \ln(d/R), \quad (2.131)$$

from which we get the result:

$$E(d) = -\frac{\partial V}{\partial d} = \frac{V_0}{\ln(R/r)} \frac{1}{d}. \quad (2.132)$$

Discussion

The d^{-1} scaling of the electric field makes the cylindrical tube suitable for charge multiplication. For example, assuming typical values $r = 20 \,\mu\text{m}$, $R = 5 \,\text{cm}$, $V_0 = 2 \,\text{kV}$, the electric field at a distance of $100 \,\mu\text{m}$ from the wire is about $20 \,\text{kV/cm}$, which is enough to trigger the formation of an avalanche with its resulting charge multiplication. As an eample, the gas multiplication factor M for a cylindrical chamber filled with P-10 gas (90% Ar, 10% CH_4) at STP can be estimated from *Diethorn* formula:

$$\ln M = \frac{V_0}{\ln(R/r)} \frac{\ln 2}{\Delta V} \ln\left(\frac{V_0}{pr \,\ln(R/r)\, K}\right) \approx 7.3 \quad \Rightarrow \quad M = 1.5 \times 10^3, \quad (2.133)$$

where p is the gas pressure and K and ΔV are gas-specific parameters, see e.g. Table 6.1 of Ref. [7] for a few examples.

Suggested Readings

An introduction to the physics of electronic avalanches in gas can be found in Ref. [1, 7]. For a more comprehensive review of gaseous detectors, the reader is addressed to Ref. [8].

Bando n. 1N/R3/SUB/2005

Problem 2.52 The cathode readout can be used in wire detectors, like multiwire chambers, TPC, LST, and even RPC. What does it mean? What are the main advantages of this setup?

Discussion

Multiwire proportional chambers (MWPC), time projection chambers (TPC), limited streamer tubes (LST), and resistive plate chambers (RPC) are all examples of gaseous ionisation detectors that measure the ionisation charge left behind by particles interacting with the gas. A gaseous detector consists in a pair of electrodes kept at different electrostatic potentials and separated by a gaseous medium. The anode is usually shaped in a way as to produce intense electric fields nearby its surface. A metallic wire kept at a positive voltage bias is the solution at the basis of the MWPC, TPC, and LST technology. A plane capacitor with small inter-plane distance is an other option, which is e.g. used in RPC detectors. The cathode confines the electric field and shields the detector from the outside. The usual way of opertaing a gaseous detector is to ground the cathode and read the anode in AC-coupling, i.e. separating the bias voltage from the readout electronics by means of a capacitor, see Problem 2.53. Alternatively, one can set the cathode at a negative bias voltage, and couple the anode directly to the readout electronics.

Solution

Let's consider the case where the anode consists in a set of parallel wire with small inter-distance, stretched along the coordinate x, and let y be the orthogonal coordinate. The passage of a ionising particle induces the formation of an electron avalanche in the neighbourhood of the anode. The positive ions drift towards the cathode inducing a signal (a time-dependent voltage pulse) between the two electrodes. If the detector is operated in anode readout, only the y coordinate can be measured with good resolution. If the cathode is segmented along y, like in the form of parallel strips, then a cathode readout, i.e. a measurement of the pulse induced at the cathode, offers the possibility of measuring also the x coordinate. If the cathode readout is analogic, a centre-of-gravity method allows to measure the x position with high precision (indeed, only limited by the noise of the electronics). If the readout is digital-only, the x resolution is instead determined by the granularity of the cathode.

Suggested Readings

An introduction to gaseous detectors and to their readout can be found in Sect. 6.6 of Ref. [1]. A more advanced and complete reference on the subject is provided by

Fig. 2.12 AC and DC couplings for a generic detector

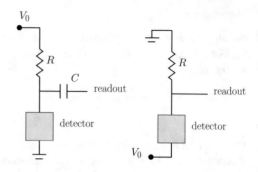

Ref. [8]. A stimulating discussion on this subject can be also found in the Nobel lecture by G. Charpak (1992), the inventor of the MWPC.

Bando n. 18211/2016

Problem 2.53 Does a radiation detector AC-coupled to its electronics have a larger noise compared to a DC-coupled detector with the same electronics?

Discussion

The readout electrode of a charge-sensitive detector, like a microstrip silicon detector, an RPC, a MWPC, etc., can be either set to a large bias voltage or be grounded. In the former case, the front-end electronics, which usually starts with a pre-amplifier, needs to be decoupled from the bias voltage by a capacitance (*AC-coupling*). In the latter case, the electrode can be directly accessed by the pre-amplifier (*DC-coupling*), see Fig. 2.12.

Solution

AC-coupling offers the advantage of having the opposite electrode (e.g. the cathode, for wire detectors) grounded, resulting in a convenient configuration to insulate the detector. However, it provides an extra decoupling capacitance in input to the readout chain, thus increasing the electronic noise compared to a DC-coupling. Indeed, for a capacitive sensor, the charge-equivalent noise Q_n can be parametrised as:

$$Q_n^2 = i_n^2 \, F_i \, T_S + \left(\frac{e_n^2 \, F_v}{T_S} + F_{vf} \, A_f \right) C^2, \qquad (2.134)$$

where C is the sum of all capacitances shunting the input, i_n^2 and e_n^2 are the quadratic current and voltage noise densities, T_S is the characteristic shaping time, and $F_{i,v,vf}$ are devise-specific constants [2].

Suggested Readings

For an introductory discussion on the readout of silicon detectors, the reader is addressed to Sect. 10.9 of Ref. [1]. More details on low-noise electronics for capacitive detectors can be found in the dedicated PDG review [2].

Bando n. 1N/R3/SUB/2005

Problem 2.54 A discriminator is operated with a threshold $V_{th} = 0.4$ V and receives in input signals that have a constant rise-time equal to $T_S = 10$ ns, but an amplitude variation between $V_{min} = 0.5$ V and $V_{max} = 1$ V. Estimate the lower bound on the time resolution due to the variable amplitude. Which technique would you use to reduce such an effect?

Discussion

A *discriminator* is a device that produces a digital signal when an analogical input pulse overcomes a predefined threshold. A discriminator in combination with a TDC device can be used for timing measurements of signals. When the input signals differ in amplitude and/or rise-times, the time measurement performed by a discriminator with fixed threshold is affected by event-by-event fluctuations on the pulse shape, giving rise to the so-called *time walk*. A number of *time-pickoff methods* can be deployed to mitigate the walk effect. A common method is based on the *constant fraction triggering* (CFT), which consists in analysing the zero-crossing of a signal obtained by a linear combination of the pulse V, delayed by a fixed time τ_d, with $-k\,V$, where k is an attenuation coefficient. The triggering time t_R is defined as the time at which:

$$V(t_R - \tau_D) - k\,V(t_R) = 0. \tag{2.135}$$

Since Eq. (2.135) is homogeneous in V, signals with the same time-shape, but different amplitude, will give the same triggering time t_r, see Fig. 2.13. This method is however affected by a residual walk effect if the pulse shape differ from one event to another. In this case, one can try to reduce the delay time τ_D as to trigger on the rising edge of the signals, where event-by-event changes are smaller, a technique known as *amplitude and risetime compensation* (ARC).

Solution

The time derivative of the signal is distributed in the range

$$\frac{dV}{dt} \in \left[\frac{V_{min}}{T_S}, \frac{V_{max}}{T_S} \right] = [0.05, 0.1]\,\frac{V}{ns}. \tag{2.136}$$

Signals with time derivatives at the edge of the interval of Eq. (2.136) will trigger the discriminator at times:

Fig. 2.13 Application of the CFT technique to a pair of Gaussian-like signals with different amplitude

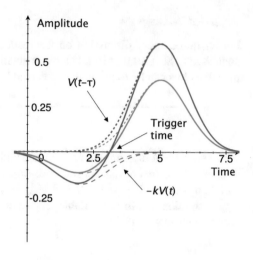

$$t_{min} = \frac{V_{th}}{V_{max}/T_S} = \frac{0.4\,V}{0.1\,V/ns} = 4\,ns$$

$$t_{max} = \frac{V_{th}}{V_{min}/T_S} = \frac{0.4\,V}{0.05\,V/ns} = 8\,ns \qquad (2.137)$$

The time walk Δt is therefore given by

$$\Delta t = t_{max} - t_{min} = 4\,ns. \qquad (2.138)$$

A technique to eliminate the time walk is for example the constant fraction triggering, which is appropriate for this case since the signals feature the same rise-time.

Suggested Readings

Discriminators are briefly discussed in Sect. 14.0 of Ref. [1], while a few time-pickoff methods are described in Sect. 17.2 of the same reference. The reader is also addressed to Ref. [23] for application of discriminators in experiments.

Appendix 1

The computer program below illustrates the numerical evaluation of the information I_{ε_π} from Problem 2.28. The algorithm approximates the Rieman integral by the finite sum of rectangles:

$$\int dx\, f(x) \approx \sum_i f\left(\frac{x_{i+1} - x_i}{2}\right) \cdot \Delta x$$

The integral to be approximated is given by:

$$I_{\varepsilon_\pi} = \mathrm{E}\left[-\frac{\partial^2 f(x, \varepsilon_\pi)}{\partial^2 \varepsilon_\pi}\right] \equiv \int_{-\infty}^{+\infty} dt\, f(t, \varepsilon_\pi) \left[-\frac{\partial^2 \ln f(x, \varepsilon_\pi)}{\partial^2 \varepsilon_\pi}\right], \qquad (2.139)$$

with:

$$\frac{\partial \ln f(x, \varepsilon_\pi)}{\partial \varepsilon_\pi} = \frac{\mathcal{N}(t;\, t_\pi, \sigma_t) - \mathcal{N}(t;\, t_K, \sigma_t)}{f(t, \varepsilon_\pi)} \qquad (2.140)$$

$$\frac{\partial^2 \ln f(x, \varepsilon_\pi)}{\partial^2 \varepsilon_\pi} = -\frac{[\mathcal{N}(t;\, t_\pi, \sigma_t) - \mathcal{N}(t;\, t_K, \sigma_t)]^2}{f(t, \varepsilon_\pi)^2} \qquad (2.141)$$

```
import math

# gaussian function
def gaus(x, m, s):
    return 1./math.sqrt(2*math.pi)/s * math.exp(-math.pow(x - m,2)/2/s/s)

m_pi = 6.68 $ TOF\index{Time of flight@Time-of-flight} for pi m_k
= 6.87 # TOF\index{Time of flight@Time-of-flight} for K sigma =
0.2 # std of TOF\index{Time of flight@Time-of-flight} measurement

def integrate(x_l=6.0, x_h=7.5, step=0.01, f_pi=0.5):
    integ = 0.0
    n_step = int((x_h-x_l)/step)
    for s in xrange( n_step ):
        t = x_l + (s+0.5)*step
        g_pi = gaus(t, m_pi, sigma)
        g_k = gaus(t, m_k, sigma)
        val = math.pow(g_pi - g_k, 2)/(f_pi*g_pi + (1. - f_pi) * g_k )
        integ += val*step
    return integ

###############################
for f_pi in [0.1, 0.3]:
    res = integrate(x_l=5., x_h=10, step=0.001, f_pi=f_pi)
    print f_pi"==>", res
```

References

1. W.R. Leo, *Techniques for Nuclear and Particle Physics Experiments*, 2nd edn. (Springer, Berlin, 1993)
2. C. Patrignani et al., (Particle Data Group). Chin. Phys. C **40**, 100001 (2016)
3. J.D. Jackson, *Classical Electrodynamics*, 3rd edn. (Wiley, New Jersey, 1999)
4. Yung-Su Tsai, Rev. Mod. Phys **46**, 815 (1974)
5. http://pdg.lbl.gov/2016/AtomicNuclearProperties

6. National Institute of Standards and Technology. XCOM: Photon Cross Sections Database (2012), http://nist.gov/pml/data/xcom/index.cfm
7. G.F. Knoll, *Radiation Detection and Measurement*, 4th edn. (Wiley, New Jersey, 2010)
8. F. Sauli, *Gaseous Radiation Detectors : Fundamentals and Applications* (Cambridge University, Cambridge, 2014)
9. A. Thompson et al., *X-ray Data Booklet*, LBNL/PUB-490 Rev.3 (2009)
10. D.E. Groom et al., Atomic Data Nucl. Data Tab. **78**, 183 (2001), https://doi.org/10.1006/adnd.2001.0861
11. A.A. Kochanov et al., Astropart. Phys. **30**, 219233 (2008), https://doi.org/10.1016/j.astropartphys.2008.09.008
12. ALEPH Collaboration Nucl, Instr. Methods A **360**, 481 (1995)
13. M. Peskin, D.V. Schroeder, *An introduction to Quantum Field Theory* (Advanced Book Program, Westview Press, 1995)
14. CMS Collaboration, Accepted by JINST, CMS-PRF-14-001, CERN-EP-2017-110, arXiv:1706.04965 (2017)
15. H. Primakoff, S. Rosen, Rep. Prog. Phys. **22**, 121–166 (1959)
16. *Stopping-power and range tables for electrons*, ICRU Report No. 37 (1984)
17. T. Tabarelli de Fatis, Eur. Phys. J. C **65**, 359 (2010), https://doi.org/10.1140/epjc/s10052-009-1207-8
18. K.S. Hirata et al., Phys. Rev. D **38**, 448–457 (1988), https://doi.org/10.1103/PhysRevD.38.448
19. R. Cahn, G. Goldhaber, *The Experimental Foundations of Particle Physics*, 2nd edn. (Cambridge University, Cambridge, 1989)
20. C.W. Fabjan, F. Gianotti, Rev. Mod. Phys. **75**, 1243–1286 (2003), https://doi.org/10.1103/RevModPhys.75.1243
21. L. Rossi et al., *Pixel Detectors* (Spinger, Berlin, 2006)
22. W. Blum, W. Riegler, L. Rolandi, *Particle Detection with Drift Chambers* (Springer, Berlin, 2008)
23. R. Frühwirth et al., *Data Analysis Techniques for High-Energy Physics*, 2nd edn. (Cambridge Press, Cambridge, 2000)
24. ATLAS Collaboration, JINST **11** P04008 (2016), https://doi.org/10.1088/1748-0221/11/04/P04008
25. C.M.S. Collaboration, JINST **8**, P04013 (2013), https://doi.org/10.1088/1748-0221/8/04/P04013
26. L. Rolandi, F. Ragusa, New J. Phys. **9**, 336 (2007), https://doi.org/10.1088/1367-2630/9/9/336

Chapter 3
Accelerators and Experimental Apparatuses

Abstract The subject of the third chapter is the motion of charged particles induced by electromagnetic fields. The first section focuses on the kinematics of charged particles inside static fields and its application for particle tracking. The second section is dedicated to the physics of accelerators. The last section is devoted to the concept of luminosity and event rates at colliders.

3.1 Tracking of Charged Particles

Tracking a charged particle is the general problem of determining the particle trajectory by interpolating a collection of position measurements. In some circumstances, one can assume that the particle trajectory gets sampled by the detector with negligible impact on the particle kinematics: the equation of motion of the particle is therefore the same as if there were no detector at all. While this is in general the case for high-energy and minimum ionising particles and for sufficiently thin detectors, this assumption breaks down when the particles have a high probability of interacting inside the detector. In general, the interaction of the particle with the detector has to be accounted for, and a variety of techniques, either analytical or Monte Carlo-based, exist for the purpose of estimating the kinematics of the particle cleared from detector effects. Two main cases should be considered when dealing with the tracking of charged particles. In the first situation, the particle momentum is assumed to be known by other means, or perhaps is not relevant at all, and one is rather interested on the track direction and/or position at an arbitrary location in space, given a set of measurements. In the second situation, tracking is performed in a static magnetic field at the purpose of extracting the particle momentum.

Linear Tracking

Let us assume that the tracking system consists of $N + 1$ measuring stations located at the positions $x_0, \ldots, x_N$, equally spaced across the spectrometer length L (also known as *lever arm*), and characterised by the same spatial resolution σ. In the absence of a magnetic field, the particle trajectory is a straight line. The measured points $\mathbf{y} = (y_1, \ldots, y_N)$ can be therefore fitted to a linear function:

© Springer International Publishing AG 2018
L. Bianchini, *Selected Exercises in Particle and Nuclear Physics*,
UNITEXT for Physics, https://doi.org/10.1007/978-3-319-70494-4_3

Fig. 3.1 Examples of linear
tracking using $N + 1$ equally
spaced measuring stations
over a total lever arm L

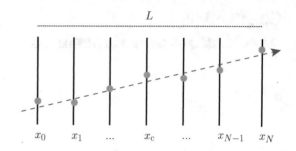

$$y(x) = a + b x, \tag{3.1}$$

see Fig. 3.1. Since the vector **y** of measurements depends linearly on the unknown
parameters a and b, the theory of χ^2 estimators can be applied analytically to yield
the estimators of the intercept ($\hat{a}$) and slope of the line ($\hat{b}$). The variance of the
estimators are given by [1]:

$$\text{Var}\left[\hat{b}\right]_{\text{meas.}} = \frac{\sigma^2}{L^2} \frac{12\,N}{(N+1)(N+2)}, \quad \text{Var}\left[\hat{a}\right]_{\text{meas.}} = \frac{\sigma^2}{N+1} + x_c^2\,\text{Var}\left[\hat{b}\right]_{\text{meas.}}, \tag{3.2}$$

where $x_c = (x_N + x_0)/2$ is the coordinate of the middle point of the spectrometer.
Furthermore, if $x_c = 0$, the two estimators are uncorrelated. Given $N + 1$ measuring
stations and a total lever arm L, the configuration leading to the smallest possible
value of $\text{Var}\left[\hat{b}\right]$ corresponds to half stations clustered at the front and half at the rear
of the spectrometer, giving

$$\text{Var}\left[\hat{b}\right]_{\text{meas.}}^{\text{opt.}} = \frac{\sigma^2}{L^2} \frac{2}{(N+1)} \tag{3.3}$$

The effect of multiple scattering through small angles (MS), on the slope, see
Sect. 2.1, is given by the second of Eq. (2.8), namely:

$$\text{Var}\left[\hat{b}\right]_{\text{MS}} = \frac{\langle y \rangle^2}{L^2} = \frac{\langle \theta_y^2 \rangle}{3}, \tag{3.4}$$

where $\langle \theta_y^2 \rangle$ can be computed according to Eq. (2.7) taking into account the full material budget in the the $N + 1$ stations.

Tracking in Magnetic Field

In the presence of a static magnetic field, the trajectory of a charged particle is
no longer a straight line. In the simplest case, the magnetic field can be assumed
uniform inside the spectrometer, $\mathbf{B}(\mathbf{r}) = \mathbf{B}_0$. The trajectory is then given by a helix,

see Problem 3.2, whose projection onto the plane orthogonal to the magnetic field is a circle of radius R:

$$(x - x_0)^2 + (y - y_0)^2 = R^2, \quad \text{with} \quad R = \frac{|\mathbf{p}| \cos \lambda}{q \, |\mathbf{B}|}, \tag{3.5}$$

where λ is the angle between the particle momentum and the plane orthogonal to the $\mathbf{B}$ field direction (*dip angle*). The momentum component onto this plane is usually denoted as *transverse momentum* $p_T \equiv |\mathbf{p}| \cos \lambda$. For large transverse momenta, the radius R is usually much larger than the lever arm L, so that Eq. (3.5) can be approximated to first order to give:

$$y(x) = a + bx + cx^2, \tag{3.6}$$

with $c = (2R)^{-1}$. Again, the measurement vector $\mathbf{y}$ depends linearly on the unknown parameters, so that a χ^2 estimator can be obtained analytically to yield the estimators $\hat{a}$, $\hat{b}$, and $\hat{c}$, from which the particle momentum can be inferred.

The momentum resolution $\delta p_T / p_T$ achievable by tracking a charged particle in magnetic field depends on the layout of the spectrometer, i.e. the number and location of the measuring layers, the position resolution of the measuring stations, the lever arm, the material budget, the magnetic field, etc. However, a simplified treatment of tracking in magnetic field for energetic particles, i.e. such that the bending power of Eq. (3.43) is small, allows to derive the dependence of $\delta p_T / p_T$ upon the main dimensionful quantities involved in the problem, namely the lever arm of the spectrometer L, the magnetic field intensity B, the position resolution δy, and the material budget, conveniently measured in radiation lengths X_0. To this purpose, let's consider the trajectory of a charged particle on the plane transverse to the magnetic field. As shown in Problem 3.2, the trajectory is an arc of length L and radius R given by Eq. (3.45). The bending angle $\theta = L/R$ is assumed to be small. The maximal distance of the arc from its cord is the so-called *sagitta*, and is usually denoted by s. From its definition, it follows that:

$$s = R \left(1 - \cos \frac{\theta}{2} \right) \approx R \frac{\theta^2}{8} = \frac{L^2}{8R}. \tag{3.7}$$

Regardless of the details of its design, the spectrometer will sample the track in a number of points, thus allowing to indirectly measure s. The uncertainty on the sagitta will be of the order of the single-point uncertainty δy, i.e. $\delta s \sim \delta y$. For small uncertainties δs one has:

$$\left(\frac{\delta p_T}{p_T} \right)_{\text{meas.}} = \frac{\delta R}{R} = \frac{\delta s}{s} \sim \frac{\delta y}{s} = \frac{8 \, \delta y \, R}{L^2} = \frac{8 \, \delta y}{q \, B \, L^2} p_T$$

$$= \frac{8 \delta y}{0.3 \, z \, B \, L^2} p_T, \tag{3.8}$$

where the last equality assumes that the units GeV/c, T, and m are used, see Problem 3.3. Therefore, the relative momentum resolution from the position measurements alone depends linearly on the momentum, and is inversely proportional to $B\,L^2$. At low momenta, another effect usually dominates the momentum resolution, namely the multiple scattering (MS) inside the spectrometer. Again, this effect should be evaluated by considering the full spectrometer design. However, the main features and the order-of-magnitude of the MS-induced uncertainty can be determined by considering a simple configuration where the particle track receives a deflection angle $\delta\theta = \sqrt{\langle\theta_{\mathrm{T}}^2\rangle}$ as given by Eq. (2.8). This angle will induce a shift in the measured radius R such that:

$$\delta\theta = \left|\delta\left(\frac{L}{R}\right)\right| = \frac{L}{R^2}\,\delta R,$$

$$\left(\frac{\delta p_{\mathrm{T}}}{p_{\mathrm{T}}}\right)_{\mathrm{MS}} = \frac{\delta R}{R} = \frac{R}{L}\,\delta\theta = \left(\frac{p_{\mathrm{T}}}{0.3\,z\,B}\right)\frac{1}{L}z\frac{0.0136}{\beta p_{\mathrm{T}}}\sqrt{\frac{L}{X_0}}$$

$$= \frac{1}{0.3\,B}\frac{0.0136}{\beta}\sqrt{\frac{1}{L\,X_0}}, \tag{3.9}$$

and again the last equality assumes that the units GeV/c, T, and m are used. Therefore, the relative momentum resolution from MS alone does not depend on the momentum, and is inversely proportional to $B\sqrt{L}$. It also depends on the traversed material as $\sim X_0^{-1/2}$, so the effect is more relevant in condensed and high-Z materials, and less in gas and low-Z materials. Since the uncertainty from the position measurement and the MS inside the detector can be assumed to be independent, the overall momentum resolution is generally parametrised as:

$$\left(\frac{\delta p_{\mathrm{T}}}{p_{\mathrm{T}}}\right)_{\mathrm{tot}} = \left[\left(\frac{\delta p_{\mathrm{T}}}{p_{\mathrm{T}}}\right)_{\mathrm{meas.}} \oplus \left(\frac{\delta p_{\mathrm{T}}}{p_{\mathrm{T}}}\right)_{\mathrm{MS}}\right] = a\,p_{\mathrm{T}} \oplus b, \tag{3.10}$$

where a and b are constants. Differently from calorimeters, where the relative energy resolution improves with energy, see Eq. (2.93), the relative momentum resolution of tracking detectors gets worse at larger momenta.

Three configurations are worth being considered because of the possibility to determine analytically the corresponding momentum resolution. They are illustrated in Fig. 3.2. In all cases, the $N+1$ measuring stations are assumed to be characterised by the same spatial resolution σ. In the first configuration, the measuring stations are uniformly spaced. It can be shown [1] that the theory of χ^2 estimators provides the following relative momentum resolution:

$$\left(\frac{\delta p_{\mathrm{T}}}{p_{\mathrm{T}}^2}\right)_{\mathrm{meas.}} = \frac{\sigma\sqrt{A_N}}{0.3\,|\mathbf{B}|\,L^2}, \tag{3.11}$$

where A_N is a function of N given by [1]:

$$A_N = \frac{720\,N^3}{(N-1)(N+1)(N+2)(N+3)} \approx \frac{720}{N+5}. \tag{3.12}$$

Numerically, $\sqrt{A_N}/8 \approx 0.8 \div 1.2$ up to $N \approx 10$, thus showing that the estimate of Eq. (3.8) applies to this configuration to better than 20% for typical values of N. The effect of MS has been calculated as well [1], giving

$$\left(\frac{\delta p_T}{p_T^2}\right)_{MS} = \frac{1}{0.3\,|\mathbf{B}|}\frac{0.0136}{\beta}\sqrt{\frac{C_N}{L\,X_0}}, \tag{3.13}$$

with $C_N = 1.3 \div 1.4$.

Given a fully magnetised spectrometer with lever arm L and $N+1$ available stations, one may wonder what is the configuration that yields the smallest momentum resolution. The answer is provided by the second layout of Fig. 3.2, where half of the stations are clustered around x_c, and the other half are equally distributed at the rear and at the front of the magnetised volume. It can be shown that the momentum resolution from measurement uncertainty only achievable by such configuration is given by [1]:

$$\left(\frac{\delta p_T}{p_T^2}\right)_{meas.}^{opt.} = \frac{\sigma\,\sqrt{B_N}}{0.3\,|\mathbf{B}|\,L^2}, \quad \text{with} \quad B_N = \frac{256}{N+1}, \tag{3.14}$$

thus giving a factor of about 1.4 smaller relative resolution for $N \approx 10$, compared to Eq. (3.11).

Finally, we consider the case where the lever arm is L, but the magnetised volume has a length of $\ell < L$. It can be shown [2] that a configuration like the one illustrated in the third cartoon of Fig. 3.2, with half stations uniformly distributed at the front and rear of the spectrometer and the other half at the centre, i.e. near to the magnet, yields a relative momentum resolution:

$$\left(\frac{\delta p_T}{p_T^2}\right)_{meas.} = \frac{8\sigma}{\sqrt{N+1}}\frac{1}{0.3\,|\mathbf{B}|\,\ell\,L}. \tag{3.15}$$

A simple proof is proposed in Problem 3.8.

Problems

Problem 3.1 Determine the equation of motion of a classical point-like particle of charge q, mass m, and initial velocity $\mathbf{v}_0$, moving inside a static and uniform electric $\mathbf{E}$ and magnetic field $\mathbf{B}$.

Fig. 3.2 Examples of particle tracking using $N+1$ measuring stations over a total lever arm L. In the first and second case, the spectrometer is immersed in a uniform magnetic field providing a total bending power of $|\mathbf{B}|L$. Two layouts of the measuring stations are deployed: uniformly spaced (top) and clustered at the front, middle, and rear of the spectrometre (centre). In the last case, the spectrometer is not fully magnetised and the bending power is only $|\mathbf{B}|\ell$ (bottom); the measuring stations are equally distributed at the front and rear of the magnetised volume, and at the front and rear of the spectrometer as to achieve a lever arm L

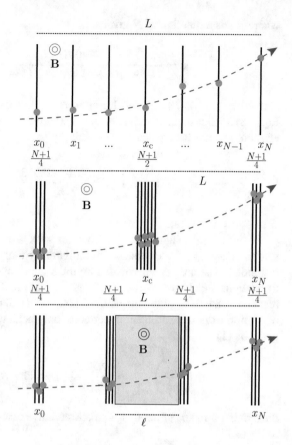

Solution

A classical particle of charge q and mass m moving inside a uniform electric field $\mathbf{E}$ superimposed to a uniform magnetic field $\mathbf{B}$ is subject to the classical force

$$\mathbf{F} = q\,(\mathbf{E} + \mathbf{v} \times \mathbf{B}). \tag{3.16}$$

The second term in the right-hand side of Eq. (3.16) described the *Lorentz* force. The equation of motion $\mathbf{r}(t)$ is determined by Newton's law:

$$\frac{d\mathbf{p}}{dt} = \mathbf{F} = q\,(\mathbf{E} + \mathbf{v} \times \mathbf{B}), \tag{3.17}$$

which is a system of three coupled ODE. We choose the reference frame so that the z-axis is aligned along $\mathbf{B}$, i.e. $\mathbf{B} = (0, 0, |\mathbf{B}|)$, and the y-axis is parallel to $\mathbf{E} \times \mathbf{B}$, so that $\mathbf{E} = (E_\perp, 0, E_\parallel)$ with $E_\perp > 0$, and. We further define the origin such that at time $t = 0$ the particle is at the origin of the reference frame. Equation (3.17) and the boundary conditions can be written by components as:

$$\begin{cases} \ddot{x} = \frac{eE_\perp}{m} + \frac{e|\mathbf{B}|}{m}\dot{y}(t) \equiv \mathscr{E}_\perp + \omega_B\,\dot{y} \\ \ddot{y} = -\frac{e|\mathbf{B}|}{m}\dot{x}(t) \equiv -\omega_B\,\dot{x} \\ \ddot{z} = \frac{eE_\parallel}{m} \equiv \mathscr{E}_\parallel \end{cases} , \quad \begin{cases} \dot{x}(0) = v_x^0 \\ \dot{y}(0) = v_y^0 \\ \dot{z}(0) = v_z^0 \end{cases} , \quad \begin{cases} x(0) = 0 \\ y(0) = 0 \\ z(0) = 0 \end{cases}$$

(3.18)

where we have defined:

$$\omega_B \equiv \frac{e\,|\mathbf{B}|}{m}, \qquad \mathscr{E}_\perp \equiv \frac{e\,E_\perp}{m}, \qquad \mathscr{E}_\parallel \equiv \frac{e\,E_\parallel}{m}.$$

(3.19)

The first of Eq. (3.19) is also known as *cyclotron frequency*. Integrating once the three equations in the system (3.18), we get:

$$\begin{cases} \dot{x} = \mathscr{E}_\perp\,t + \omega_B\,y + v_x^0 \\ \dot{y} = -\omega_B\,x + v_y^0 \\ \dot{z} = \mathscr{E}_\parallel\,t + v_z^0 \end{cases}$$

(3.20)

Inserting the second of Eq. (3.20) into the first, we have an equation in $x(t)$ alone:

$$\ddot{x} = -\omega_B^2\,x + (\mathscr{E}_\perp + \omega_B\,v_y^0) = -\omega_B^2\left(x + \frac{E_\perp/|\mathbf{B}| + v_y^0}{\omega_B}\right),$$

(3.21)

which can be integrated to yield:

$$x(t) = \left(-\frac{1}{\omega_B}\sqrt{\left(\frac{E_\perp}{|\mathbf{B}|} + v_y^0\right)^2 + v_x^0}\right)\cos(\omega_B\,t + \alpha) + \frac{E_\perp/|\mathbf{B}| + v_y^0}{\omega_B},$$

(3.22)

where α is defined such that

$$\tan\alpha = \frac{v_x^0}{v_y^0 + E_\perp/|\mathbf{B}|}.$$

(3.23)

Inserting this result into the second of Eq. (3.20), we get:

$$y(t) = \left(\frac{1}{\omega_B}\sqrt{\left(\frac{E_\perp}{|\mathbf{B}|} + v_y^0\right)^2 + v_x^0}\right)\sin(\omega_B\,t + \alpha) - \frac{E_\perp}{|\mathbf{B}|}t - \frac{v_x^0}{\omega_B}.$$

(3.24)

Finally, the last component can be obtained by integrating the third equation of the system (3.20) to give:

$$z(t) = \frac{\mathscr{E}_\parallel}{2} t^2 + v_z^0 t. \tag{3.25}$$

Therefore, the equation of motion of the particle on the x-y plane is a uniform circular motion with angular frequence ω_B centred around the point:

$$\mathbf{r}_c(t) = \left(\frac{E_\perp/|\mathbf{B}| + v_y^0}{\omega_B}, \; -\frac{E_\perp}{|\mathbf{B}|} t - \frac{v_x^0}{\omega_B}, 0 \right), \tag{3.26}$$

which drifts with constant velocity along the direction of $\mathbf{E} \times \mathbf{B}$. Along the direction of the magnetic field, the motion is instead uniformly accelerated. The velocities along the y-axis at the time when the motion along x reverts its direction are given by:

$$0 = \dot{x}(t_n) = \sqrt{\left(\frac{E_\perp}{|\mathbf{B}|} + v_y^0\right)^2 + v_x^0} \; \sin(\omega_B t_n + \alpha) \quad \Rightarrow \quad t_n = k\pi - \alpha$$

$$\dot{y}(t_n) = \pm\sqrt{\left(\frac{E_\perp}{|\mathbf{B}|} + v_y^0\right)^2 + v_x^0} - \frac{E_\perp}{|\mathbf{B}|}. \tag{3.27}$$

Hence, the motion along the y direction reverts direction if at least one among $|v_x^0|$ and $|v_y^0|$ is larger than zero, otherwise the trajectory goes through cusps. For the special case $E_\parallel = 0$, and assuming that the particle is initially at rest, i.e. $\mathbf{v}_0 = 0$, the mean velocity $\langle \mathbf{v} \rangle$ is given by the time-average of $\mathbf{v}(t)$, namely:

$$\langle \mathbf{v} \rangle = \begin{pmatrix} \langle \dot{x} \rangle \\ \langle \dot{y} \rangle \\ \langle \dot{z} \rangle \end{pmatrix} = \begin{pmatrix} 0 \\ -\frac{|\mathbf{E}|}{|\mathbf{B}|} \\ 0 \end{pmatrix} = \frac{\mathbf{E} \times \mathbf{B}}{|\mathbf{B}|^2}. \tag{3.28}$$

Discussion

The motion of charged particles inside a simultaneous static electric and magnetic field is a common situation in particle physics experiments. A typical example is provided by the Time Projection Chamber discussed in Problem 2.4, where the electric field is needed to drift the ionisation electrons to the multiplication region, while the magnetic field allows to measure the track curvature, hence the particle momentum, see Problem 3.2. In a TPC, the fields are parallel, i.e. $E_\perp = 0$, a configuration that helps reducing the lateral diffusion of the charge, see Problem 2.42. Another canonical example is a microstrip detector (see Problem 2.43) inversely polarised and immersed in an orthogonal magnetic field, i.e. $E_\parallel = 0$. In both cases, however, the equation of motion are not given by the solution found here because the electrons do not move in vacuum, but interact with the medium (the filling gas for TPC, the silicon crystal for the microstrip). The problem should be therefore treated with a proper kinetic theory of the interactions with the traversed material. A simple friction model predicts that the charged particles (e.g. electrons) will drift with velocity

$$\mathbf{v} = \frac{\mu_e}{1 + \omega_B^2 \tau^2} \left(\mathbf{E} + \frac{\omega_B \tau}{|\mathbf{B}|} \mathbf{E} \times \mathbf{B} + \frac{\omega_B^2 \tau^2}{|\mathbf{B}|^2} (\mathbf{E} \cdot \mathbf{B}) \mathbf{B} \right), \qquad (3.29)$$

where $\mu_e = e\tau/m_e$ is the electron mobility, see Eq. (2.112), and τ is the mean time between two collisions with the medium constituents. See Chap. 34.6 of Ref. [3] for more details. In particular, for $\mathbf{E}$ orthogonal to $\mathbf{B}$, the drift velocity makes an angle Θ_L with respect to the electric field direction such that:

$$\tan \Theta_L = \omega_B \tau = \mu_e |\mathbf{B}|. \qquad (3.30)$$

In semiconductive devices, like pixels or microstrips, this drift angle is also called *Lorentz angle* and has important implications on the position resolution of the detector. Using a typical value for electron mobility in silicium $\mu_e \sim 10^3 \, \text{V}^{-1} \, \text{cm}^2 \, \text{s}^{-2}$ and magnetic field intensities of order ~ 1 T, the Lorentz angle turn out to be sizable, i.e. $\Theta_L \sim 10^{-1}$.

Suggested Readings

The notation has been adapted from Chap. 4.8 of Ref. [4]. See also Chap. 34.6 of Ref. [3] for more details on the motion of charged particles in gas uncer the combined effect of electric and magnetic fields.

Problem 3.2 Show that the trajectory of a charged particle of momentum $\mathbf{p}$ and charge q moving inside a uniform magnetic field $\mathbf{B}$ is a helix. What is the relation between the particle momentum and the radius of curvature of the trajectory?

Discussion

When dealing with relativitic particles, the laws of kinematics change, but the time-evolution law of Eq. (3.17) maintains its form. This can be for example proved by deriving the equation from the relativistic generalisation of the Euler-Lagrange equation for a point-like charge q and mass m moving in a classical electromagnetic field:

$$\frac{d}{dt} \frac{\partial L}{\partial \dot{\mathbf{r}}} - \frac{\partial L}{\partial \mathbf{r}} = 0, \quad \text{with} \quad L(\mathbf{r}, \dot{\mathbf{r}}) = -mc^2 \sqrt{1 - \frac{|\dot{\mathbf{r}}|^2}{c^2}} - q \, \phi(\mathbf{r}) + q \, \dot{\mathbf{r}} \cdot \mathbf{A}(\mathbf{r}),$$
$$(3.31)$$

where $\dot{\mathbf{r}} \equiv \mathbf{v}$, and $\phi(\mathbf{r})$ and $\mathbf{A}(\mathbf{r})$ are the scalar and vector potential of the electric and magnetic field, respectively, such that $E = -\nabla \phi$ and $\mathbf{B} = \nabla \times \mathbf{A}$. Since $\mathbf{B}$ is constant, we can always assume $\partial_t \mathbf{A} = 0$. Computing explicitly the derivatives in Eq. (3.31), we get:

$$\frac{d}{dt} (p_i + q A_i) = \frac{dp_i}{dt} + \frac{\partial A_i}{\partial t} + v_j \partial_i A_j = -q \left(\partial_i \phi - v_j \partial_i A_j \right),$$

$$\frac{dp_i}{dt} = -q \left(\partial_i \phi - v_j \left[\partial_i A_j - \partial_j A_i \right] \right) = -q \left(\partial_i \phi - v_j \, \varepsilon_{ijk} \left[\nabla \times \mathbf{A} \right]_k \right) =$$

$$= -q \left(\partial_i \phi - \left[\mathbf{v} \times (\nabla \times \mathbf{A}) \right]_i \right) = q \, E_i + q \, (\mathbf{v} \times \mathbf{B})_i \,,$$

$$\frac{d\mathbf{p}}{dt} = q \, \mathbf{E} + q \, \mathbf{v} \times \mathbf{B}. \tag{3.32}$$

Hence, the relativistic extension of the time-evolution law for a point-like charge is identical to the classical expression of Eq. (3.17), provided that the relativistic definition of momentum is used.

Solution

From Eq. (3.17) with $\mathbf{E} = 0$, one can easily see that the momentum magnitude $|\mathbf{p}|$ is constant in time, since

$$\frac{d|\mathbf{p}|^2}{dt} = 2 \, \mathbf{p} \cdot \frac{d\mathbf{p}}{dt} \propto \mathbf{p} \cdot (\mathbf{p} \times \mathbf{B}) = 0. \tag{3.33}$$

Hence, Eq. (3.17) becomes:

$$m\gamma \frac{d^2\mathbf{r}}{dt^2} = q \frac{d\mathbf{r}}{dt} \times \mathbf{B}, \quad m\gamma \, |\mathbf{v}|^2 \frac{d^2\mathbf{r}}{ds^2} = q|\mathbf{v}| \frac{d\mathbf{r}}{ds} \times \mathbf{B}, \quad \frac{d^2\mathbf{r}}{ds^2} = \frac{q}{|\mathbf{p}|} \frac{d\mathbf{r}}{ds} \times \mathbf{B}, \tag{3.34}$$

where we have changed variable from the laboratory time t to the trajectory length $s = |\mathbf{v}| t$, so that we can directly get a geometrical parametrisation of the curve. We choose the reference frame like in Problem 3.1, so that the system (3.34) can be written by components as:

$$\begin{cases} x'' = \frac{qB}{|\mathbf{p}|} y' \equiv \frac{y'}{K} \\ y'' = -\frac{qB}{|\mathbf{p}|} x' \equiv -\frac{x'}{K} \\ z'' = 0 \end{cases}, \quad \begin{cases} x'(0) = -\cos \lambda \sin \Phi_0 \\ y'(0) = -\cos \lambda \cos \Phi_0 \\ z'(0) = \sin \lambda \end{cases}, \quad \begin{cases} x(0) = x_0 \\ y(0) = y_0 \\ z(0) = z_0 \end{cases} \tag{3.35}$$

Here, Φ_0 can be interpreted *a posteriori* as the angle in the x–y plane formed by the radial vector that connects the centre of the circumference to the initial position, measured with respect to the x-axis, see Fig. 3.3. The constant K is defined as $K \equiv |\mathbf{p}|/q|\mathbf{B}|$, while $\lambda \in [0, \pi/2]$ is the dip angle with respect to the x–y plane. The boundary condition on $\mathbf{r}_0'$ is subject to the constraint $|\mathbf{r}_0'| = 1$.

Integrating once the second of the system (3.35), we have:

$$y' = -\frac{x'}{K} + C \quad \Rightarrow \quad x'' = -\frac{1}{K^2} (x + KC), \tag{3.36}$$

Fig. 3.3 Representation of the trajectory **r**(s) of a charged particle moving inside a uniform magnetic field **B**. The angle λ is called *dip angle*, whereas Φ_0 is the angle of the radial vector joining the centre of the circumference to $\mathbf{r}_0$, with respect to the x-axis

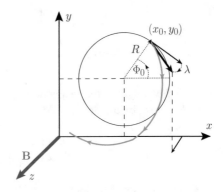

with C a constant of integration, hence $x(s)$ satisfies the equation for a displaced harmonic oscillator. Taking into account the boundary conditions in (3.35), we get:

$$x(s) = x_0 + R\left[\cos\left(\frac{s\,\cos\lambda}{R} - \Phi_0\right) - \cos\Phi_0\right]. \tag{3.37}$$

where $R \equiv K\,\cos\lambda$. Substituting this expression into the second of (3.35), and taking into account the boundary conditions for y:

$$y' = \frac{R}{\cos\lambda}x'' = -\cos\lambda\,\cos\left(\frac{s\,\cos\lambda}{R} - \Phi_0\right),$$

$$y(s) = y_0 - R\left[\sin\left(\frac{s\,\cos\lambda}{R} - \Phi_0\right) + \sin\Phi_0\right]. \tag{3.38}$$

The last component can be trivially integrated to give:

$$z(s) = z_0 + \sin\lambda\, s. \tag{3.39}$$

The projection of the curve onto the x–y plane is a circle of radius

$$|R| = \frac{|\mathbf{p}|\cos\lambda}{|q|\,|\mathbf{B}|} = \frac{p_T}{|q|\,|\mathbf{B}|}, \tag{3.40}$$

where p_T is the transverse momentum, while the z-component increments linearly with the curve length: the curve is therefore a helix of pitch $p = 2\pi R\tan\lambda$. In particular, for $|s| \ll |R|$, i.e. for large momenta, the trajectories in the x–z and y–z planes can be approximated by straight lines:

$$x(z) \approx x_0 + \frac{\sin\Phi_0}{\tan\lambda}(z - z_0)$$

$$y(z) \approx y_0 - \frac{\cos\Phi_0}{\tan\lambda}(z - z_0) \tag{3.41}$$

The radius of curvature ρ of the track along the trajectory is given by:

$$\rho(s) = \left| \frac{d^2\mathbf{r}}{ds^2} \right|^{-1} = \left| \frac{q}{|\mathbf{p}|} \frac{d\mathbf{r}}{ds} \times \mathbf{B} \right|^{-1} = \frac{|R|}{\cos^2 \lambda}. \tag{3.42}$$

The *bending angle* of the magnetic field is defined as the angle by which a charged particle is bent by the field all along its path, i.e.:

$$\Delta\theta = \int \frac{ds}{\rho} = \int ds \left| \frac{d^2\mathbf{r}}{ds^2} \right| = \frac{q}{|\mathbf{p}|} \int ds \left| \frac{d\mathbf{r}}{ds} \times \mathbf{B}(\mathbf{r}) \right| \equiv \frac{\mathscr{B}}{(|\mathbf{p}|/q)}. \tag{3.43}$$

The integral at the right-hand side of Eq. (3.43) is called *bending power*. Geometrically, it corresponds to the integral along the path of the magnetic field component orthogonal to the trajectory. The bending angle is therefore proportional to the bending power and inversely proportional to the ratio $|\mathbf{p}|/q$, which in accelerator physics is also known as *beam rigidity*, i.e. a measure of the particle resilence to be bent by a magnetic field. In general, the bending power depends on the initial particle direction, and it has to be computed numerically. For uniform magnetic fields, like those generated at the inside of a long solenoid of internal radius R_{coil} and intensity $|\mathbf{B}|$, the bending power from the interaction point up to the position of entrance in coil $\mathbf{r}_{coil}$ is given by:

$$\mathscr{B} = \left| \int_0^{\mathbf{r}_{coil}} ds \frac{d\mathbf{r}}{ds} \right| |\mathbf{B}| \cos \lambda = |\mathbf{r}_{coil}| |\mathbf{B}| \cos \lambda = R_{coil} |\mathbf{B}|. \tag{3.44}$$

For example, the solenoid of the CMS experiment has an inner radius $R_{coil} \approx 3$ m and generates a magnetic field of intensity $|\mathbf{B}| \approx 4$ T, giving a bending power $\mathscr{B} \approx 12$ T m uniformly with respect to the pseudorapidity η, as defined in Eq. (1.161). At large values of η, however, the magnetic field is no longer uniform and the bending power features a dependence on the direction η. See Ref. [5] for a more exact evaluation of $\mathscr{B}$ in the ATLAS toroid and CMS solenoid.

Suggested Readings

The formalism used in this exercise and its application in real experiments can be found in Ref. [5].

Problem 3.3 Show that the transverse momentum p_T [GeV/c] of a particle of charge $q = z e$, moving inside a uniform magnetic field of intensity B [T], along a circular path of radius R [m] on the transverse plane, is given by the formula:

$$p_T = 0.3 z B R. \tag{3.45}$$

Solution

From Eq. (3.40), we see that the radius of curvature R of a particle moving in the plane orthogonal to $\mathbf{B}$ is related to the transverse momentum p_T by the relation:

$$p_T = q \, |\mathbf{B}| \, R. \tag{3.46}$$

Dividing the two sides of Eq. (3.46) by the units in which we wish to express the dimensionful quantities, we get:

$$\frac{p_T}{\text{GeV}/c} = \frac{1}{\text{GeV}/c} (z \cdot e) \cdot \left(\frac{|\mathbf{B}|}{\text{T}} \cdot \text{T} \right) \cdot \left(\frac{R}{\text{m}} \cdot \text{m} \right) =$$

$$= z \left(\frac{|\mathbf{B}|}{\text{T}} \right) \cdot \left(\frac{R}{\text{m}} \right) \frac{2.998 \times 10^8 \, \text{m/s} \, e \, \text{T} \, \text{m}}{10^9 \, e \, \text{V}} =$$

$$= 0.3 \, z \left(\frac{|\mathbf{B}|}{\text{T}} \right) \left(\frac{R}{\text{m}} \right), \tag{3.47}$$

where we have used the fact that $\text{T} = \text{V} \, \text{m}^{-2} \, \text{s}$. The last approximation in Eq. (3.47) is valid to better than the permill level, which is usually fine for most of the applications.

Problem 3.4 A drift chamber is immersed in a magnetic field with intensity $|\mathbf{B}| = 0.8 \, \text{T}$. A photon converts into a $e^+ e^-$ pair. Two tracks on the plane orthogonal to the magnetic field, and with radius of curvature $R = 20 \, \text{cm}$, are observed. The tracks start parallel to each other. Determine the energy of the incoming photon.

Solution

The electron and positron have momentum $|\mathbf{p}_e|$ given by Eq. (3.45), namely

$$|\mathbf{p}_e| = 0.3 \, |\mathbf{B}| \, R = 0.3 \cdot 0.8 \cdot 0.2 \, \text{MeV} = 48 \, \text{MeV}. \tag{3.48}$$

Since the tracks are intially parallel, the momentum transfered to the nucleus is of order $|\mathbf{q}| \sim m_e^2 / E_\gamma \ll m_e$, and the kinetic energy $|\mathbf{q}|^2 / 2 m_N$ of the recoiling nucleus is negligible, see Problem 1.48. To good approximation, the photon energy is therefore given by:

$$E_\gamma \approx 2 \, E_e \approx 2 \, |\mathbf{p}_e| = 96 \, \text{MeV}. \tag{3.49}$$

We can thus verify a posteriori that the most probable interaction process for the incoming photon is indeed pair-production.

Problem 3.5 The LEP dipoles allowed for a maximum magnetic field $|\mathbf{B}| = 0.135$ T, and covered up to 2/3 of the 27 km-long storage ring. What was the largest electron energy attainable?

Solution

The effective radius of curvature at LEP was about $(0.66 \cdot 27/2\pi)$ km ≈ 2.9 km. The maximum electron energy is therefore given by Eq. (3.45):

$$E_{max} = |\mathbf{p}_{max}| = (0.3 \cdot 0.135 \cdot 2.9 \times 10^3)\,\text{GeV} \approx 117\,\text{GeV}. \qquad (3.50)$$

The actual limit achieved at LEP was a bit smaller than this, namely $E_{max} = 104.5$ GeV, and was dictated by the limitations of the RF accelerators. This limit had important implications on the reach for the Higgs boson later discovered at the LHC [6, 7], since the maximum centre-of-mass energy of $2\,E_{max} \approx 209$ GeV allowed the experiments to probe Higgs boson masses below about 115 GeV, thus outside the experimental value of 125 GeV established years later at the LHC [8].

Suggested Readings

The energy record at LEP was achieved in 2000, see e.g. the original CERN announcement [9].

Problem 3.6 A beam of momentum $|\mathbf{p}| = 2$ GeV contains both kaons and pions. The beam passes between the plates of an electrostatic separator of length L. The electric field between the plates has uniform intensity $|\mathbf{E}| = 10$ kV/cm. Determine the opening angle between the two beams emerging from the separator.

Solution

The kinetic energy acquired by the beam due to the work done by the electric field is negligible, so we can assume $|\mathbf{p}|$ to be approximately constant during the motion. Let's denote the coordinate parallel to the electric field by y, such that $y(0) = 0$. From Eq. (3.32), we get:

$$e\,|\mathbf{E}| = m\,\frac{d(\beta_y\gamma)}{dt} \approx m\,\gamma\,\dot{\beta}_y, \qquad \ddot{y} = \frac{e\,|\mathbf{E}|}{m\,\gamma}, \qquad y(t) = \frac{1}{2}\frac{e\,|\mathbf{E}|}{m\,\gamma}t^2,$$

$$y(x) = \frac{e\,|\mathbf{E}|}{2|\mathbf{p}|}\frac{x^2}{\beta}, \qquad y'(x) = \frac{e\,|\mathbf{E}|}{|\mathbf{p}|}\frac{x}{\beta}. \qquad (3.51)$$

The opening angle between the two beams at the exit of the separator is therefore given by:

$$|\Delta y'(L)| = \frac{e\,|\mathbf{E}|\,L}{|\mathbf{p}|}\left(\sqrt{1 + \frac{m_K^2}{|\mathbf{p}|^2}} - \sqrt{1 + \frac{m_\pi^2}{|\mathbf{p}|^2}}\right) \approx \frac{e\,|\mathbf{E}|\,L}{2|\mathbf{p}|^3}(m_K^2 - m_\pi^2), \qquad (3.52)$$

where we have expanded the square root to first order. Numerically, we get:

$$|\Delta y'(L)| = \frac{10^6 \, \text{eV/m} \cdot 1\,\text{m}}{2 \cdot (2\,\text{GeV})^3} \cdot (0.23\,\text{GeV}^2) = 1.4 \times 10^{-5}\,\text{rad}. \qquad (3.53)$$

Discussion

Equation (3.52) shows that a static electric field can be used to separate particles of different mass but same momentum, which may have been for example selected by a magnetic field. However, due to the $\sim |\mathbf{p}|^{-3}$ scaling of the opening angle, this method becomes inefficient for momenta in excess of a few GeV, as the numerical calculation also shows. Indeed, electric fields more intense than a few tens of kV/cm can give rise to the formation of discharges (in air at standard pressure, the break-down voltage is about 30 kV/cm). For larger beam momenta, separation techniques based on the use of RF cavities become more efficient.

Suggested Readings

This topic is briefly discussed in Chap. 11.4 of Ref. [10].

Problem 3.7 A mass spectrometer consists of a pair of electrodes kept at a potential difference $|V_0| = 1\,\text{MV}$, followed by a magnetic analyzer of length $L = 1\,\text{m}$ and uniform magnetic field with intensity $|\mathbf{B}| = 1\,\text{kG}$. Positive ions are produced at rest at the grounded electrode and accelerated by the electric field towards the cathode. The position of arrival y at the opposite side of the spectrometer is measured with an uncertainty σ_y. The beam to be composed of $^{12}\text{C}^+$ and $^{13}\text{C}^+$ ions in unknown proportion. What is the minimum resolution σ_y necessary to identify the correct mass number to better than 3σ per incident ion?

Discussion

The use of accelerated beams and magnetic analysers for mass spectroscopy is a well established technology known under the name of *accelerated mass spectroscopy* (AMS). An application of AMS is radiocarbon dating for assessing the age of organic samples. See e.g. Ref. [11] for more informations on the subject.

Solution

The kinetic energy of the ion at the entrance of the spectrometer is $T = qV_0 = |\mathbf{p}|^2/2m$. We can assume classical kinematics since $qV_0 = 1\,\text{MeV}$ is small compared to the mass of carbon ions. The radius of curvature R is then given by Eq. (3.40):

$$R = \frac{|\mathbf{p}|}{qB} = \frac{\sqrt{2mqV_0}}{qB} \qquad (3.54)$$

Fig. 3.4 Sketch of a mass spectrometer

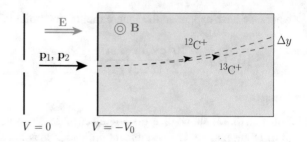

Replacing the symbols by their values, we get for the $^{12}C^+$ population:

$$R(^{12}C^+) = \sqrt{\frac{2 \cdot 10^6 \, V}{1.6 \times 10^{-19} \, C} \frac{12 \times 10^{-3} \, kg}{6.0 \times 10^{23}} \frac{1}{0.1 \, T}} = 5.0 \, m. \qquad (3.55)$$

In Eq. (3.55), we have use the fact that one mole of ^{12}C weighs *exactly* 12 g. The apparatus measures the lateral displacement y with respect to the initial beam position, see Fig. 3.4. Since $L/R \approx 0.2$, we can approximate the ion trajectory with a parabola to yield:

$$y \approx \frac{L^2}{2R} = \sqrt{\frac{q}{2m V_0}} \frac{|B| L^2}{2}, \qquad (3.56)$$

which is accurate to the percent level. For $^{12}C^+$, $y \approx 10$ cm, which is smalle compared to L, thus justifying the use of Eq. (3.56). Since $\Delta m = 1$ a.u. is small compared to the ^{12}C mass, we can approximate:

$$\frac{\Delta y}{y} \approx \frac{1}{2} \frac{\Delta m}{m} \quad \Rightarrow \quad \Delta y \approx \frac{y}{2} \frac{\Delta m}{m}. \qquad (3.57)$$

In order to separate the two mass numbers to better than N_σ standard deviations per ion, the position resolution needs to satisfy:

$$\sigma_y < \frac{\Delta y}{N_\sigma} = \frac{1}{4 N_\sigma} \sqrt{\frac{q}{2m V_0}} |B| L^2 \frac{\Delta m}{m} =$$

$$= \frac{1}{4 \cdot 3} \sqrt{\frac{1.6 \times 10^{-19} \, C}{2 \cdot 2 \times 10^{-26} \, kg \cdot 10^6 \, V}} \cdot 0.1 \, T \cdot 1 \, m^2 \cdot \frac{1}{12} = 0.1 \, cm. \qquad (3.58)$$

Bando n. 13705/2010

Problem 3.8 A tracking system consists of a pair of multiwire chambers separated by a distance $d = 1$ m, and with spatial resolution $\sigma = 150 \, \mu$m, followed by a dipolar magnet with bending power $\mathscr{B} = 2\,\mathrm{T\,m}$, and by an identical pair of chambers. Determine the relative resolution on the momentum transverse to the magnetic field as a function of the transverse momentum itself. Assume that both the dip angle of the impinging particle and the angular deflection induced by the magnet are small. Neglect the effect of multiple scatterring inside the chambers.

Solution

The spectrometer design corresponds to the last of Fig. 3.2. We can therefore apply the result from Eq. (3.15) giving:

$$\frac{\delta p_T}{p_T} = \frac{8\sigma}{\sqrt{N+1}} \frac{1}{0.3\,\mathscr{B}\,L} p_T = \frac{8 \cdot 150 \times 10^{-6}}{\sqrt{4}} \frac{p_T}{0.3 \cdot 2 \cdot 2} = 5 \times 10^{-4}\,p_T, \quad (3.59)$$

where L is the total length of the spectrometer, i.e. $L = 2\,d = 2\,\mathrm{m}$, and p_T is measured in GeV. One can convince himself of this result by noticing that the system measures two segments, before and after the magnet, respectively. The uncertainty on the slope $\hat{b}$ of each segment is given by the first of Eq. (3.2), namely $\sigma_{\hat{b}} = \sqrt{2}\sigma/d = 2\sqrt{2}\sigma/L$. The uncertainty on the bending angle is therefore $\sqrt{2}\sigma_{\hat{b}} = 4\sigma/L$. This angle is related to the momentum and bending power via Eq. (3.43), thus giving $\delta p_T/p_T^2 = 4\sigma/(0.3\,\mathscr{B}\,L)$, in agreement with Eq. (3.59).

Suggested Readings

For more information on the topic, the reader is addressed to dedicated textbooks, in particular Chap. 8 of Ref. [12] and Chap. 3.3 of Ref. [2].

Bando n. 1N/R3/SUB/2005

Problem 3.9 Discuss how the transverse momentum resolution $\delta p_T/p_T$ for a charged particle measured by a magnetic spectrometer depends on the transverse momentum p_T if the particle moves in air or inside iron. Assume a typical value of $100\,\mu\mathrm{m}$ and $1\,\mathrm{m}$ for the position resolution and lever arm, respectively.

Solution

The relative momentum resolution from the measurement error and from multiple scattering can be parametrised as in Eq. (3.10): it is roughly constant at low momenta, while it grows linearly with p_T at large momenta. The effect of MS is more relevant in iron than in air because of the larger particle density and Z number. The radiation lengths for air and iron can be found in Table 2.3. Using these values, we get:

$$\left(\frac{\delta p_T}{p_T}\right)_{MS}^{\mathrm{air}} / \left(\frac{\delta p_T}{p_T}\right)_{MS}^{\mathrm{Fe}} = \sqrt{\frac{X_0^{\mathrm{Fe}}}{X_0^{\mathrm{air}}}} = \sqrt{\frac{1.8\,\mathrm{cm}}{300\,\mathrm{m}}} = 8 \times 10^{-3}. \quad (3.60)$$

The momentum at which the measurement error and multiple-scattering become of the same size is given by:

$$p_T = \frac{z}{8\,\delta y}\,\frac{0.0136}{\beta}\,\frac{L^{3/2}}{\sqrt{X_0}}. \tag{3.61}$$

Using typical values $\delta y = 100\,\mu m$ and $L = 1\,m$, we get:

$$p_T \approx \begin{cases} 100\,\text{GeV} & \text{Fe} \\ 1\,\text{GeV} & \text{air} \end{cases} \tag{3.62}$$

Therefore, multiple scattering is negligible in air for momenta in excess of a few GeV, so that $\delta p_T/p_T \sim a\,p_T$, whereas momentum resolution in iron is approximately constant, $\delta p_T/p_T \sim b$, up to momenta of about 100 GeV.

Bando n. 13705/2010

Problem 3.10 A tracking system for measuring high-momentum muons consists of three chambers, separated by a distance $L = 1\,m$ and with spatial resolution $\sigma_0 = 100\,\mu m$, $\sigma_1 = 50\,\mu m$, and $\sigma_2 = 100\,\mu m$. The chambers are located inside a uniform magnetic field of intensity $|\mathbf{B}| = 1\,T$. Consider muons of $p_T = 1\,TeV$ that impinge almost normally to the chambers: what is their the charge misidentification probability?

Solution

Let the initial muon direction be aligned along the x-axis, and denote the orthogonal coordinate by y. We further chose the direction of the y-axis such that the concavity of the true muon trajectory is positive, see Fig. 3.5.

A given event will produce three spatial coordinates $\mathbf{y} = (y_0, y_1, y_2)$. For later use, we define the new coordinate

$$y_1^{\text{ext}} = \frac{(y_2 + y_0)}{2}. \tag{3.63}$$

Fig. 3.5 Illustration of the measuring system considered in the Problem

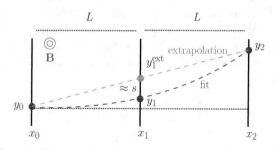

It is easy to convince oneself that a sign-flip corresponds to $s = (y_1^{\text{ext}} - y_2) < 0$, where s coincides with the sagitta of the curve modulo terms of order $L^2/2R$. Since s is given by the difference between two gaussian and independent variables, it will be also normally distributed with mean and standard deviation:

$$\mu_s = \langle y_1^{\text{ext}} \rangle - \langle y_1 \rangle = \frac{(2L)^2}{4R} - \frac{L^2}{2R} = \frac{0.3\,|\mathbf{B}|\,L^2}{2\,p_T} = \frac{0.3 \cdot 1 \cdot (1)^2}{2 \cdot 10^3} = 150\,\mu\text{m},$$

$$\sigma_s = \sigma_2^2 + \frac{\sigma_0^2 + \sigma_2^2}{4} = 87\,\mu\text{m}, \tag{3.64}$$

where we have used the second of Eq. (3.2) to relate the variance of y_1^{ext} to the variance of y_0 and y_2. Therefore:

$$\text{Prob}\,[s < 0] = \frac{1}{\sqrt{2\pi}\,\sigma_s} \int_{-\infty}^{0} ds\,\exp\left[-\frac{(s - \mu_s)^2}{2\,\sigma_s^2}\right] = \frac{1}{2}\left[1 + \text{erf}\left(-\frac{\mu_s}{\sqrt{2}\,\sigma_s}\right)\right], \tag{3.65}$$

where $\text{erf}(x)$ is the error function defined by:

$$\text{erf}(x) = \frac{2}{\sqrt{\pi}} \int_{0}^{x} dt\,e^{-t^2}, \tag{3.66}$$

which is available in almost all standard libraries, e.g. the math library in Python. Replacing the symbols in Eq. (3.65) by their numerical values, we get a misidentification probability:

$$\text{Prob}\,[s < 0] = \frac{1}{2}\left[1 + \text{erf}\left(-\frac{150\,\mu\text{m}}{\sqrt{2} \cdot 87\,\mu\text{m}}\right)\right] = 4.2\% \tag{3.67}$$

The analytical result can be cross-checked by a simple Monte Carlo simulation of the experimental setup, see Appendix 3.4. For example, by using 1M toy events, we get $\text{Prob}\,[s < 0]_{\text{MC}} = (4.18 \pm 0.02)\%$, in agreement with the analytical result of Eq. (3.67).

Bando n. 1N/R3/SUB/2005

Problem 3.11 A spectrometer measures the momentum of charged particles of unitary charge and momentum of a few GeV. It consists of three parallel planes of position detectors with spatial resolution $\sigma_x = 100\,\mu\text{m}$, separated by a distance $a = 20$ cm. The planes are immersed in a uniform magnetic field $\mathbf{B}$ with intensity of 1 T, parallel to the planes and orthogonal to the measured position x. Particles enter the spectrometer almost perpendicular to the detector panes. Estimate the transverse momentum resolution at $|\mathbf{p}| = 2$ GeV.

Solution

The spectrometer configuration corresponds to the first of Fig. 3.2 with $N = 2$. We can therefore use the analytical solution of Eq. (3.11) giving:

$$\frac{\delta p_T}{p_T} = \frac{\sigma \sqrt{A_N}}{0.3\, |\mathbf{B}|\, L^2} p_T = \frac{100^{-6} \cdot 8 \cdot 1.22}{0.3 \cdot 1 \cdot (0.4)^2} \cdot 2 = 4.1\% \qquad (3.68)$$

As we have seen, a simple argument like the one used to derive Eq. (3.8) would have given a result in agreement with Eq. (3.68) within 20%. Actually, for $N = 2$ we could have also easily guessed the extra factor of 1.22. Indeed, if we denote the three measured positions by y_i, $i = 1, 2, 3$, the sagitta s and its uncertainty δs can be estimated by:

$$s \approx y_2 - \frac{y_1 + y_3}{2} \quad \Rightarrow \quad \delta s = \sqrt{\frac{3}{2}}\sigma \approx 1.22\,\sigma, \qquad (3.69)$$

in agreement with the Eq. (3.11).

Bando n. 1N/R3/SUB/2005

Problem 3.12 A magnetised iron slab of thickness $d = 60$ cm is saturated by a constant magnetic field whose force lines can be assumed to be fully contained inside the slab. Let the other two dimensions of the slab be large compared to the thickness and let's assume that the boundary conditions impose the magnetic fiel **B** to be given by $\mathbf{B} = B(y)\, \mathbf{e}_z$, where $B(y)$ is an even function of $y - d/2$, such that $y = 0$ ($y = d$) correspond to the upper (lower) faces, and $\mathbf{e}_z$ is orthogonal to the transverse plane. Furthermore, a numerical simulation predicts that $B(y)$ is approximately constant around $B_0 = 2$ T within a tolerance of 10%. We wish to measure the bending power of the magnet to better than 1% by using cosmic muons with p_T in excess of 5 GeV. The muon direction at the entrance and at the exit of the slab is measured by two identical stations of drift tubes with angular resolution $\sigma_\phi = 2$ mrad, where ϕ measures the angle on the transverse plane. The muon momentum downstream of the slab is measured by an independent spectrometer of perfect resolution. Determine how many vertical muons are needed to achieve the desired accuracy on the bending power by assuming a cosmic muon spectrum $dN_\mu/dE \propto E^{-\alpha}$ with $\alpha = 2.7$. Does the answer change if the alignment between the two stations cannot be constrained to better than 5 mrad?

Solution

Given the approximate field map, the bending angle (3.43) is about

$$\Delta\phi \lesssim \frac{0.3 \cdot 2 \cdot 0.6}{5} = 72 \text{ mrad}, \qquad (3.70)$$

hence the total track length inside the slab for vertical muons has length d to better than 0.3%, which is an acceptable approximation. The same arguments allows to approximate the trajectory by a parabola. From Eqs. (3.34) and (3.47), it follows that:

$$\frac{d\mathbf{r}}{ds}(s) - \frac{d\mathbf{r}}{ds}(0) = \frac{0.3 \cdot z}{|\mathbf{p}|} \int d\mathbf{r} \times \mathbf{B}(\mathbf{r}), \tag{3.71}$$

where the momentum is measured in GeV, lengths in b, and the magnetic field in T. The z factor can be ± 1 for positive/negative muons. Notice that $|\mathbf{p}|$ is assumed to be constant. We first consider the ideal case that the energy loss is negligible compared to the muon momentum, and then generalise to the case where the curvature decreases throughout the trajectory because of a continuous energy loss.

Since $\mathbf{B} = B(y)\,\mathbf{e}_z$, for almost vertical muons $d\mathbf{r} \times \mathbf{B}(\mathbf{r}) = dy\,B(y)\,\mathbf{e}_\phi$, where $\mathbf{e}_\phi$ is the unit vector lying on the transverse plane and orthogonal to the projection of $d\mathbf{r}$ onto the transverse plane. Projecting both sides along $\mathbf{e}_\phi$, and neglecting terms of order $(\Delta\phi)^2$ we get:

$$p_T\,\Delta\phi = 0.3 \cdot z \int dy\,B(y). \tag{3.72}$$

Equation 3.72 implies that for all muons the product $p_T\,\Delta\phi$ is proportional to the bending power $\int dy\,B(y)$. We can therefore combine the measurements from N independent muon tracks into an estimator of the bending power. Each angular measurement comes with an uncertainty

$$\sigma_{p_T\,\Delta\phi} = p_T\sqrt{2\sigma_\phi^2 + \left(\frac{0.0136}{\beta p_T}\right)^2 \frac{d}{X_0}} \equiv p_T\sqrt{c_1 + \frac{c_2}{p_T^2}} \tag{3.73}$$

where the first term on the right-hand side accounts for the measurement error from each DT station, and the second accounts for multiple-scattering across the iron slab. Given N independent normal measurements with same mean $\langle p_T\Delta\phi\rangle$ and standard deviations σ_i, the variance of the maximum-likelihood estimator is given by

$$\sigma^2_{\langle p_T\Delta\phi\rangle} = \left(\sum_i \frac{1}{\sigma_i^2}\right)^{-1}, \tag{3.74}$$

see Eq. (4.83). For $N \gg 1$, we can approximate the finite sum at the right-hand side of Eq. (3.74) by

$$\sigma^2_{\langle p_T\Delta\phi\rangle} \approx \left(N \int dp_T\,f(p_T)\frac{1}{\sigma^2_{p_T\,\Delta\phi}(p_T)}\right)^{-1}, \tag{3.75}$$

where $f(p_T) \propto dN_\mu/dE$ is the differential distribution of muon energies that arrive at the detector, satisfying the requirement:

$$\int_{p_T^L}^{p_T^H} dp_T \, f(p_T) = 1, \tag{3.76}$$

where p_T^L and p_T^H define the muon momentum acceptance in the sample. We assume here that the detector acceptance is unity for all values of p_T: although this is hardly true in practice, the muon spectrum is so steeply falling that high-p_T events will have a negligible weight in Eq. (3.77). By assuming the power law $dN/dp_T \propto p_T^{-\alpha}$, Eq. (3.77) becomes:

$$\sigma^2_{\langle p_T \Delta\phi \rangle} \approx \left(\frac{N(\alpha - 1)}{p_T^{L\,(1-\alpha)} - p_T^{H\,(1-\alpha)}} \int_{p_T^L}^{p_T^H} dp_T \, \frac{p_T^{-\alpha}}{c_2 + c_1 p_T^2} \right)^{-1}. \tag{3.77}$$

The integral at the right-hand side of Eq. (3.77) can be evaluated numerically, see Appendix 3.5. Using the numerical inputs from the problem ($p_T^L = 5\,\text{GeV}$, $p_T^H = \infty$, $\sigma_\phi = 2\,\text{mrad}$, $X_0 = 1.8\,\text{cm}$), we get:

$$\frac{\sigma_{\langle p_T \Delta\phi \rangle}}{\langle p_T \Delta\phi \rangle_{\text{map}}} \leq 10^{-2} \quad \Rightarrow \quad N \gtrsim 700. \tag{3.78}$$

The statistical scaling of the ML estimator (3.77) is valid if the only uncertainty comes from DT measurement and from MS. In the presence of a systematic misalignment $\varepsilon_\phi = 5\,\text{mrad}$, however, the uncertainty on $\langle p_T \Delta\phi \rangle$ is limited by $p_T^L \varepsilon_\phi/0.3 = 0.08\,\text{T m}$, which alone would give a relative uncertainty on $\mathscr{B}$ of about 7%, hence larger than the target accuracy of 1%. In order to remove the misalignement bias from the measurements, we notice that the combination $\langle p_T \Delta\phi \rangle$ has a sign that depends on the muon charge. Therefore:

$$\frac{\langle p_T \Delta\phi \rangle_+ - \langle p_T \Delta\phi \rangle_-}{2} = \frac{(0.3 \int dy\, B(y) + \langle p_T \varepsilon_\phi \rangle_+) - (-0.3 \int dy\, B(y) + \langle p_T \varepsilon_\phi \rangle_-)}{2}$$
$$= 0.3 \int dy\, B(y) \tag{3.79}$$

where we have assumed that the μ^+ and μ^- spectra are identical. By dividing the sample into a sub-sample of $N/2$ positive muons and one of $N/2$ negative muons, the variance of the new estimator (3.79) is equal to Eq. (3.77) and is now free from the systematic uncertainty due to the alignment.

A word of caution for Eq. (3.71) is required. The assumption that $|\mathbf{p}|$ can be treated as constant during the motion is certainly a valid approximation for tracking in air, but for our case a quick computation shows that the muon energy loss after traversing the slab is a non-negligible fraction of the initial muon energy. Indeed, from Eq. (2.3) with the values from Table 6.1 of Ref. [3] we find:

$$\Delta E \approx 1.45\,\mathrm{MeV\,g^{-1}\,cm^2} \cdot 7.9\,\mathrm{g\,cm^{-3}} \cdot 60\,\mathrm{cm} = 0.69\,\mathrm{GeV}, \tag{3.80}$$

so that neglecting the energy loss may result in a bias on the bending power in excess of the target accuracy of 1%. Since the energy loss per unit length is to a good approximation constant throughout the trajectory, the generalisation of Eq. (3.72) becomes:

$$p_T\,\Delta\phi = 0.3 \cdot z \int dy\, \frac{B(y)}{1 - y\frac{C}{p_T}}, \tag{3.81}$$

where $C = dE/dx$ can be safely assumed to be constant for a given value of p_T. Because of Eq. (3.81), the factor $y\,C/p_T$ can be treated as a small perturbation and the integral can be expanded around $y = 0$ to yield:

$$p_T\,\Delta\phi = 0.3 \cdot z \int_0^d dy\, B(y) \left(1 + y\frac{C}{p_T} + \left(y\frac{C}{p_T} \right)^2 + \dots \right). \tag{3.82}$$

Given that $B(y)$ is constant within 10%, we can estimate the size of the $\mathcal{O}(y^2)$ term:

$$\frac{\int_0^d dy\, B(y) \left(y\frac{C}{p_T} \right)^2}{\int_0^d dy\, B(y)} \approx \frac{(B_0 d)\frac{1}{3}\left(\frac{d\,C}{p_T} \right)^2}{B_0 d} \approx 0.6\%, \tag{3.83}$$

hence we can neglect it since it's below the target resolution. Even with this approximation, we see that the product $p_T\,\Delta\phi$ is no longer proportional to the bending power alone, but involves a combination of the bending power and of the first moment of the magnetic field. Following the assumption of the problem, we can parametrise the field as:

$$B(y) = B_0 + \frac{B_0''}{2} \left(y - \frac{d}{2} \right)^2 + \dots \tag{3.84}$$

Since $\delta B/B_0 \lesssim 10\%$, the fourth and higher-order terms will give increasingly negligible contributions, and we can therefore neglect them. Under this assumption, one can easily verify that

$$\int_0^d dy\, B(y) \left(1 + y\frac{C}{p_T} \right) = \left(\int_0^d dy\, B(y) \right) \left(1 + \frac{1}{2}\frac{C\,d}{p_T} \right). \tag{3.85}$$

The correction at the right-hand side of Eq. (3.85) is about 7% at $p_T = 5\,\mathrm{GeV}$, hence it gives a non-negligible contribution to the bending angle and should not be neglected. Therefore, Eq. (3.72) can be now generalised to:

$$\frac{p_T \, \Delta\phi}{1 + \frac{1}{2}\frac{C\,d}{p_T}} = 0.3 \cdot z \int dy \, B(y). \tag{3.86}$$

Suggested Readings

This exercise is inspited to the work of Ref. [13], where a similar analysis technique has been adopted to calibrate the magnetic field map in the return yoke of the CMS spectrometer.

3.2 Accelerators

Accelerating particles into beams with the desired energy and space-time structure is a primary need in experimental particle physics. Besides being used for fundamental research, the majority of accelerators in use nowadays are devoted to industrial and medical applications. Although numerous sources of natural radiation exist (e.g. radioactive decays, cosmic radiation), the latter are often to weak to be relevant for most of the applications requiring particles. This limitation is even more sever for fundamental research, which is often conducted at the energy and intensity frontier.

Particles accelerators can be classified in two main families depending whether the electric field that provides the acceleration is static or time-dependent. Static accelerators, like the Cockroft-Walton, Ladderton, X-tube, Van de Graaf, etc., are naturally associated with the concept of linear acceleration: particles of charge q accelerate between two terminals kept at a voltage difference V_0 and acquire an extra kinetic energy $E_f = |q\,V_0|$, see Fig. 3.6

The same accelerating field cannot be used again for the same particle, since any static electric field is irrotational:

$$\nabla \times \mathbf{E} = -\frac{\partial \mathbf{B}}{\partial t} \quad \Rightarrow \quad \oint d\mathbf{s} \cdot \mathbf{E} = 0 \;\Leftrightarrow\; \text{no "circular" acceleration if } \mathbf{B} \text{ is static.}$$

$$\tag{3.87}$$

Static accelerators are thus inherently limited in energy by the maximum electric fields V_0 achievable in safe conditions in the laboratory before phenomena like spark formations and disruptive discharges occur. Time-dependent fields evade the bounds imposed by Eq. (3.87), and open the way to high-energy acceleration. This can be realised in both linear and circular fashion. Linear accelerators operated with radio-frequency fields (LINAC) work like a chain of static accelerators of equivalent voltage $V_{\text{r.f.}}$, whose accelerating gradient, as experienced by the particle, is replicated n times in cascade, so that $E_f = n\,V_{\text{r.f.}}$. In circular accelerators, each particle visits several times the same point of the apparatus, receiving a kick $q\,V_{\text{r.f.}}$ at each turn. If the particle motion and the accelerating gradient are maintained with the appropriate

Fig. 3.6 Examples of a Van de Graaf (left) and Cockroft-Walton (right) accelerator hosted at the LNF site

synchronisation, the energy transfer sums up constructively resulting in an overall energy increment. The cyclotron, betatron, and synchrotron are based upon this concept.

The purpose of this section is to make the reader familiar with the different solutions of accelerators, and thus most of the exercises have a rather qualitative solution. Among the few quantitative tests that a non-expert of accelerator physics should certainly know, we propose the calculation of radiation loss in circular and linear colliders.

Problems

Bando n. 13705/2010

Problem 3.13 Compare the energy lost by an electron initially at rest accelerated by a LINAC of length $L = 10$ km long, in which the accelerating field is 20 MV/m, with the energy lost by the same electron when, kept at an energy of 100 GeV, it makes a full round of a circular accelerator of radius 10 km. Does the result change if the particle were a proton?

Discussion

A particle of charge q that is accelerated by external forces (e.g. a magnetic field, a RF wave, etc.) loses energy in the form of electromagnetic radiation. For a classical particle, the power loss per unit solid angle is described by Larmor's formula

$$\frac{dP}{d\Omega} = \frac{e^2}{4\pi c^3}|\dot{\mathbf{v}}|^2 \sin^2\theta, \quad P = \int d\Omega \frac{dP}{d\Omega} = \frac{2}{3}\frac{e^2|\dot{\mathbf{v}}|^2}{c^3}, \quad (3.88)$$

where θ is the polar angle with respect to the particle acceleration $\dot{\mathbf{v}}$. Here, e^2 has units of [kg m^3 t^{-2}], i.e. we assume the Heaviside–Lorentz units such that the first of Maxwell equation reads $\nabla \cdot \mathbf{E} = 4\pi\rho$. The relativistic generalisation of the second of Eq. (3.88) proceeds through the replacement [14]:

$$P = \frac{2}{3}\frac{e^2|\dot{\mathbf{v}}|^2}{c^3} = \frac{2}{3}\frac{e^2}{m^2 c^3}\left(\frac{d\mathbf{p}}{dt}\cdot\frac{d\mathbf{p}}{dt}\right) \rightarrow -\frac{2}{3}\frac{e^2}{m^2 c^3}\left(\frac{dp_\mu}{d\tau}\frac{dp^\mu}{d\tau}\right) = \quad (3.89)$$

$$= \frac{2}{3}\frac{e^2}{m^2 c^3}\left[\left(\frac{d\mathbf{p}}{d\tau}\right)^2 - \beta^2\left(\frac{d|\mathbf{p}|}{d\tau}\right)^2\right] = \frac{2}{3}\frac{e^2}{c}\gamma^6\left[\dot{\boldsymbol{\beta}}^2 - (\boldsymbol{\beta}\times\dot{\boldsymbol{\beta}})^2\right], \quad (3.90)$$

where we have used the fact that $dE = |\mathbf{v}|\,d|\mathbf{p}|$. The last equality has been proved in Problem 1.4. We can now consider two cases: a linear acceleration, i.e. $d\mathbf{p}/dt \propto \mathbf{p}$, and a circular acceleration, i.e. $d\mathbf{p}/dt \cdot \mathbf{p} = 0$.

For a linear accelerator, Eq. (3.89) becomes:

$$P_{\text{lin.}} = \frac{2}{3}\frac{e^2}{m^2 c^3}\gamma^2\left(\frac{d\mathbf{p}}{dt}\right)^2[1-\beta^2] = \frac{2}{3}\frac{e^2}{m^2 c^3}\left(\frac{d\mathbf{p}}{dt}\right)^2 = \frac{2}{3}\frac{e^2}{m^2 c^3}\left(\frac{dE}{dx}\right)^2,$$
$$(3.91)$$

where we have used again the relation $dE = |\mathbf{v}|\,d|\mathbf{p}| \Rightarrow dE/dx = d|\mathbf{p}|/dt$. Since dE/dx is proportional to the gradient of the accelerating field, it is independent of the particle energy and depends only on the external field. The ratio between the power loss and the power supplied by the external accelerating field is therefore:

$$\frac{P_{\text{lin.}}}{(dE/dt)} = \frac{2/3\,(e^2/m^2 c^3)\,(dE/dx)^2}{(dE/dx)\,|\mathbf{v}|} = \frac{2}{3}\frac{e^2(dE/dx)}{m^2 c^4\beta} = \frac{2}{3}\frac{(dE/dx)}{\beta(mc^2/r_c)},$$
$$(3.92)$$

where $r_c = e^2/mc^2$ is the classical particle radius. For an electron, $r_c = 2.8 \times 10^{-13}$ cm, and the denominator at the right-hand side of Eq. (3.92) as about 1.8×10^{12} MeV/cm. An accelerating field able to supply an energy per unit length comparable to this value, the field intensity should of order 10^{12} MV/cm, which is far above the maximum limit attainable in laboratories. Therefore, energy loss in linear accelerators is always negligible.

For a circular accelerator, $d|\mathbf{p}|/dt$ is approximately zero, while $|d\mathbf{p}/dt| = \omega|\mathbf{p}|$, where ω is the revolution frequency Eq. (3.89) becomes:

$$P_{\text{circ.}} = \frac{2}{3} \frac{e^2}{m^2 c^3} \gamma^2 \omega^2 |\mathbf{p}|^2 = \frac{2}{3} \frac{e^2 c}{R^2} \beta^4 \gamma^4, \tag{3.93}$$

where R is the radius of curvature of the circular orbit. The energy lost per orbit is given by

$$\Delta E_{\text{orbit}} = P_{\text{circ.}} \cdot \frac{2\pi}{\omega} = P_{\text{circ.}} \cdot \frac{2\pi R}{\beta c} = \frac{4\pi}{3} \frac{e^2}{R} \beta^3 \gamma^4. \tag{3.94}$$

For an electron and proton beam, Eq. (3.94) takes the numerical values:

$$\begin{aligned}
\frac{\Delta E_{\text{orbit}}}{\text{MeV}} &\approx \frac{4\pi}{3} \frac{1}{137} \frac{197 \,\text{MeV fm}}{R} \left(\frac{E}{\text{GeV}}\right)^4 \frac{1}{(mc^2/\text{GeV})^4} = \\
&= \begin{cases} \frac{8.8 \times 10^{-2}}{(R/m)} \left(\frac{E}{\text{GeV}}\right)^4 & \text{electrons} \\ \frac{7.8 \times 10^{-15}}{(R/m)} \left(\frac{E}{\text{GeV}}\right)^4 & \text{protons} \end{cases}
\end{aligned} \tag{3.95}$$

Solution

Assuming a constant accelerating field of intensity $E = 20 \,\text{MV/m}$ over a distance $L = 10 \,\text{km}$, we see that the electron becomes relativistic already after a few tens of centimeters, so that we can assume $\beta = 1$ since the beginning. Equation (3.91) then gives:

$$\begin{aligned}
\Delta E_{\text{LINAC}} &= P_{\text{lin.}} \cdot \frac{L}{c} = \frac{2}{3} \frac{e^2 L}{m_e^2 c^4} \left(\frac{dE}{dx}\right)^2 = \frac{2}{3} \frac{r_e L}{m_e c^2} \left(\frac{dE}{dx}\right)^2 = \\
&= \frac{2}{3} \frac{2.8 \times 10^{-15} \,\text{m} \cdot 10^4 \,\text{m}}{0.511 \,\text{MeV}} \cdot 400 \frac{\text{MeV}^2}{\text{m}^2} = 1.5 \times 10^{-8} \,\text{MeV} \tag{3.96}
\end{aligned}$$

For the case of circular collider, we can use directly Eq. (3.95) to give:

$$\Delta E_{\text{orbit}} = \frac{8.8 \times 10^{-2}}{(R/m)} \left(\frac{E}{\text{GeV}}\right)^4 \,\text{MeV} = \frac{8.8 \times 10^{-2}}{10^4} (100)^4 = 880 \,\text{MeV} \tag{3.97}$$

Hence, we find that the ratio between the two energy losses is equal to

$$\frac{\Delta E_{\text{LINAC}}}{\Delta E_{\text{orbit}}} \approx 2 \times 10^{-11}. \tag{3.98}$$

If the accelerated were a proton, Eqs. (3.95) and (3.96) would give

$$\begin{cases} \Delta E_{\text{LINAC}} = 8.0 \times 10^{-10} \,\text{MeV} \\ \Delta E_{\text{orbit}} = 7.8 \times 10^{-11} \,\text{MeV} \end{cases} \Rightarrow \frac{\Delta E_{\text{LINAC}}}{\Delta E_{\text{orbit}}} \approx 10, \tag{3.99}$$

hence the LINAC acceleration would result in a larger energy loss.

Suggested Readings

The master reference for this problem is the classical textbook on electrodynamics by Jackson [14].

Bando n. 1N/R3/SUB/2005

Problem 3.14 What is the ratio between the power radiated by a LHC proton and en electron at LEP-I? And between the latter and an electron at DAΦNE?

Solution

Refering to Eq. (3.93), the ratio between the power emitted by beam/accelerator A and the power emitted by a different beam/accelerator B is given by:

$$\frac{P_A}{P_B} = \left(\frac{E_A}{E_B}\right)^4 \left(\frac{m_B}{m_A}\right)^4 \left(\frac{R_B}{R_A}\right)^2 \tag{3.100}$$

The CERN accelerators LEP-I and the LHC are hosted by the same tunnel of length $L \approx 27$ km. The former used to collide electrons/positrons of energy $E \approx 45$ GeV, while the latter collides protons with a maximum beam energy of 7 TeV. The DAΦNE collider located at the Frascati Laboratories collides electrons and positrons at an energy of about $E = m_\phi/2 = 510$ MeV and consists of two circular rings of length $L \approx 100$ m. Therefore:

$$\text{power ratio} = \begin{cases} \left(\frac{45\,\text{GeV}}{7\,\text{TeV}}\right)^4 \left(\frac{938\,\text{MeV}}{0.511\,\text{MeV}}\right)^4 \approx 2 \times 10^4 & \text{LEP-I/LHC} \\ \left(\frac{45\,\text{GeV}}{510\,\text{MeV}}\right)^4 \left(\frac{100\,\text{m}}{27\,\text{km}}\right)^2 \approx 830 & \text{LEP-I/ DAΦNE} \end{cases} \tag{3.101}$$

Energy supply and heat dissipation at LEP were one of the main challenge and ultimately limited the energy reach of LEP-II, see also Problem 3.5.

Suggested Readings

More information on the electron synchrotron DAΦNE, hosted by the LNF, can be found in Ref. [15]

Bando n. 18211/2016

Problem 3.15 How does the power emitted by a relativistic particle of energy E and mass m moving in a circular orbit depend on the ratio E/m? Discuss an application or a consequence of this energy emission.

Solution

The power emitted by a charged particle moving along a circular trajectory in the form of syncrotron radiation is given by Eq. (3.93). In particular, the emitted power scales like $\sim (E/m)^4$. Because of this scaling law, for a fixed radius R and beam energy E, it is more expensive to maintain closed orbits for light particles, like electrons or positrons, rather than for heavier ones, like protons or ions. The challenge is twofold: on the one hand, the lost power needs to be supplied at each turn as to maintain the particle orbit at the ring radius R. This is normally achieved by RF cavities, whose accelerating gradients are however limited in practice by power consumption. On the other hand, the radiation emitted by the particle needs to be removed by a cooling system. The latter can become challenging for accelerators operated at cryogenic temperatures (e.g. the LHC).

On the positive side, the enhanced emission of syncrotron radiation by lighter particles finds technological applications, e.g. for the production of syncrotron light as a diagnostic tool (like X-ray spectroscopy). Another advantage of syncrotron radiation is the natural development of a beam polarisation orthogonal to the orbital plane [14] and a reduced beam energy spread, which make electron beams appealing for a number of research applications.

Suggested Readings

The synchrotron radiation is discussed in detail in Chap. 14 of Ref. [14]. For a reference on beam polarisation at LEP and its applications, see e.g. Ref. [16].

Bando n. 5N/R3/TEC/2005

Problem 3.16 What is the role of a magnetic quadrupole in an accelerator?

Discussion

A magnetic quadrupole can be obtained by four dipoles, rotated by $90°$ one from the other, and with alternated polarities (e.g. N-S-N-S). If the dipoles are aligned along the bisectrix of the $x-y$ plane, symmetry dictates the magnetic field and Lorentz force experienced by a charged particle moving along the z-direction at the centre of the quadrupole to given by:

$$\begin{cases} B_x(x, y) = \frac{\partial B_y}{\partial x} y \\ B_y(x, y) = \frac{\partial B_y}{\partial x} x \end{cases} \Rightarrow \begin{cases} F_x(x, y) = -\frac{q|\mathbf{p}|}{m\gamma} \frac{\partial B_y}{\partial x} x \\ F_y(x, y) = +\frac{q|\mathbf{p}|}{m\gamma} \frac{\partial B_y}{\partial x} y \end{cases} \tag{3.102}$$

where the derivatives are computed at the centre of the quadrupole. The *quadrupole strength* is defined as

$$K \equiv \frac{1}{|\mathbf{B}|\rho} \frac{\partial B_y}{\partial x}, \tag{3.103}$$

which is the field derivative normalised to the beam rigidity $|\mathbf{B}|\rho$. Equation (3.102) implies that the force exerted by a quadrupole is focusing in one direction and defocusing on the orthogonal direction, depending on the sign of the derivative, or, equivalently, of the quadrupole strength. It can be proved that a series of identical quadrupoles of length ℓ, but of alternated polarity, i.e. rotated by $90°$, guarantees

Fig. 3.7 Example of a $FODO$ cell hosted at the LNF site: a magnetic dipole is followed by a pair of quadrupoles with alternated polarity, and by a RF cavity

stability to the beam provided that the distance between two quadrupoles is less than $2f$, where $f = (\ell K)^{-1}$ is the focal length of the quadrupole (*strong focusing*). The basic element of this lattice is called a $FODO$ cell, since it consists of a focusing (F) and defocusing (D) quadrupole, separated by a region of no drift (O), for example provided by a dipole, see Fig. 3.7.

Modern synchrothrons are built upon the principle of strong focusing.

Solution

Owing to the focusing/defocusing property of quadrupole fields, see Eq. (3.102), the latter find application as the building bricks of the magnetic lattice needed to maintain the beams in stable circular orbits. Another application of quadrupoles is as the last beam focusing elements in the proximity of the interaction point, where the minimum beam size is usually desired.

Suggested Readings

For an introduction to the physics of accelerators, the reader is addressed to Ref. [17]. See also Sect. 30 of Ref. [3].

Bando n. 5N/R3/TEC/2005

Problem 3.17 Define the emittance of a beam. What is the normalised emittance ε_n? What is the relation between emittance and brightness?

Discussion

The motion of a particle of momentum $|\mathbf{p}|$ and charge q, moving in a circular collider whose lattice consists of a series of focusing/defocusing and dipole elements, can be described the terms of the displacements x, y, and z with respect to the ideal trajectory (a circle of radius ρ). Since these displacements are very small compared to the radius ρ, the problem can be linearised, resulting in a system of differential equations, which are instances of Hills equation. The co-rotating reference frame moves along the ideal trajectory and the position of the centre is parametrised by the path length variable s. The x-coordinate is usually taken as the outgoing radial direction of the co-rotating frame, the y-coordinate as the vertical direction, and the z-direction is tangent to the ideal trajectory. The linearised equations read as [3]

$$\begin{cases} x'' + K_x(s)\, x = 0, & K_x(s) = +\frac{1}{|\mathbf{B}|\rho}\frac{\partial B_z}{\partial x} + \frac{1}{\rho^2} \\ y'' + K_y(s)\, y = 0, & K_y(s) = -\frac{1}{|\mathbf{B}|\rho}\frac{\partial B_z}{\partial x} \\ z' + x/\rho = 0 \end{cases} \qquad (3.104)$$

where the quadrupole strength K of Eq. (3.103) has been used to account for the focusing/defocusing effect of quadrupoles, ρ is the radius of curvature of the ideal trajectory, and $|\mathbf{B}|$ is the field intensity of the dipoles. The first two equations in (3.104) can be proved as follows. Consider for example the y displacement. When the particle moves inside the quadrupole along an infinitesimal length ds, the bending angle (3.43) along y is given by:

$$d\theta_y = d\left(\frac{dy}{ds}\right) = \frac{(\partial_x B_z\, y)\cdot ds}{|\mathbf{B}|\,\rho} \equiv -K_y\, y \quad \Rightarrow \quad y'' + K_y\, y = 0 \qquad (3.105)$$

where the magnetic field component in the y direction is given by Eq. (3.102). For the x displacement, a similar equation hold with inverted sign of K, as to reflect the opposite behaviour of quadrupoles along orthogonal directions. Even in the absence of quadrupoles, the transverse motion along x features equilibrium oscillations around the ideal trajectory with wavenumber ρ. For example, consider the case that the trajectory differs from the ideal one by a constant displacement δ: the x displacement as a function of s is a projection of the displacement along a rotating radial vector: $x = |\delta|\cos(s/\rho + \phi_0)$, or in terms of infinitesimal variations, $d(dx/ds) = -(x\,ds)/\rho^2$. The equation for the longitudinal motion can be instead proved by noticing that a displacement x from the ideal orbit will make the particle position along z move forward, or, backward, by $-(x/\rho)\,ds$, hence $z' = -x/\rho$.

The transverse motion consists of two decoupled equations. The solution can be written in the form

$$x(s) = \sqrt{\varepsilon_x\,\beta_x(s)}\,\cos(\psi_x(s)), \qquad (3.106)$$

$$\text{with}: \quad \psi_x' = \frac{1}{\beta_x}, \quad \text{and} \quad 2\beta_x\,\beta_x'' - \beta_x'^2 + 4\beta_x^2\, K_x(s) = 4, \qquad (3.107)$$

and similar for y. Here, ε_x is constant and is fixed by the initial conditions. The *amplitude functions* $\beta_{x,y}$ are periodic functions of s of period $2\pi\rho$. The play both the role of an amplitude envelope along the circumference (the maximum displacement is bound by $|x| \leq \max_s \sqrt{\varepsilon_x \, \beta_x(s)})$, and of local wavelength. The phase ψ_x at the position s along the circumference is given by

$$\psi_x(s) = \psi_x^0 + \int_0^s \frac{ds}{\beta_x}, \tag{3.108}$$

After completing an integer number or turns, i.e. $s = k \, 2\pi\rho$, the phase has advanced by a number which is in general not a multiple of 2π, thus implying that the particle won't necessarily find itself in the same displaced position after a full turn. After some algebra, one can easily show that

$$\left[\frac{1 + (-\beta_x'/2)^2}{\beta_x} \right] x^2 + 2 \left(-\frac{\beta_x'}{2} \right) x \, x' + \beta_x \, x'^2 = \varepsilon_x \tag{3.109}$$

In terms of the phase-space variables (x, x'), the motion is takes place along an ellipse of area $\pi\varepsilon$, whose principal axes are continuously stretched and rotated as s changes. In particular, when $\beta_x' = 0$, the equation takes the nice form:

$$\beta_x' = 0 \quad \Rightarrow \quad \frac{x^2}{\beta_x} + \beta_x \, x'^2 = \varepsilon_x, \tag{3.110}$$

i.e. the semi-axes have lengths $\Delta x = \sqrt{\varepsilon_x \, \beta_x}$ and $\Delta x' = \sqrt{\varepsilon_x/\beta_x}$. This condition is for example realised at the interaction points of circular colliders as to ensure the smallest beam sizes in the transverse plane.

Solution

When the beam is made of $N \gg 1$ non-interacting particles, each particle will obey Eq. (3.104) with a value of the invariant $\varepsilon_{x,y}$ and of the initial phase $\phi_{x,y}^0$ sampled from a statistical distribution. IUnder the assumption that the beam dynamics is described by a conserved Hamiltonian, which is the case if the beam particles are weakly interacting and if the beam radiation is negligible, the Hamiltonian flow conserves the phase-space volume $\int p_{x,y} \, dq_{x,y}$, where (p, q) are coniugate variables (Liouville theorem). Since the latter is proportional to $\int dx \, x'$ and $\int dy \, y'$, respectively, the area $\varepsilon_0^{x,y}$ in the displacement-divergence plane that contain a certain fraction of the beam particles, is also a time-invariant. The latter is called *beam emittance*, and is measured in units of (mm mrad). If the particle distribution in these planes is gaussian, one can define the transverse beam sizes $\sigma_{x,y}$ as the lengths that contain a certain number of standard deviations of the phase-space distribution. For example, for a 87% fraction of the beam particles, one has:

$$\varepsilon_0^{x,y} = \frac{(2\,\sigma_{x,y}^2)}{\beta_{x,y}}, \tag{3.111}$$

where now $\sigma_{x,y}$ give the beam sizes in the displacement spaces (x, y).

In the relativistic limit, Liouville theorem states that $\int p \, dq \sim \beta\gamma \int dx \, x'$ is conserved, therefore the beam emittance decreases like γ^{-1} for $\gamma \gg 1$. The *normalised emittance* is the true time-invariant along the accelerator chain, and can be defined as the beam emittance for $\gamma = 1$, i.e.:

$$\varepsilon_n = \gamma \, \varepsilon_0. \qquad (3.112)$$

When a beam is created and injected into the accelerating stage, its emittance is ε_n. After the acceleration, the emittance gets shrunk by a factor γ. The *brightness B* of a beam is defined as the particle multiplicity per unit of emittance, i.e.:

$$B = \frac{N}{\varepsilon_0}. \qquad (3.113)$$

Suggested Readings

For an overview on the physics of accelerators, the reader is addressed to Ref. [17]. A summary of the relevant notation can be found in the dedicated PDG review [3].

Problem 3.18 Determine the functional form of the amplitude function $\beta(s)$ in a region free of magnetic fields around the interaction point of a circular collider.

Solution

As discussed in Problem 3.17, in order to maximise the luminosity, the beam optics is adjusted such that the amplitude function is at a minimum at the interaction point (IP). We chose the coordinate system such that $s = 0$ corresponds to the IP position, so that $\beta'(0) = 0$. From Eq. (3.106), we can write the ODE for $\beta(s)$ in a region free of magnetic field by setting $K(s) = 0$. For simplicity of notation, we suppress the coordinate index. We then have:

$$2\beta\beta'' - \beta'^2 - 4 = 0. \qquad (3.114)$$

To solve this ODE, we make the *ansatz*:

$$\beta(s) = \beta^* + \alpha^* s^2, \qquad (3.115)$$

where $\beta^* = \beta(0)$. Actually, the α^* parameter is not independent from β^*, since Eq. (3.114) with the boundary condition $\beta'(0) = 0$ implies that $\beta''(0) = 2/\beta^*$, hence $\alpha^* = 1/\beta^*$. Equation (3.114) then becomes:

$$2\left(\beta^* + \frac{s^2}{\beta^*}\right)\frac{2}{\beta^*} - \left(\frac{2s}{\beta^*}\right)^2 - 4 = 0. \qquad (3.116)$$

Hence, the solution is indeed given by:

$$\beta(s) = \beta^* + \frac{s^2}{\beta^*},\qquad(3.117)$$

which is a parabola in s, whose constant terms and curvature at the origin are not independent.

Discussion

We can notice that at large distance from the IP, the amplitude function grows quadratically with s, while the beam RMS increases linearly, see Eq. (3.111). The increase of the transverse beam size away from the IP is called *hourglass effect*. By reducing the amplitude function at the interaction point (β^*), the beam size at the centre will shrink, while the size at the boundary of the luminous region will increase proportionally, thus reducing its contribution to the overall luminosity. The net effect depends on how the longitudinal beam size compares to β^*. An an example, the design β^* at the LHC interaction points is about 0.5 m.

Problem 3.19 Consider a LHC proton beam at $\sqrt{s} = 14\,\text{TeV}$. Estimate the size of the proton transverse momentum at the interaction point of the ring by assuming $\varepsilon_n = 3.75\,\mu\text{m}$ and $\beta^* = 0.55\,\text{m}$. How does this number compare with the typical quark transverse momentum of order $\hbar/2r_p \sim 2.5\,\text{GeV}$? And with the transverse momentum due to a beam-beam crossing angle $\mu = 300\,\mu\text{rad}$?

Solution

The interaction points correspond to local minima of the amplitude function $\beta(s)$. From Eq. (3.110), it follows that the phase-space points of the beam particles in the displacement-divergence plane lie along ellipses with semiaxes $\sqrt{\varepsilon\,\beta^*}$ and $\sqrt{\varepsilon/\beta^*}$, respectively. The divergence at the IP is maximal and is given by

$$x' = \sqrt{\frac{\varepsilon}{\beta^*}} = \sqrt{\frac{\varepsilon_n}{\gamma\,\beta^*}} \approx \sqrt{\frac{3.75\,\mu\text{m}}{7\times10^3\cdot0.55\,\text{m}}} = 3\times10^{-5},\qquad(3.118)$$

corresponding to $p_T = x'|\mathbf{p}| \approx 200\,\text{MeV}$, i.e. approximately one order of magnitude smaller than the typical transverse momentum of bounded quarks inside the proton. When the two beams have a finite crossing angle ϕ (assuming an angle of $\phi/2$ of each beam direction with respect to the z-axis), the proton transverse momentum is given by $|\mathbf{p}|\phi/2 \approx 1\,\text{GeV}$.

Bando n. 5N/R3/TEC/2005

Problem 3.20 What are the betratron oscillations?

Solution

The betatron oscillation, thus called because they were first observed in betatrons, see Problem 3.22, are the oscillation of beam particles in the transverse plane around the reference trajectory, i.e. the trajectory of the ideal particle that undergoes a perfectly circular and closed path, see Problem 3.17. These oscillations are described by the amplitude functions $\beta_{x,y}(s)$, which provides an envelope to the particle displacements from the ideal trajectory and plays the role of a local wavelength of the transverse motion. The number of cycles in the two transverse directions per turn,

$$Q_{x,y} = \frac{1}{2\pi} \oint \frac{ds}{\beta_{x,y}(s)}, \tag{3.119}$$

are called *beam tunes*. If the beam tunes are rational numbers, i.e. $p\, Q_{x,y} = q$ with p and q integers, then the particles find themselves in the same position after a certain number of turns. This condition is to be avoided to prevent local defects of the lattice to sum up constructively, a situation that can cause beam instabilities.

Bando n. 13153/2009

Problem 3.21 What is the cyclotron frequency?

Solution

Th cyclotron frequency for a charged particle of mass m and electric charge e moving on a plane orthogonal to a uniform magnetic field of intensity $|\mathbf{B}|$ is defined as $\omega_B = e\,|\mathbf{B}|/m$. From the equation of motion of Eq. (3.32), it follows that such a particle completes a full revolution in a time $T = 2\pi/\omega$, where the revolution frequency ω is given by:

$$\frac{d\mathbf{p}}{dt} = \frac{e}{m\gamma} \mathbf{p} \times \mathbf{B} = \frac{e\,|\mathbf{B}|}{m\,\gamma} |\mathbf{p}|\, \mathbf{e}_r,$$

$$\frac{d\mathbf{p}}{dt} \cdot \mathbf{e}_r \equiv \omega\,|\mathbf{p}| = \frac{e\,|\mathbf{B}|}{m\,\gamma} |\mathbf{p}| \quad \Rightarrow \quad \omega = \frac{e\,|\mathbf{B}|}{m\,\gamma} = \frac{\omega_B}{\gamma}. \tag{3.120}$$

Therefore, the cyclotron frequency ω_B coincides with the revolution frequency ω of a charged particle moving in an orthogonal magnetic field, provided that the motion is not relativistic, i.e. $\gamma \approx 1$. At relativistic energies, the revolution frequency decreases. This can be intuitively understood by noticing that the the the length of the circumference scales linearly with the particle momentum, see Eq. (3.40), but the velocity saturates to c, so that the revolution period scales like $\sim|\mathbf{p}|$.

Bando n. 5N/R3/TEC/2005

Problem 3.22 Explain how the magnetic field intensity, the radiofrequency, and the orbital radius change during the acceleration of a proton in:

1. a cyclotron;
2. a betatron;
3. a synchrotron.

Discussion

The cyclotron, betatron, and synchrotron are three different kinds of non-static circular accelerators, where charged particles are progressively accelerated and maintained in orbit by means of guide magnetic fields. Historically, the cyclotron has been the first accelerator used for particle physics, although the idea behind the betatron is even older (Wideröe, 1919). The synchrotron represents the historical evolution of the cyclotron and is, as of today, the adopted solution for high-energy accelerators.

Solution

In a cyclotron, particles are accelerated by time-varying electric fields between the faces of two D-shaped magnets, where an orthogonal and constant magnetic field bends the particles by 180°. The polarity of the electric field in inverted at each half-turn. The revolution frequency is given by Eq. (3.120), hence it is approximately independent of the particle energy for $\beta \ll 1$: for non-relativistic particle, the frequency of the accelerating potential is constant. At higher energies, the frequency needs to decrease in order to be always in phase with the arrival of the particle. Furthermore, since the radius of curvature changes proportionally to the momentum, the radius of the D-shaped magnets determines the maximum energy. Therefore, we can summarise:

- B: constant;
- R: time-dependent, it grows linearly with the particle momentum;
- f: constant for $\beta \ll 1$, time-dependent otherwise.

In a betatron, time-varying magnetic fields provide both the guide and the acceleration. In its simplest implementation, it consists of a yoke encircled by a toroidal beam tube of fixed radius. Particles inside the tube are accelerated by a rotational electric field generated by the time varying magnetic field in the yoke:

$$\nabla \times \mathbf{E} = -\frac{\partial \mathbf{B}_a}{\partial t} \quad \Rightarrow \quad |\mathbf{E}| = \frac{\dot{\Phi}(t)}{2\pi R} = \frac{\pi R^2 \langle |\dot{\mathbf{B}}_a| \rangle}{2\pi R} = \frac{R}{2} \langle |\dot{\mathbf{B}}_a| \rangle, \qquad (3.121)$$

$$\text{with} \quad \langle |\dot{\mathbf{B}}_a| \rangle = \frac{1}{\pi R^2} \int d\boldsymbol{\sigma} \cdot \dot{\mathbf{B}}_a \qquad (3.122)$$

A guide field $\mathbf{B}_g$ orthogonal to the beam plane, will maintain the beam at a fixed radius R provided that

$$\dot{\mathbf{p}} = q\,|\dot{\mathbf{B}}_g|\,R = q\,|\mathbf{E}| = q\,\frac{R}{2}\,\langle|\dot{\mathbf{B}}_a|\rangle \quad \Rightarrow \quad |\mathbf{B}_g| = \frac{1}{2}\left(\frac{1}{\pi\,R^2}\int d\boldsymbol{\sigma}\cdot\mathbf{B}_a\right) \quad (3.123)$$

This implies a proportion of 2 : 1 between the accelerating and guide fields. There-fore, we can summarise:

- B: time-dependent;
- R: constant;
- f: constant.

In a synchrotron, a toroidal beam tube is immersed in a guide magnetic field generated by dipoles located all around the tube. During the acceleration stage, the beam energy is progressively incremented by radiofrequency (RF) cavities synchronised with the arrival of the particles. The guide field is incremented as to catch up with the increasing beam energy. When the ultimate beam energy is reached, the field is maintained at a constant value. Therefore, we can summarise:

- B: time-dependent, it grows linearly with the particle momentum until the maximum beam energy is reached;
- R: constant;
- f: time-dependent, it must be a multiple of the beam revolution frequency. Approximately constant for $\gamma \gg 1$.

Suggested Readings

For an introduction to the physics of accelerators, the reader is addressed to Ref. [17].

Bando n. 18211/2016

Problem 3.23 What are the main differences between a cyclotron and a synchrotron? What factors do limit the maximum energy achievable by each of them?

Solution

The main features of cyclotrons and synchrotrons have been discussed in Problem 3.22. In a cyclotron, the radius of the accelerated beam changes proportionally with the beam momentum, the magnetic field is constant, and the radiofrequency is approximately constant for non-relativistic particles. The maximum beam energy is limited by the outer radius of the magnets and by the necessity of maintaining the RF in phase with the revolution frequency. In a synchrotron, the radius if fixed, while the magnetic field and the accelerating RF change with time until the largest energy is achieved. For a given radius R, the latter is mostly limited by the maximum intensity of the guide field. For electron/positron acceleration, the limiting factor may come instead from the synchrotron radiation, see Problem 3.13.

Suggested Readings

For an introduction to the physics of accelerators, the reader is addressed to Ref. [17].

Bando n. 18211/2016

Problem 3.24 Discuss a way to excite wakes in plasma acceleration? What would be the advantages of a plasma accelerator compared to traditional acceleration techniques?

Discussion

Wake-field acceleration is a technique for charged particle acceleration consisting in using the electric field induced by the passage of a moving charge (wakefield) as driving force to accelerate a beam of particles.

Solution

Laser plasma acceleration consists in the acceleration of charged particles by the wakefield induced by a resonant short-pulse (pulse duration of order of the plasma frequency, see Eq. (2.22)), high-intensity laser source in an underdense plasma ($n_e \sim 10^{17}$ cm^{-3}). The phase velocity of the plasma wave is approximately given by the group velocity of the laser. The main advantage of LPA is the intense accelerating field, which can exceed the maximum electric fields of traditional RF cavities by about three orders of magnitude. For example, for an ambient electron density of 10^{17} cm^{-3}, the accelerating plasma electric field is of order

$$E = 96 \, (n_e/\text{cm}^{-3})^{1/2} \, \text{V/m} \approx 30 \, \text{GV/m}, \tag{3.124}$$

to be compared with the RF electric fields of about 30 MV/m.

The largest accelerating gradients achievable by LPA would considerably reduce the length and costs of the accelerating facilities. The main challenge are represented by the maximum luminosity achievable and by the overall power efficiency (luminosity as a function of electrical input power).

Suggested Readings

The state of the art in LPA technology has been summarised in the IFCA White Paper [18].

Bando n. 5N/R3/TEC/2005

Problem 3.25 The bremsstrahlung spectrum emitted by a X-ray tube for medical applications features a Gaussian-like shape, with a variety of peaks superimposed. How can you explain these peaks?

Solution

X-ray tubes are devices where electrons are accelerated through an electric potential and impinge on a metallic anode emitting bremsstrahlung radiation. The spectrum of the emitted radiation is a continuous distribution, modulated at low frequency from the absorption by dead material, featuring a number of peaks superimposed. The latter correspond to excitation of the atomic levels of the atoms at the anode (K, L, M levels).

Bando n. 18211/2016

Problem 3.26 What is the highest energy of an α particle produced by a 5 MV tandem accelerator?

Discussion

The concept of Tandem accelerator was invented in order to achieve higher beam energies compared to single-ended accelerators at the same maximum terminal voltage V. In a Tandem accelerator, negative ions of charge $-e$ are produced at one electrode, usually grounded or connected to a negative potential $-V_I$, and are accelerated towards the high-voltage terminal. Here, they encounter a charge stripper (usually made of gas or of a thin carbon foil), where they get ionised, becoming positive ions of charge $+z\,e$. At this point, they accelerate again towards another grounded electrode where they are finally extracted (usually through the use of a magnet analyser). The kinetic energy acquired by the ion when reaching the final electrode is therefore

$$T = e\,V_I + e\,V + e\,z\,V = e\,V_I + e\,(1+z)\,V, \qquad (3.125)$$

where $-V_I$ is the potential of the ion source.

Solution

For a He source, the maximum ion charge after the stripper is $z = +2$. From Eq. (3.125) with $V = 5$ MV and $V_I = 0$, we therefore obtain a maximum energy

$$T_{\max} = 3\,e\,V = 15 \text{ MeV}, \qquad (3.126)$$

hence three times large than by using a single-ended accelerator at the same terminal voltage.

Suggested Readings

An instructive overview on electrostatic accelerators and on the concept of Tandem acceleration is presented in the lectures [19]. For more detailed information on negative ion beams, charge strippers, and technological aspects of electrostatic accelerators, the reader is addressed to Ref. [11].

3.3 Luminosity and Event Rates

The concept of luminosity is of cardinal importance in accelerator and collider physics. As discussed in Sect. 1.3, the luminosity of a collider determines the event rate. More specifically, the luminosity is defined as the coefficient of proportionality between the measured event rate and the cross section. Assuming that the interacting particles are prepared in beams of fixed momentum, as it is usually the case at colliders, Eq. (1.288) implies:

$$\mathscr{L}(t) = v_{\rm rel} \int d\mathbf{r} \, [n_1(\mathbf{r}, t) \, n_2(\mathbf{r}, t)] \tag{3.127}$$

where $v_{\rm rel}$ is the relative velocity, see Problem 1.11, and $n_{1,2}(\mathbf{r}, t)$ are the particle densities of the colliding beams. The integration is performed over the volume around the interaction point. We can specialise Eq. (3.127) to a few special cases.

Continuous Beam Against a Fixed Target

Let's assume that a direct current beam (1) with particle velocity $\mathbf{v}_1$ is directed onto a fixed target (2), see Fig. 3.8. We chose the reference frame such that the beam direction is aligned along the z-axis, while the x, y axes span the transverse plane. The beam distribution along z is assumed to be uniform. The target is at rest and is characterised by a particle distribution $n_2(\mathbf{r})$. The relative velocity is $v_{\rm rel} = |\mathbf{v}_1|$. Equation (3.127) becomes

$$\mathscr{L} = |\mathbf{v}_1| \int dx\,dy\,dz\, n_1(x, y)\, n_2(x, y, z) = |\mathbf{v}_1| \int dx\,dy\, n_1(x, y)\, \delta_2(x, y) =$$
$$= \int d\Phi_1(x, y)\, \delta_2(x, y), \tag{3.128}$$

where $\delta_2(x, y)$ is the surface density of the target at the point (x, y) while $d\Phi_1$ is the beam flux (number of particles per unit time) across an infinitesimal surface element $dx\,dy$ centred around (x, y). For the special case of uniform target density and uniform beam flux over a limited area, Eq. (3.134) becomes:

$$\mathscr{L} = \Phi_1\, \delta_2, \tag{3.129}$$

Fig. 3.8 Illustration of a continuous beam of density $n_1(\mathbf{r})$ and velocity $\mathbf{v}_1$ colliding against a fixed target of density $n_2(\mathbf{r})$

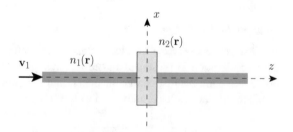

Fig. 3.9 Illustration of a bunched beam of bunch density $n_1(\mathbf{r})$, velocity $\mathbf{v}_1$, colliding with a fixed target of density $n_2(\mathbf{r})$ with a collision frequency f_{coll}

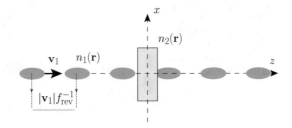

where Φ_1 is the beam flux across the surface of minimum area between the beam and the target transverse size. This result is in agreement with Eq. (1.290).

Bunched Beam Against a Fixed Target

Let's assume that a bunched beam (1) with bunch velocity $\mathbf{v}_1$ is directed against a fixed target (2), see Fig. 3.9. We further assume that the bunches are equally spaced in time, so that the frequency f_{coll} of bunch collision with the target is a constant. Equation (3.127) becomes

$$\mathcal{L}(t) = |\mathbf{v}_1| \int dx\, dy\, dz\, n_1(x, y, z, t)\, n_2(x, y, z) \qquad (3.130)$$

Strictly speaking, the luminosity here is a function of time. For example, it's zero before any bunch collision, and non-zero during the collision. We can however re-define the luminosity as the time average of Eq. (3.127) over many bunches N_b, i.e.

$$\mathcal{L} \equiv \frac{f_{\text{coll}}}{N_b} \int_0^{N_b\, f_{\text{coll}}^{-1}} dt\, \mathcal{L}(t) = f_{\text{coll}} |\mathbf{v}_1| \int dx\, dy\, dz\, dt\, n_1(x, y, z, t)\, n_2(x, y, z) =$$
$$= f_{\text{coll}} \int dx\, dy\, dz\, ds\, n_1(x, y, z, s)\, n_2(x, y, z). \qquad (3.131)$$

The second equality holds if the beam structure is periodic with time period f_{coll}^{-1}: in this case, it suffices to integrate the particle density over a time large enough to contain the full bunch crossing through the target. In the last equality, the variable $s = |\mathbf{v}_1|\, t$ has been introduced to make the integration variables homogeneous. Since the motion takes place along the z-axis, the density n_1 must be a function of $z - s$, so that:

$$\mathcal{L} = f_{\text{coll}} \int dx\, dy\, \left(\int ds\, n_1'(x, y, z - s) \right) \left(\int dz\, n_2(x, y, z) \right)$$
$$= f_{\text{coll}} \int dx\, dy\, \delta_1(x, y)\, \delta_2(x, y) \qquad (3.132)$$

For the special case of uniform target and bunch density over a limited area, Eq. (3.140) becomes:

Fig. 3.10 Illustration of two bunched beams of bunch density $n_1(\mathbf{r})$ and $n_2(\mathbf{r})$, and velocities $\mathbf{v}_1$ and $\mathbf{v}_2$, colliding head-on with a collision frequency f_{coll}

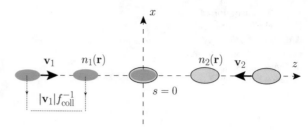

$$\mathscr{L} = f_{\mathrm{coll}} \, N_1 \, \delta_2, \tag{3.133}$$

where N_1 is the number of particles per bunch. Again, if the transverse size of the target is smaller than the bunch, one should consider only those particles inside the smallest of the two areas.

Bunched Beam Against a Bunched Beam

Let's assume that two bunched beams (1) and (2) with bunch velocity $\mathbf{v}_1$ and $\mathbf{v}_2$ collide one against the other, see Fig. 3.10. We further assume that the bunches are equally spaced in time, so that f_{coll} is a constant. Since the beam structure is periodic, we shall consider the time integral of $\mathscr{L}(t)$ for two bunches, and divide the result by the time interval between two subsequent bunch crossings. Writing explicitly v_{rel} in terms of the beam velocities:

$$\mathscr{L} = f_{\mathrm{coll}} \sqrt{(\mathbf{v}_1 - \mathbf{v}_2)^2 - \frac{(\mathbf{v}_1 \times \mathbf{v}_2)^2}{c^2}} \int dx \, dy \, dz \, dt \, n_1(\mathbf{r}, t) \, n_2(\mathbf{r}, t), \tag{3.134}$$

We can further simplify the expression by introducing the *crossing angle* ϕ, defined such that the angle between the two velocities is $\pi - \phi$. We further assume that the beam are relativistic, so that $|\mathbf{v}_1| = |\mathbf{v}_2| \approx c$. With this choice:

$$v_{\mathrm{rel}} = c \sqrt{\left(2 \cos \frac{\phi}{2}\right)^2 - \sin^2 \phi} = c \, (1 + \cos \phi) = 2c \cos^2 \frac{\phi}{2} \tag{3.135}$$

For back-to-back collisions, $\phi = 0$ and we obtain the expected result $v_{\mathrm{rel}} = 2c$. Putting everything together, we have:

$$\mathscr{L} = 2 \cos^2 \frac{\phi}{2} \, f_{\mathrm{coll}} \int dx \, dy \, dz \, ds \, n_1(\mathbf{r}, s) \, n_2(\mathbf{r}, s), \tag{3.136}$$

where we have again introduced the variable $s = c\,t$.

Integrated Luminosity

Since the statistical significance of a measurement grows like $\sqrt{N}$, at the end of the day what matters for the physics is the *integrated luminosity* defined as:

$$\mathscr{L}_{\text{int}} = \int_0^{T_{\text{run}}} dt \, \mathscr{L}(t) \equiv T_{\text{run}} \langle \mathscr{L} \rangle, \tag{3.137}$$

where the integration is performed over the full run time T_{run}. The latter is limited by other contingent factors, like the availability of the accelerating facility for the desired research program, power supply limitations, turn-around vs beam life times, etc. For this reason, the maximisation of $\langle \mathscr{L} \rangle$ to reduce the overall run time is an important achievement for an accelerator, and much effort is usually devoted to this task.

This section collects a number of exercises that aim at familiarising with the concept of luminosity. Particular emphasis is devoted to the luminosity of circular collider, where a number of exact calculations can be performed for several cases of interest.

Problems

Problem 3.27 Prove that the luminosity of a circular collider where two equally spaced and bunched beams with N_b bunches per beam, revolution frequency f_{rev}, bunch populations N_1 and N_2, Gaussian beam profile in three dimensions with standard deviations σ_x, σ_y, σ_z, and colliding head-on along the z-direction, is given by the formula:

$$\mathscr{L} = \frac{N_1 \, N_2 \, f_{\text{rev}} \, N_b}{4\pi \, \sigma_x \, \sigma_y}. \tag{3.138}$$

Generalise this expression to the case where the standard deviations of the two beams are different.

Solution

The luminosity can be computed from Eq. (3.136). To parametrise the bunch densities, we chose the reference frame such that the bunches move along the z-axis, while the transverse plane is spanned by the x- and y-axes. We further assume that the centre of each bunch passes through the origin at the time $s = 0$. With this choice:

$$\begin{cases} \rho_1(x, y, z, s) = \frac{(2\pi)^{-\frac{3}{2}}}{\sigma_x \sigma_y \sigma_z} \exp\left[-\frac{x^2}{2\sigma_x^2} - \frac{y^2}{2\sigma_y^2} - \frac{(z-s)^2}{2\sigma_z^2} \right] \\ \rho_2(x, y, z, s) = \frac{(2\pi)^{-\frac{3}{2}}}{\sigma_x \sigma_y \sigma_z} \exp\left[-\frac{x^2}{2\sigma_x^2} - \frac{y^2}{2\sigma_y^2} - \frac{(z+s)^2}{2\sigma_z^2} \right] \end{cases} \tag{3.139}$$

Assuming $\phi = 0$ and relativistic bunches, the integral at the right-hand side of Eq. (3.136) can be computed explicitly yielding:

$$\iint ds \, d\mathbf{r} \, [\rho_1(\mathbf{r}, s) \, \rho_2(\mathbf{r}, s)] =$$

$$= \frac{(2\pi)^{-3}}{\sigma_x^2 \, \sigma_y^2 \, \sigma_z^2} \int dx \, \exp\left[-\frac{x^2}{\sigma_x^2} \right] \int dy \, \exp\left[-\frac{y^2}{\sigma_y^2} \right] \int dz \, \exp\left[-\frac{z^2}{\sigma_z^2} \right] \int ds \, \exp\left[-\frac{s^2}{\sigma_z^2} \right]$$

$$= \frac{(2\pi)^{-3}}{\sigma_x^2 \, \sigma_y^2 \, \sigma_z^2} (\sqrt{\pi}\sigma_x)(\sqrt{\pi}\sigma_y)(\sqrt{\pi}\sigma_z)^2 = \frac{1}{8\pi \, \sigma_x \, \sigma_y} \tag{3.140}$$

Equation (3.136) then becomes:

$$\mathscr{L} = 2\, N_b\, f_{rev}\, N_1\, N_2 \left(\frac{1}{8\pi\, \sigma_x\, \sigma_y} \right) = \frac{N_1\, N_2\, f_{rev}\, N_b}{4\pi\, \sigma_x\, \sigma_y}. \tag{3.141}$$

It is interesting to notice that under these assumptions, the luminosity is independent of the bunch resolution along the direction of collision. Using Eqs. (3.111) and (3.113), we can equivalently write the luminosity in a form that makes explicit its dependence on the accelerator parameters:

$$\mathscr{L} = \gamma \frac{N_1\, N_2\, f_{rev}\, N_b}{4\pi\, \sqrt{\beta_x^*\, \varepsilon_x^n}\, \sqrt{\beta_y^*\, \varepsilon_y^n}}. \tag{3.142}$$

Let us now consider the more generic case that the beam dimensions are different. The integral of Eq. (3.140) now becomes:

$$\iint ds\, d\mathbf{r}\, [\rho_1(\mathbf{r}, s)\, \rho_2(\mathbf{r}, s)] = \frac{(2\pi)^{-3}}{(\sigma_{1x}\sigma_{2x})(\sigma_{1y}\sigma_{2y})(\sigma_{1z}\sigma_{2z})} \times$$

$$\int dx\, \exp\left[-\frac{1}{2} \left(\frac{1}{\sigma_{x1}^2} + \frac{1}{\sigma_{x2}^2} \right) x^2 \right] \int dy\, \exp\left[-\frac{1}{2} \left(\frac{1}{\sigma_{y1}^2} + \frac{1}{\sigma_{y2}^2} \right) y^2 \right] \times$$

$$\iint dz\, ds\, \exp\left[-\frac{1}{2} \left(\frac{1}{\sigma_{z1}^2} + \frac{1}{\sigma_{z2}^2} \right) \left(z^2 - 2zs \left(\frac{\sigma_{2z}^2 - \sigma_{1z}^2}{\sigma_{2z}^2 + \sigma_{1z}^2} \right) + s^2 \right) \right] =$$

$$= \frac{(2\pi)^{-\frac{3}{2}}}{\sqrt{\sigma_{1x}^2 + \sigma_{2x}^2}\sqrt{\sigma_{1y}^2 + \sigma_{2y}^2}\sqrt{\sigma_{1z}^2 + \sigma_{2z}^2}} \times$$

$$\int ds\, \exp\left[-\frac{1}{2} \frac{\sigma_{1z}^2 + \sigma_{2z}^2}{\sigma_{1z}^2 \sigma_{2z}^2} \left(1 - \frac{(\sigma_{1z}^2 - \sigma_{2z}^2)^2}{(\sigma_{1z}^2 + \sigma_{2z}^2)^2} \right) s^2 \right] =$$

$$= \frac{1}{2^{3/2}\pi^{3/2}\sqrt{\sigma_{1x}^2 + \sigma_{2x}^2}\sqrt{\sigma_{1y}^2 + \sigma_{2y}^2}\sqrt{\sigma_{1z}^2 + \sigma_{2z}^2}} \sqrt{\frac{\pi}{2}} \sqrt{\sigma_{1z}^2 + \sigma_{2z}^2} =$$

$$= \frac{1}{4\pi\sqrt{\sigma_{1x}^2 + \sigma_{2x}^2}\sqrt{\sigma_{1y}^2 + \sigma_{2y}^2}}. \tag{3.143}$$

Putting everything together, we get:

$$\mathscr{L} = \frac{N_1\, N_2\, f_{rev}\, N_b}{2\pi\sqrt{\sigma_{1x}^2 + \sigma_{2x}^2}\sqrt{\sigma_{1y}^2 + \sigma_{2y}^2}}, \tag{3.144}$$

which reduces to Eq. (3.142) when $\sigma_{1x} = \sigma_{2x}$ and $\sigma_{1y} = \sigma_{2y}$. Again, we find that the luminosity is independent of the bunch RMS along z. This could have been predicted a priori, since

$$\mathscr{L} \sim \frac{1}{\sigma_{1z}\sigma_{2z}} \iint dz\, ds\, f\left(\frac{z-s}{\sigma_{1z}}\right) f\left(\frac{z+s}{\sigma_{2z}}\right) = \frac{1}{2}\left(\int d\xi\, f(\xi)\right)^2, \quad (3.145)$$

where we have just performed the change of variables $\xi = (z-s)/\sigma_{1z}$ and $\xi = (z+s)/\sigma_{2z}$, with Jacobian $(\sigma_{1z}\sigma_{2z})/2$. The integral at the right-hand side of Eq. (3.145) is clearly independent of σ_{iz}.

Suggested Readings

The lecture notes [20] represent a valid starting point for introducing the concept of luminosity. The technical note [21] contains a number of useful formula.

Problem 3.28 Determine the luminosity of a circular collider where two equally spaced and bunched beams with N_b bunches per beam, revolution frequency f_{rev}, bunch populations N_1 and N_2, and rectangular shape in the (x, y, z) dimensions of size a, b, and c, collide head-on along the z-direction.

Solution

The luminosity is given by Eq. (3.136) with the choice:

$$\begin{cases} \rho_1(x, y, z, s) = \frac{1}{abc} I(x)_{[-\frac{a}{2}, +\frac{a}{2}]} I(y)_{[-\frac{b}{2}, +\frac{b}{2}]} I(z)_{[-\frac{c}{2}+s, +\frac{c}{2}+s]} \\ \rho_2(x, y, z, s) = \frac{1}{abc} I(x)_{[-\frac{a}{2}, +\frac{a}{2}]} I(y)_{[-\frac{b}{2}, +\frac{b}{2}]} I(z)_{[-\frac{c}{2}-s, +\frac{c}{2}-s]} \end{cases} \quad (3.146)$$

where I is the index function. The integral at the right-hand side of Eq. (3.136) can be computed explicitly yielding:

$$\iint ds\, d\mathbf{r}\, [\rho_1(\mathbf{r}, s)\, \rho_2(\mathbf{r}, s)] =$$

$$= \frac{1}{(abc)^2} \int dx\, I(x)_{[-\frac{a}{2}, +\frac{a}{2}]} \int dy\, I(y)_{[-\frac{b}{2}, +\frac{b}{2}]} \iint ds\, dz\, I(z)_{[-\frac{c}{2}+s, +\frac{c}{2}+s]} I(z)_{[-\frac{c}{2}-s, +\frac{c}{2}-s]} =$$

$$= \frac{1}{abc^2} \int_{-\frac{c}{2}}^{0} ds\, \left[\left(s + \frac{c}{2}\right) - \left(-s - \frac{c}{2}\right)\right] + \int_{0}^{-\frac{c}{2}} ds\, \left[\left(-s + \frac{c}{2}\right) - \left(s - \frac{c}{2}\right)\right] =$$

$$= \frac{1}{abc^2}\left(\frac{c^2}{2}\right) = \frac{1}{2ab}. \quad (3.147)$$

We can express the beam sizes in terms of the RMS along the x and y components through $\sigma_x = a/\sqrt{12}$ and $\sigma_y = b/\sqrt{12}$, yielding:

$$\mathscr{L} = \frac{2N_1 N_2 f_{rev} N_b}{2(\sqrt{12})^2 \sigma_x \sigma_y} = \frac{N_1 N_2 f_{rev} N_b}{12\, \sigma_x \sigma_y} \quad (3.148)$$

A comparison with Eq. (3.142) shows that the luminosity for identical rectangular beams is again independent of the bunch size along z, and differs from the result of Gaussian bunches by about 4.5%.

Discussion

The fact that the luminosity formula (3.142), that was obtained under the assumption of Gaussian profiles, agrees numerically with the rectangular distribution, is not a mere accident. Indeed, it can be proved analytically [22] that this result holds for several functional forms, see Problem 4.4. In particular, the value of the functional $\mathscr{L}\sqrt{\langle x^2\rangle}$ is at a minimum for a parabolic distribution centred around $x = 0$ and defined on a compact support. Since most of the reasonable assumptions for $\rho(x)$ are of this functional form (e.g. a Gaussian, a triangle, a rectangle, etc.), one can expect that the value of $\mathscr{L}\sqrt{\langle x^2\rangle}$ will not differ much from the result obtained under the assumption of a parabolic distribution, which is $3/5^{\frac{3}{2}} = 0.268$, and indeed one finds variations not larger than 5%, see e.g. Table 10 of Ref. [22] for some numerical results.

To summarise, Eq (3.142) represents a valid approximation for reasonable distributions in the transverse plane, provided that σ is replaced by the appropriate RMS.

Suggested Readings

The goodness of the RMS as a measure of the beam size in the calculation of luminosities has been studied in detail in Ref. [22].

Problem 3.29 Consider the case of two identical bunched beams with Gaussian profile that collide with a crossing angle angle $\phi \ll 1$. Determine how the luminosity gets reduced with respect to the zero crossing angle. Evaluate the correction factor using the LHC design parameters $\phi = 285\,\mu\mathrm{rad}$, $\sigma_x = 16.7\,\mu\mathrm{m}$, and $\sigma_z = 8.8\,\mathrm{cm}$.

Solution

Without loss of generality, we can assume that that the angle ϕ between the two beams is in the xz-plane, see Fig. 3.11. The luminosity is given by Eq. (3.136) with the choice:

$$\begin{cases} \rho_1(x, y, z, s) = \frac{(2\pi)^{-\frac{3}{2}}}{\sigma_x \sigma_y \sigma_z} \exp\left[-\frac{x^2_{+\phi/2}}{2\sigma_x^2} - \frac{y^2}{2\sigma_y^2} - \frac{(z_{+\phi/2}-s)^2}{2\sigma_z^2} \right] \\ \rho_2(x, y, z, s) = \frac{(2\pi)^{-\frac{3}{2}}}{\sigma_x \sigma_y \sigma_z} \exp\left[-\frac{x^2_{-\phi/2}}{2\sigma_x^2} - \frac{y^2}{2\sigma_y^2} - \frac{(z_{-\phi/2}+s)^2}{2\sigma_z^2} \right] \end{cases} \tag{3.149}$$

Fig. 3.11 Illustration of two bunched beams of the same bunch density $n(\mathbf{r})$ colliding at an angle $\pi - \phi$

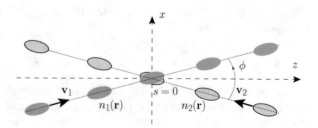

Here, we have indicated by $x_{\pm\phi/2}$ and $z_{\pm\phi/2}$ the rotated of (x, z) by $\pm\phi/2$ around the y-axis, i.e.

$$\begin{cases} z_{+\phi/2} = +z\cos\frac{\phi}{2} + x\sin\frac{\phi}{2} \\ x_{+\phi/2} = -z\sin\frac{\phi}{2} + x\cos\frac{\phi}{2} \end{cases}, \qquad \begin{cases} z_{-\phi/2} = +z\cos\frac{\phi}{2} - x\sin\frac{\phi}{2} \\ x_{-\phi/2} = +z\sin\frac{\phi}{2} + x\cos\frac{\phi}{2} \end{cases} \qquad (3.150)$$

The integrand at the right-hand side of Eq. (3.136) can be simplified to:

$$\exp\left[-\frac{1}{2\sigma_x^2}\left[\left(x\cos\frac{\phi}{2} - z\sin\frac{\phi}{2}\right)^2 + \left(x\cos\frac{\phi}{2} + z\sin\frac{\phi}{2}\right)^2\right] - \frac{y^2}{\sigma_z^2}\right] \times$$

$$\exp\left[-\frac{1}{2\sigma_z^2}\left[\left(z\cos\frac{\phi}{2} + x\sin\frac{\phi}{2} - s\right)^2 + \left(z\cos\frac{\phi}{2} - x\sin\frac{\phi}{2} + s\right)^2\right]\right] =$$

$$= \exp\left[-x^2\left(\frac{\cos^2\frac{\phi}{2}}{\sigma_x^2} + \frac{\sin^2\frac{\phi}{2}}{\sigma_z^2}\right) - z^2\left(\frac{\sin^2\frac{\phi}{2}}{\sigma_x^2} + \frac{\cos^2\frac{\phi}{2}}{\sigma_z^2}\right) - \frac{s^2}{\sigma_z^2} + \frac{2xs\sin\frac{\phi}{2}}{\sigma_z^2}\right].$$

$$(3.151)$$

The integration over s gives:

$$\int ds\, \exp\left[-\frac{s^2}{\sigma_z^2} + \frac{2xs\sin\frac{\phi}{2}}{\sigma_z^2}\right] = \sqrt{\pi}\sigma_z \exp\left[\frac{x^2\sin^2\frac{\phi}{2}}{\sigma_z^2}\right]. \qquad (3.152)$$

Restoring all the constants, and integrating over x, y, and z:

$$\mathcal{L} = 2N_1 N_2 f_{\text{rev}} N_b \cos^2\frac{\phi}{2}\frac{(2\pi)^{-3}}{\sigma_x^2\sigma_y^2\sigma_z^2}\sqrt{\pi}\sigma_y\sqrt{\pi}\sigma_z\times$$

$$\times \int dx\,dz\, \exp\left[-x^2\left(\frac{\cos^2\frac{\phi}{2}}{\sigma_x^2}\right) - z^2\left(\frac{\sin^2\frac{\phi}{2}}{\sigma_x^2} + \frac{\cos^2\frac{\phi}{2}}{\sigma_z^2}\right)\right] =$$

$$= \frac{N_1 N_2 f_{\text{rev}} N_b \cos\frac{\phi}{2}}{4\pi\sigma_y\sqrt{\sigma_z^2\sin^2\frac{\phi}{2} + \sigma_x^2\cos^2\frac{\phi}{2}}} = \frac{N_1 N_2 f_{\text{rev}} N_b}{4\pi\sigma_x\sigma_y}\frac{1}{\sqrt{1 + \frac{\sigma_z^2}{\sigma_x^2}\tan^2\frac{\phi}{2}}}. \qquad (3.153)$$

If we now make the assumption that $\phi \ll 1$ and that $\sigma_z \gg \sigma_x$, which is typically the case at colliders, then the result can be further simplified to give:

$$\mathcal{L} = \frac{N_1 N_2 f_{\text{rev}} N_b}{4\pi\sigma_x\sigma_y}\left[1 + \left(\frac{\sigma_z}{\sigma_x}\frac{\phi}{2}\right)^2\right]^{-\frac{1}{2}}. \qquad (3.154)$$

Even though the crossing angle can be rather small in absolute scale (for example, at the LHC, $\phi \approx 3 \times 10^{-4}$ rad), the other angle that sets the reference scale is the ratio

σ_z/σ_x, which can be large. The adimensional quantity $\left(\frac{\phi}{2}\frac{\sigma_z}{\sigma_x}\right)$ is often referred to as *Piwinski angle*. For example, using the LHC design parameters, we have:

$$\left[1 + \left(\frac{7.7\,\text{cm}}{16.7\,\mu\text{m}}\frac{285 \times 10^{-6}}{2}\right)^2\right]^{-\frac{1}{2}} = 0.835, \qquad (3.155)$$

i.e. the luminosity gets reduced by about 16% compared to the case of perfectly head-on collisions.

Suggested Readings

See the lecture notes [20] for an overview on the subject.

Problem 3.30 Consider the beam setup of Problem 3.29, and assume that the beams have an offset Δ with respect to the z-axis, as illustrated in Fig. 3.12. How does the luminosity change as a function of the beam offset?

Solution

We can refer to Problem 3.29 for much of the discussion and for a good fraction of the calculations. We need to modify Eq. (3.150) to include two offsets d_1 and d_2 along the axes orthogonal to the two beam directions, namely:

$$\begin{cases} z_{+\phi/2} = +z\cos\frac{\phi}{2} + x\sin\frac{\phi}{2} \\ x_{+\phi/2} = d_1 - z\sin\frac{\phi}{2} + x\cos\frac{\phi}{2} \end{cases}, \qquad \begin{cases} z_{-\phi/2} = +z\cos\frac{\phi}{2} - x\sin\frac{\phi}{2} \\ x_{-\phi/2} = d_2 + z\sin\frac{\phi}{2} + x\cos\frac{\phi}{2} \end{cases}$$
$$(3.156)$$

Clearly, if $d_1 = d_2$ nothing should happen since this case corresponds to a rigid translation of the two beams in the xz-plane. We can therefore expect the result to depend only on the relative offset $\Delta = |d_1 - d_2|$ Referring to Eq. (3.151), we can easily see that the integration over s and y is unaffected. After collecting all terms that depend on x and z, the integrand is proportional to:

Fig. 3.12 Illustration of two bunched beams of the same bunch density $n(\mathbf{r})$ colliding at an angle $\pi - \phi$ and with an offset Δ

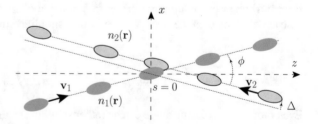

$$\exp\left[-\left(x\frac{\cos\frac{\phi}{2}}{\sigma_x}+\frac{d_1+d_2}{2\sigma_x}\right)^2-\frac{(d_2-d_1)^2}{4\sigma_x^2}\right]\times$$

$$\exp\left[-z^2\left(\frac{\sin^2\frac{\phi}{2}}{\sigma_x^2}+\frac{\cos^2\frac{\phi}{2}}{\sigma_z^2}\right)-z\frac{\sin\frac{\phi}{2}}{\sigma_x^2}(d_2-d_1)\right] \qquad (3.157)$$

The integration over x is straightforward and gives the same contribution as in Eq. (3.153). The integration over z gives the offset-dependent term in addition to the Piwinski term. We can therefore use the result Eq. (3.154) and multiply the right-hand side by the offset-dependent corrections to give:

$$\mathscr{L}=\frac{N_1\,N_2\,f_{\text{rev}}\,N_b}{4\pi\sigma_x\,\sigma_y}\left[1+\left(\frac{\sigma_z}{\sigma_x}\tan\frac{\phi}{2}\right)^2\right]^{-\frac{1}{2}}\exp\left[\frac{\Delta^2\sin^2\frac{\phi}{2}}{4\sigma_x^4\left(\frac{\sin^2\frac{\phi}{2}}{\sigma_x^2}+\frac{\cos^2\frac{\phi}{2}}{\sigma_z^2}\right)}-\frac{\Delta^2}{4\sigma_x^2}\right].$$

$$(3.158)$$

In particular, for $\phi\ll1$, the luminosity depends on the displacement like:

$$\mathscr{L}=\mathscr{L}_0\exp\left[-\frac{\Delta^2}{4\sigma_x^2}\right], \qquad (3.159)$$

where $\mathscr{L}_0$ is the luminosity for perfectly head-on and aligned collisions.

Discussion

From Eq. (3.159), it follows that the counting rate for any reaction occurring when the two beams interact will be given by

$$\frac{dN}{dt}=\frac{dN_0}{dt}\exp\left[-\frac{\Delta^2}{4\sigma_x^2}\right], \qquad (3.160)$$

Hence, one can directly measure the transverse RMS σ_x by recording the counting rates at different values of Δ during a beam scan. This method, known as *Van der Meer scan*, was first suggested by S. van der Meer at the CERN Intersecting Storage Ring [23]. As of today, it represents the principal method for an absolute luminosity calibration at hadron colliders.

Problem 3.31 In a fixed target collider, a monochromatic and continuous beam with flux Φ_{in} passes through a target at rest of unknown particle density. Interactions between the beam and the target remove beam particles. Let the cross section for such destructive absorbtion be σ_0. Determine the luminosity of the collider from the beam flux measured before and after the target.

Solution

Let's denote the beam velocity by $\mathbf{v}_1$ and the beam density by $n_1(x, y, z)$. The assumption of stationary density is clearly valid if the linear density along the beam direction is constant. Equation (3.127) can be written as

$$\mathscr{L} = |\mathbf{v}_1| \int dx\, dy\, dz\, n_1(x, y, z)\, n_2(x, y, z). \tag{3.161}$$

The number of particles removed from the beam in an infinitesimal volume centred around the point (x, y, z) by the interaction with the target is:

$$dn_1 = -n_1(x, y, z)\, \underbrace{n_2(x, y, z)\sigma_0}_{1/\lambda_0}\, dz. \tag{3.162}$$

Plugging Eq. (3.162) into (3.161), we get:

$$\mathscr{L} = -\frac{1}{\sigma_0} \int dx\, dy\, d(|\mathbf{v}_1|n_1) = -\frac{1}{\sigma_0} \int d\Phi_1 = \frac{\Phi_{\text{in}} - \Phi_{\text{out}}}{\sigma_0}, \tag{3.163}$$

where we have used the fact that $J_1(\mathbf{r}) = |\mathbf{v}_1|n_1(\mathbf{r})$ is the beam flux density, and $J_1 \cdot (dx\, dy)$ is the infinitesimal flux across an area $dx\, dy$. In particular, if the interaction responsible for the beam absorbtion is the same that produces the signal events, we get the pretty intuitive result:

$$\frac{dN}{dt} = \mathscr{L}\sigma_0 = \Phi_{\text{in}} - \Phi_{\text{out}}, \tag{3.164}$$

i.e. the observed event rate is equal to the flux absorbed from the beam. For the case of a uniform target of thickness d, the flux decreases as a function of z according to an exponential law:

$$\Phi(z) = \Phi_{\text{in}} \exp[-n_2\, \sigma_0\, z], \tag{3.165}$$

see e.g. Problem 2.14. The luminosity is therefore given by:

$$\mathscr{L} = \frac{\Phi_{\text{in}}}{\sigma_0} \left(1 - e^{-n_2\, \sigma_0\, d}\right) \approx \Phi_{\text{in}}\, n_2\, d \tag{3.166}$$

where the last approximation is valid for d much smaller than the interaction length, and agrees with Eq. (3.135).

Bando n. 1N/R3/SUB/2005

Problem 3.32 Prove that the LHC luminosity is $\mathscr{L} = 10^{34}$ cm^{-2}s^{-1} knowing that the beam current is $I = 0.5$ A and assuming 3000 bunches per beam. How many protons per bunch are there? What is the beam cross section in the interaction region?

Solution

The beam current is related to the total number of protons per beam by the relation:

$$I = e\,N_b\,N\,f_{rev} \quad \Rightarrow \quad (N\,f_{rev}\,N_b) = \frac{I}{e}. \tag{3.167}$$

From Eqs. (3.167) and (3.138), we thus have:

$$\mathscr{L} = \frac{(N\,f_{rev}\,N_b)^2}{4\pi\,\sigma_x\,\sigma_y\,f_{rev}\,N_b} = \frac{(I/e)^2}{4\pi\,\sigma_x\,\sigma_y\,f_{rev}\,N_b}. \tag{3.168}$$

The luminosity can be estimated if the collision frequency and the transverse dimensions of the bunches at the IP are known. The LHC revolution frequency is $f_{rev} = 11,245$ kHz. Using the design value $\sigma_{x,y} = 16.7\,\mu$m [24] for the RMS, we have:

$$= \frac{(0.5\,\text{A}/1.6 \times 10^{-19}\,\text{C})^2}{4\pi \cdot (16.7 \times 10^{-4}\,\text{cm})^2 \cdot 11.245\,\text{kHz} \cdot 3000} = 0.9 \times 10^{34}\,\text{cm}^{-2}\,\text{s}^{-1}, \tag{3.169}$$

in agreement with the design value $10^{34}\,\text{cm}^{-2}\,\text{s}^{-1}$. Notice that the collision frequency differs from the well-known value of 40 MHz. This is due to the fact that the LHC beams contain empty sections, so that the overall collision frequency is less than the expectation from 25 ns spaced bunches. The number of protons per bunch can be obtained directly from Eq. (3.167):

$$N = \frac{I/e}{f} = \frac{0.5\,\text{A}/1.6 \times 10^{-19}\,\text{C}}{11.245\,\text{kHz} \cdot 3000} = 0.9 \times 10^{11}. \tag{3.170}$$

A more refined calculation ($N_b = 2801$, $I = 0.584$ A) would give the well-known result of 1.15×10^{11} protons/bunch.

Problem 3.33 A circular *pp* collider is characterised by a luminosity $\mathscr{L}_0$ at the start of the collision run. The colliding beams consist of N_b equally populated bunches (initial bunch population N_0) with revolution frequency f_{rev}, and that intersect in two points located along the ring. Let the total *pp* cross section be σ_{pp}. How does $\mathscr{L}$ change as a function of run time?

Solution

In order to determine the time evolution for the luminosity, we first derive the time evolution of the bunch population N, which is related to the luminosity by Eq. (3.138). Let's denote the number of interaction points by n_{IP}. During a time interval dt, assumed large compared to the interval between two subsequent collisions, so that several collisions take place and $dN \gg 1$, the number of particles removed from the bunch is given by

$$dN = -dt \, N^2 \left(\frac{f_{\mathrm{rev}} \, n_{\mathrm{IP}}}{4\pi \, \sigma_x \, \sigma_y} \right) \sigma_{pp}, \qquad \frac{dN}{N_0} = -dt \left(\frac{N}{N_0} \right)^2 \frac{\mathscr{L}_0 \, \sigma_{pp} \, n_{\mathrm{IP}}}{N_b \, N_0},$$

$$-\frac{dn}{n^2} = \frac{dt}{\tau_{\mathrm{nucl}}}, \qquad \text{with} \quad n(t) = N(t)/N_0, \quad n(0) = 1, \quad \tau_{\mathrm{nucl}} \equiv \frac{N_b \, N_0}{\mathscr{L}_0 \, \sigma_{pp} \, n_{\mathrm{IP}}}$$

$$\text{(3.171)}$$

This last ODE can be trivially integrated to give:

$$N(t) = \frac{N_0}{(1 + t/\tau_{\mathrm{nucl}})} \quad \Rightarrow \quad \mathscr{L}(t) = \frac{\mathscr{L}_0}{(1 + t/\tau_{\mathrm{nucl}})^2}. \tag{3.172}$$

We can evaluate numerically τ_{nucl} using the LHC design parameters (for two experiments):

$$\tau_{\mathrm{nucl}} = \frac{2808 \cdot 1.15 \times 10^{11}}{10^{34} \text{ cm}^{-2} \text{ s}^{-1} \cdot 100 \text{ mbarn} \cdot 2} = 44.8 \text{ h} \tag{3.173}$$

In analogy with the exponential decay, we can define the beam lifetime as the run time at which the luminosity gets reduced to a factor of $1/e$ from its initial value. With this convention, $\tau = \tau_{\mathrm{nucl}}(\sqrt{e} - 1) \approx 29$ h.

Discussion

Besides nuclear collisions, other effects, characterised by time scales τ_i, contribute to beam losses. The time evolution of the luminosity can be often approximated by an exponential law $\mathscr{L}(t) = \mathscr{L}_0 \, e^{-t/\tau}$, where $\tau = \left(\sum_i \tau_i^{-1} \right)^{-1}$ is the total beam lifetime, including collision loss.

Suggested Readings

For an overview of the LHC design parameters, the reader is addressed to the technical report [24].

Problem 3.34 Given a total operation time T_{tot}, determine the optimal run time of a collider, i.e. the run duration T_{run} that provides the largest total integrated luminosity, given a beam lifetime τ and a turn-around time T_{cycl}. Estimate the maximum total integrated luminosity at the LHC using $T_{\mathrm{run}} = 4800$ h (equivalent to 200 full-days), $\tau = 15$ h.

Solution

Given a total operation time T_{tot} and a turn-around time of the machine T_{cycl}, the optimal run time T_{run} is defined as the run duration that maximises the integrated luminosity over the entire operation time. Using the exponential approximation, see Problem 3.33, T_{run} can be determined from the condition:

$$0 = \frac{\partial}{\partial T_{run}} \underbrace{\left[\underbrace{\frac{T_{tot}}{T_{run} + T_{cycl}}}_{\text{number of runs}} \underbrace{\int_0^{T_{run}} dt \, \mathscr{L}(t)}_{\text{int. lum. per run}} \right]}_{T_{run}^{opt}} = \frac{\partial}{\partial T_{run}} \left[\frac{T_{tot} \, \tau \, \mathscr{L}_0(0)(1 - e^{-T_{run}/\tau})}{T_{run} + T_{cycl}} \right]_{T_{run}^{opt}},$$

$$0 = \frac{T_{run}^{opt}}{\tau} - \ln\left(1 + \frac{T_{run}^{opt} + T_{cycl}}{\tau}\right). \tag{3.174}$$

The last equation can be solved numerically. For example, by using the LHC parameters $T_{cycl} = 1.2$ h, which is dictated by the injection cycle PS $\to$ SPS $\to$ LHC and by the energy ramping in the LHC, $\tau = 15$ h (to be compared to the value τ_{nucl} of Eq. (3.173)), a numerical scan of Eq. (3.174) shows that the zero of the equation is obtained for $T_{run}^{opt} \approx 5.5$ h. With this value, we can estimate the integrated luminosity at the end of the operation time:

$$\mathscr{L}_{int} = \frac{4800 \text{ h}}{5.5 \text{ h} + 1.2 \text{ h}} \cdot 10^{34} \text{ cm}^{-2} \text{ s}^{-1} \cdot 15 \text{ h} \cdot \left(1 - \exp\left[-\frac{5.5 \text{ h}}{15 \text{ h}}\right]\right) \approx 110 \text{ fb}^{-1}. \tag{3.175}$$

This value is a crude approximation and should be taken as an order-of-magnitude estimate.

Suggested Readings

For an overview of the LHC design parameters, the reader is addressed to the technical report [24].

Bando n. 1N/R3/SUB/2005

Problem 3.35 Compute the average number of interactions per bunch crossing if the luminosity is $\mathscr{L} = 2.5 \times 10^{31}$ cm^{-2}s^{-1}, the total cross section $\sigma = 20$ mbarn and the bunches cross every $4 \, \mu$s. What is the probability of having zero interactions per bunch crossing?

Discussion

Equation (3.142) encodes the recipes for maximising the luminosity of a given collider. The beam energy γ is either limited by the accelerator or imposed by the physics program. The normalised emittance depends on the full acceleration cycle and is mostly fixed by technical limitations. The revolution frequency is determined by the beam radius. The number of bunches is limited by the minimum spacing of *buckets* along the beam, which depends for example on the RF cavities. The two easiest ways to increase the luminosity are therefore to minimise the amplitude functions at the IP, β^* and/or to maximise the number of particles per bunch. While the luminosity clearly benefits from these two operations, on the other hand the extra rate is fully concentrated in a small time window around each bunch crossing. This would not be the case if the number of bunches or the revolution frequency could be increased. The collisions events occurring simultaneously in the same bunch crossing are called *pile-up* events.

Solution

The average rate of events is given by $\mathscr{L}\sigma$. The collision events are actually clustered in short time windows with duration of order $T_{coll} \sim \sigma_z/c \sim$ ns, separated by longer inter-bunch intervals with no activity. We can compute the number of events per bunch crossing as:

$$\frac{\text{\# collisions}}{\text{bunch crossing}} = \frac{\text{\# collisions / second}}{\text{\# bunch crossing / second}} \quad \Leftrightarrow \quad \mu_{PU} = \frac{\mathscr{L}\sigma}{f}, \qquad (3.176)$$

where f is the collision frequency, i.e. $f = N_b f_{rev}$ using the notation of Eq. (3.142). From the numbers given by the exercise:

$$\mu_{PU} = 10^{31} \text{ cm}^{-2}\text{s}^{-1} \cdot 20 \text{ mbarn} \cdot 4\,\mu\text{s} = 2.0 \qquad (3.177)$$

The number of pile-up events will be distributed as a Poissonian variable with mean μ_{PU}, hence:

$$\mathscr{P}(n \mid \mu_{PU}) = \frac{\exp[-\mu_{PU}]\,\mu_{PU}^n}{n!} \quad \Rightarrow \quad \mathscr{P}(0;\,\mu_{PU}) = e^{-\mu_{PU}} \approx 13.5\%. \quad (3.178)$$

In hadron colliders, where the inelastic cross sections are large, the number of pile-up events can be very large. For example, at the LHC with design conditions, $\mu_{PU} = 27$, which is expected to become as large as $\mu_{PU} \approx 140$ in the future high-luminosity LHC upgrade [25].

Bando n. 18211/2016

Problem 3.36 An antiproton beam with momentum $|\mathbf{p}| = 6$ GeV and total current $I = 0.16$ mA moves inside a circular ring of length $L = 300$ m. At each round, the beam crosses a hydrogen gas target with a surface density $\delta = 10^{14}$ cm^{-2}. Calculate the revolution frequency, the number of antiprotons, and the integrated luminosity after $T = 6$ min. What beam energy would it be needed if the same centre-of-mass energy had to be achieved by using a symmetric proton-antiproton collider?

Solution

The beam velocity is given by

$$|\mathbf{v}| = \frac{|\mathbf{p}|}{\sqrt{|\mathbf{p}|^2 + m_p^2}} = 0.988\,c, \qquad (3.179)$$

which we can approximate by c within 1%, which is fine given the accuracy on the other data. The revolution frequency is therefore:

$$f_{\text{rev}} = \frac{|\mathbf{v}|}{L} \approx \frac{3 \times 10^8 \text{ m s}^{-1}}{300 \text{ m}} = 1.0 \text{ MHz}. \tag{3.180}$$

We can use Eq. (3.167) to determine the number of circulating protons in the beam:

$$N_{\text{b}} = \frac{I/e}{f_{\text{rev}}} = \frac{0.16 \times 10^{-3} \text{ A}/1.6 \times 10^{-19} \text{ C}}{1.0 \text{ MHz}} = 1.0 \times 10^9. \tag{3.181}$$

Since the hydrogen target is at rest, the collider behaves like a fixed-target experiment. The luminosity is therefore given by Eq. (1.290):

$$\mathscr{L} = \Phi \, \delta = \frac{I}{e} \delta = \frac{0.16 \times 10^{-3} \text{ A}}{1.6 \times 10^{-19} \text{ C}} \cdot 10^{14} \text{ cm}^{-2} = 10^{29} \text{ cm}^{-2} \text{ s}^{-1}. \tag{3.182}$$

The integrated luminosity after $T = 6$ min is given by:

$$\mathscr{L}_{\text{int}} = \mathscr{L} \, T = 10^{29} \text{ cm}^{-2} \text{ s}^{-1} \cdot (6 \times 60) \text{ s} = 36 \, \mu\text{b}^{-1}, \tag{3.183}$$

where we have used the conversion $\text{cm}^{-2} = 10^{-24} \text{ barn}^{-1}$. The centre-of-mass energy is given by Eq. (1.87) with $m_1 = m_2 = m_p$. The beam energy of circular collider operating at the same value of $\sqrt{s}$ is:

$$2 \, E_{\text{circ}} = \sqrt{2m_p^2 + 2m_p(|\mathbf{p}|^2 + m_p^2)^{\frac{1}{2}}},$$

$$E_{\text{circ}} = m_p \sqrt{\frac{1}{2} + \frac{1}{2}\left(1 + \frac{|\mathbf{p}|^2}{m_p^2}\right)^{\frac{1}{2}}} = 1.81 \text{ GeV.} \tag{3.184}$$

Problem 3.37 A 10 GeV proton beam crosses an iron slab of thickness $d = 0.1$ cm. The proton flux corresponds to a current $I = 0.016$ nA. The interaction length in iron is $\lambda = 17$ cm. Estimate the rate of charged and neutral pions emerging from the slab.

Solution

The interaction length is defined as the probability of interaction per particle and per unit length, see Eq. (1.291). Since $d \ll \lambda$, we can approximate this probability by d/λ, so that the number of proton-iron interactions per second is:

$$\frac{dN}{dt} = \frac{I}{e}\left(\frac{d}{\lambda}\right) = \frac{1.6 \times 10^{-11} \text{ A}}{1.6 \times 10^{-19} \text{ C}} \frac{0.1 \text{ cm}}{17 \text{ cm}} = 0.59 \text{ MHz.} \tag{3.185}$$

As discussed in Problem 2.35, high-energy protons in matter first produce one energetic hadron, typically, a pion, sharing a sizable fraction of the incoming proton energy, e.g. $p + p \rightarrow p + n + \pi^+$, $p + n \rightarrow p + n + \pi^0$, etc., so we can assume that the total rate of pions (neutral and charged) is equal to the interaction rate of Eq. (3.185).

Bando n. 1N/R3/SUB/2005

Problem 3.38 An experiment with a direct-current beam is instrumented with a trigger system with efficiency $\varepsilon = 20\%$. The rate of events that need to be selected online is $f = 5$ kHz. The data acquisition system generates a dead time of $T = 1$ ms for each recorded event; during this latency, both the trigger and the detector are not sensitive to additional events. Determine the average data acquisition rate.

Discussion

The data acquisition system illustrated in this problem is defined *non-paralyzable* because the system is not sensitive to additional events arriving after one trigger, so that the dead time cannot be extended beyond the single-event processing time. This is in contrast with the so-called *paralyzable* systems, where the arrival of a new event before the system has finished processing the previous one extends the time window in which the system is not sensitive to additional events. Given a true event rate ν, a dead time τ, and a trigger efficiency ε (probability of accepting an event given that it has been accepted by the DAQ system), the measured rate m in a non-paralysable system can be obtained like follows. Consider a time interval $T \gg \max\left[\tau, \nu^{-1}\right]$. The number of triggers m' in this time interval will be given by the number of true events occurring in the same time window minus the number of true events occurring during the total dead time generated by the accepted triggers, i.e.:

$$m' T = \nu T - m' T \tau \nu \quad \Rightarrow \quad m = \varepsilon m' = \frac{\varepsilon \nu}{1 + \nu \tau}. \tag{3.186}$$

For a paralyzable system, we first notice that, in the assumption that the number of true events in a given time window is Poisson distributed, the time intervals between two subsequent events has an exponential distribution with decay constant ν, where ν is the true event rate. This can be easily proved by noticing that the probability of zero events after a time Δt from a given event is $e^{-\nu \Delta t}$. Only true events spaced by more than τ can generate a trigger, hence the measured rate is equal to the fraction of events such that $\Delta t > \tau$, i.e.:

$$m = \varepsilon \nu e^{-(\nu \tau)}. \tag{3.187}$$

Differently from the non-paralyzable, case Eq. (3.188) admits two solutions for ν at a given measured rate m.

Solution

From Eq. (3.186), we can estimate the average data acquisition rate as:

$$m = \frac{0.2 \cdot 5\,\text{kHz}}{1 + 1 \times 10^{-3}\,\text{s} \cdot 5 \times 10^3 \text{s}^{-1}} = 0.17\,\text{kHz}. \tag{3.188}$$

It is interesting to notice that a paralzsable system operating with the same conditions would yield a trigger rate smaller by a factor of ≈ 25 compared to a non-paralyzable one.

Suggested Readings

An introduction to the concept of paralysable and non-paralysable DAQ systems can be found in Refs. [26, 27]. A more complete overview on the topic can be found in Ref. [2].

Bando n. 1N/R3/SUB/2005

Problem 3.39 Collision events at a rate of 5 kHz are analysed to decide whether an event has to be recorded or not. The decision takes a time $T_{\text{trig}} = 20\,\mu\text{s}$ while the digitization takes $T_{\text{dig}} = 1$ ms. What rate of accepted events can be sustained if the dead time has to be maintained below $\delta = 20\%$?

Solution

Let's denote the collision rate by f and the fraction of signal events that we wish to select online by $\varepsilon_{\text{trig}}$. For example, $\varepsilon_{\text{trig}}$ can be the fraction of events that pass the online selection devised to identify signal events, i.e. $\varepsilon_{\text{trig}} = \sigma_{\text{acc}}/\sigma_{\text{tot}}$, where σ_{acc} is the cross section for producing an event that passes the selection, while σ_{tot} is the total cross section. The DAQ system gives rise to two types of dead time: the first comes from the trigger system and affects the full collision rate. The second arises from the digitizer and affects only the rate of events that have fired the trigger. Assuming both systems to be non-paralyzable (see Problem 3.38), we can visualise the rate reduction as a two-stage process:

$$f \xrightarrow{\text{Trigger}} \left(\nu_{\text{trig}} = \frac{\varepsilon_{\text{trig}}\, f}{1 + f\, T_{\text{trig}}} \right) \xrightarrow{\text{Digitizer}} \left(\nu_{\text{dig}} = \frac{\nu_{\text{trig}}}{1 + \nu_{\text{trig}}\, T_{\text{dig}}} \right). \tag{3.189}$$

If one wants to maintain a fraction of lost events less than δ, where now the dead time refers by extension to the fraction of signal events lost because of dead time in the DAQ system, the rate in output to the digitizer needs to satisfy:

$$\frac{\nu_{\text{dig}}}{\varepsilon_{\text{trig}} f} \geq 1 - \delta, \qquad \left(\frac{\varepsilon_{\text{trig}} f}{1 + f T_{\text{trig}}} \frac{1}{1 + \frac{\varepsilon_{\text{trig}} f}{1 + f T_{\text{trig}}} T_{\text{dig}}} \right) \frac{1}{\varepsilon_{\text{trig}} f} \geq 1 - \delta$$

$$\frac{1}{1 + f (T_{\text{trig}} + \varepsilon_{\text{trig}} T_{\text{dig}})} \geq 1 - \delta, \qquad \varepsilon_{\text{trig}} \leq \frac{1}{f \, T_{\text{dig}}} \frac{\delta}{1 - \delta} - \frac{T_{\text{trig}}}{T_{\text{dig}}}. \qquad (3.190)$$

With the choice $\delta = 0.2$, we thus get

$$\varepsilon_{\text{trig}} < \frac{1}{5 \times 10^3 \text{ s}^{-1} \cdot 10^{-3} \text{ s}} \frac{0.2}{1 - 0.2} - \frac{2 \times 10^{-5} \text{ s}}{10^{-3} \text{ s}} = 0.05 - 0.02 = 3\%,$$
$$(3.191)$$

and the accepted rate is:

$$\nu_{\text{trig}} = \varepsilon_{\text{trig}} f = 3\% \cdot 5 \text{ kHz} = 150 \text{ Hz}. \qquad (3.192)$$

Notice that the largest dead time of the system is achieved by chosing $\varepsilon_{\text{trig}} = 1$ (all triggering events are accepted), corresponding to a dead time

$$\delta = 1 - \frac{1}{1 + f (T_{\text{trig}} + T_{\text{dig}})} \approx 0.83. \qquad (3.193)$$

If a smaller dead time is desired for the same efficiency, possible solutions are to pipeline the events while waiting for a trigger accept, or parallelise the event digitization by using multiple CPU's.

Suggested Readings

For an introduction to the queueing theory and to the use of trigger in HEP, the reader is addressed to Ref. [2].

Bando n. 1N/R3/SUB/2005

Problem 3.40 Two scintillating counters are used to monitor the luminosity at a collider. The two counters are located before and after the interaction point. Bunches cross every $T = 10\,\mu\text{s}$ and, at each bunch crossing, two digital signals, A and B, are generated by the counters in the occurrence of an event. A logic AND of the two signals measures the coincidence of the two signals in the same bunch crossing. Background events, like those induced by beam halo particles, or collisions with the redioal gas inside the beam pipe, are superimposed to the true events, thus generating a random coincidence. The following mean rates are measured for the three signals: $f_A = 7.53 \times 10^4$ Hz, $f_B = 6.67 \times 10^4$ Hz, and $f_{A \cap B} = 5.43 \times 10^4$ Hz. Determine the mean number of collisions per bunch crossing.

Discussion

A standard technique for monitoring the luminosity at circular colliders is through the use of two symmetric counters located at both sides of the interaction point. Since the acceptance of the counters is fixed, the luminosity is proportional to the rate of coincident events. If the cross section of the scattering process in the counter acceptance is also known (e.g. Bhabba scattering for e^+e^- machines), an absolute normalisation of the luminosity becomes possible. Because of background events, false coincidences can be generated at random. By measuring independently the individual counter rates and the coincidence rate, the rate of signal events can be disentangled from the false coincidences.

Solution

Let's first consider the signal in one of two counters, say counter A, and let us assume for simplicity that any collision produces a signal in A. The probability of observing an event in A for a given bunch crossing is the sum of probabilities of two independent hypotheses:

$$\text{Prob}\,[A] = \text{Prob}\,[\text{coll.}] + (1 - \text{Prob}\,[\text{coll.}]) \cdot \text{Prob}\,\big[\text{bkg.}\big]. \tag{3.194}$$

Let us now consider $N \gg 1$ bunch crossings over a time $T \gg 1/f$, where f is the bunch crossing frequency. Multiplying both sides of Eq. (3.194) by N/T, and remembering that $\text{Prob}\,[\text{coll.}] = \nu/f$, where ν is the coll. rate, we get:

$$\frac{N \cdot \text{Prob}\,[A]}{T} = \frac{N \cdot \text{Prob}\,[\text{coll.}]}{T} + \left(1 - \frac{\nu}{f}\right) \cdot \frac{N \cdot \text{Prob}\,\big[\text{bkg.}\big]}{T},$$

$$f_A = \nu + \left(1 - \frac{\nu}{f}\right) b_A, \tag{3.195}$$

where b_A is the background rate in counter A. Similarly, we can consider the coincidence signal, and compute its probability as:

$$\text{Prob}\,[A \cap B] = \text{Prob}\,[\text{coll.}] + (1 - \text{Prob}\,[\text{coll.}]) \cdot \text{Prob}\,[A]\,\text{Prob}\,[B],$$

$$\frac{N \cdot \text{Prob}\,[A \cap B]}{T} = \frac{N \cdot \text{Prob}\,[\text{coll.}]}{T} + \left(1 - \frac{\nu}{f}\right) \cdot \frac{N \cdot \text{Prob}\,[A]}{T} \frac{N \cdot \text{Prob}\,[B]}{T} \frac{1}{N/T},$$

$$f_{A \cap B} = \nu + \left(1 - \frac{\nu}{f}\right) \frac{b_A b_B}{f}. \tag{3.196}$$

To summarise, we can write a system of three equations in three unknowns ν, b_A, and b_B:

$$\begin{cases} f_A = \nu + \left(1 - \frac{\nu}{f}\right) b_A \\ f_B = \nu + \left(1 - \frac{\nu}{f}\right) b_B \\ f_{A \cap B} = \nu + \left(1 - \frac{\nu}{f}\right) \frac{b_A b_B}{f}. \end{cases} \tag{3.197}$$

which can be used to determine the true collision frequency ν. Replacing the first two equations into the last one:

$$f_{A\cap B} = \nu + \frac{(f_A - \nu)(f_B - \nu)}{f\left(1 - \frac{\nu}{f}\right)}, \quad (f - \nu)(f_{A\cap B} - \nu) = (f_A - \nu)(f_B - \nu),$$

$$\nu = \frac{f\, f_{A\cap B} - f_A\, f_B}{f - f_A - f_B + f_{A\cap B}}. \tag{3.198}$$

Replacing the symbols by the measured values, we obtain an estimate:

$$\nu = \frac{(10^5)(5.43 \times 10^4) - (7.53 \times 10^4)(6.67 \times 10^4)}{10^5 - 7.53 \times 10^4 - 6.67 \times 10^4 + 5.43 \times 10^4}\; \mathrm{Hz} = 3.31 \times 10^4\; \mathrm{Hz}. \tag{3.199}$$

The mean number of collisions per bunch crossing as defined in Eq. (3.176) is given by:

$$\mu_{PU} = \frac{\nu}{f} = \frac{3.31 \times 10^4\; \mathrm{Hz}}{10^5\; \mathrm{Hz}} = 0.331. \tag{3.200}$$

Suggested Readings

The luminosity measurement at LEP using two luminosity monitors mounted at small angles with respect to the beam is discussed in details in Ref. [28].

Appendix 1

The computer program below implements a Monte Carlo simulation of the experimental apparatus considered in Problem 3.10. The measured points (y_0, y_1, y_2) are sampled from a Gaussian pdf centred around the true values $y_i = x_i^2/2R$ and with standard deviation equal to the detector resolution. The signed sagitta is computed from

$$s = \frac{y_2 + y_0}{2} - y_1, \tag{3.201}$$

and a sign flip corresponds to the outcome $s < 0$.

```
import math
import random

# the true particle positions; use MKS units and pT in GeV/c
def track(x=1.0, pt=1000., B=1.0):
    R = pt/(0.3*B)
    return 0.5*x*x/R

# track sagitta\index{Sagitta} from measured position
def sagitta(y0,y1,y2):
    return 0.5*(y2+y0) - y1

# the measured particle positions
def smear(mu, sigma):
    return random.gauss(mu,sigma)

# generate ntoys pseudo-measurement; use MKS units and pT in
 GeV/c def toys(ntoys=100, pt=1000., B=1.0, s0=1.0e-04, s1=0.5e-04,
s2=1.0e-04):
    outcomes = [0,0]
    for toy in xrange(ntoys):
        y0 = smear( track(x=0.0, pt=1000., B=1.0), s0 )
        y1 = smear( track(x=1.0, pt=1000., B=1.0), s1 )
        y2 = smear( track(x=2.0, pt=1000., B=1.0), s2 )
        s = sagitta(y0,y1,y2)
        outcomes[ int(s<0) ] += 1
        mis = float(outcomes[1])/float(ntoys)
        err = math.sqrt((1-mis)*mis/ntoys) # binomial pdf
    print ("Charge mis-id = (%.2f +/- %.2f)"% (mis*100., err*100))

################
toys(ntoys=1000000)
```

Appendix 2

The computer program below computes numerically the integral:

$$\sigma^2_{\langle p_T \Delta \phi \rangle} \approx \left(\frac{N(\alpha-1)}{p_T^{L\,(1-\alpha)} - p_T^{H\,(1-\alpha)}} \int_{p_T^L}^{p_T^H} dp_T \, \frac{p_T^{-\alpha}}{c_2 + c_1 p_T^2} \right)^{-1} \tag{3.202}$$

from Problem 3.12.

```
import math

# the constants
c1 = 2*math.pow(2.0e-03,2)
c2 = math.pow(0.0136,2)*(60./1.8)
BP = 2 * 0.6
acc = 1.0e-02

def integrate(x_l=5.0, x_h=100., alpha=2.7, step = 1.0 ):
    integ = 0.0
    n_step = int((x_h-x_l)/step)
    for s in xrange( n_step ):
        x = x_l + (s+0.5)*step
        val = math.pow(x, -alpha)/(c2 + c1*x*x)
        integ += val*step
    integ *= (alpha-1)/(math.pow(x_l, 1-alpha) - math.pow(x_h, 1-alpha))
    return 1./math.sqrt(integ)

#######################

x_l = 5.
x_h = 300.
res = integrate(x_l=x_l, x_h=x_h, alpha=2.7, step = 0.05 )
print"<sigma>=", res,"N=", res/BP * math.pow(acc, -2)
```

References

1. R.L. Glusckstern, Nucl. Instr. Meth. **24**, 381 (1963). https://doi.org/10.1016/0029-554X(63)90347-1
2. R. Frühwirth et al., *Data Analysis Techniques for High-Energy Physics*, 2nd edn. (Cambridge Press, Cambridge, 2000)
3. C. Patrignani et al., Particle data group. Chin. Phys. C **40**, 100001 (2016)
4. A. Fasano, S. Marmi, *Meccanica Analitica*, 2nd edn. (Bollati Boringhieri, 2002)
5. L. Rolandi, F. Ragusa, New J. Phys. **9**, 336 (2007). https://doi.org/10.1088/1367-2630/9/9/336
6. ATLAS Collaboration, Phys. Lett. B **716**, 1 (2012). https://doi.org/10.1016/j.physletb.2012.08.020
7. C.M.S. Collaboration, Phys. Lett. B **716**, 30 (2012). https://doi.org/10.1016/j.physletb.2012.08.021
8. ATLAS and CMS Collaborations, Phys. Rev. Lett. **114**, 191803 (2015). https://doi.org/10.1103/PhysRevLett.114.191803
9. http://cerncourier.com/cws/article/cern/28229
10. D.H. Perkins, *Introduction to Hig Energy Physics*, 4th edn. (Cambridge University press, Cambridge, 2000)
11. R. Hellborg (ed.), *Electrostatic Accelerators* (Springer, Berlin, 2005)
12. W. Blum, W. Riegler, L. Rolandi, *Particle Detection with Drift Chambers* (Springer, Berlin, 2008)
13. C.M.S. Collaboration, JINST **5**, T03021 (2010). https://doi.org/10.1088/1748-0221/5/03/T03021
14. J.D. Jackson, *Classical Electrodynamics*, 3rd edn. (Wiley, New York, 1999)
15. w3.lnf.infn.it/accelerators/dafne/?lang=en

16. A. Blondel, Phys. World **4**(1), 22 (1991). https://doi.org/10.1088/2058-7058/4/1/20
17. E. Wilson, *An Introduction to Particle Acceleretors* (Oxford University Press, Oxford, 2001)
18. White Paper of ICFA-ICUIL Joint Task Force, *High Power Laser Technology for Accelerators*, published in *ICFA Beam Dynamics Newslettern*, vol. 56 (2011). http://www-bd.fnal.gov/icfabd/
19. F. Hinterberger, CAS - CERN Accelerator School and KVI: Specialised CAS Course on Small Accelerators, Zeegse, The Netherlands, 24 May–2 Jun 2005, pp. 95-112 (CERN-2006-012). https://doi.org/10.5170/CERN-2006-012.95
20. W. Herr, B. Muratori, CAS - CERN Accelerator School: Intermediate Course on Accelerator Physics, Zeuthen, Germany, 15–26 Sep 2003, pp. 361–378 (CERN-2006-002). https://doi.org/10.5170/CERN-2006-002.361
21. T. Suzuki, KEK-76-3 (1976)
22. H.G. Hereward, CERN/MPS/DL 69-12 (1969)
23. S. van der Meer, ISR-PO/68-31 (1968)
24. *LHC Design Report*, CERN-2004-003 (2004)
25. G. Apollinari et al. CERN-2015-005 (2015)
26. W.R. Leo, *Techniques for Nuclear and Particle Physics Experiments*, 2nd edn. (Springer, Berlin, 1993)
27. G.F. Knoll, *Radiation Detection and Measurement*, 4th edn. (Wiley, New York, 2010)
28. E. Bravin et al. CERN SL-97-72 (1997)

Chapter 4
Statistics in Particle Physics

Abstract This chapter is devoted to the statistical treatment of experimental data. By its very nature, the outcome of a particle physics experiment is a to be considered as random process: the theory of probability thus provides the correct theoretical framework. A few concepts are of key importance in data analysis. After recalling a few outstanding theorems, the properties of the likelihood function are illustrated by considering simple cases that can be handled analytically, i.e. without the use of computer programs. The focus is then put on the combination of random variables, error propagation, and the construction of frequentist confidence intervals.

4.1 Elements of Statistics

The outcome of an experiment can be treated as a random variable. On the one hand, most of the reactions studies in particle physics are associated with quantum-mechanical transition amplitude. In this case, only the probability of occurrence is prescribed, but not the actual outcome of a given reaction. On the other hand, the presence of noise in any measurement introduces a certain degree of randomness. For these reasons, the interpretation of an experiment has to be carried out according to the paradigms of statistics. There is a vast literature of text books on probability, statistics, and data analysis applied to particle physics. No attempt is made here to be exhaustive in this respect. A few concepts and classical results, however, stand out for their pervasive presence and cardinal importance in the analysis of experimental data. This section collects a few outstanding results that are relevant for solving the proposed exercises.

The "large number" theorems

There are two main theorems one should always remember when dealing with large size data samples. For their proof, the reader is addressed to dedicated books, see e.g. Ref. [1].

Theorem 4.1 (Law of Large Numbers) *Let $\{X_1, \ldots, X_N\}$ be a sequence of N independent and identically distributed random variables with mean μ and variance $\sigma^2 < \infty$. Then, the sample mean*

© Springer International Publishing AG 2018
L. Bianchini, *Selected Exercises in Particle and Nuclear Physics*,
UNITEXT for Physics, https://doi.org/10.1007/978-3-319-70494-4_4

$$S_N = \frac{\sum_{i=1}^{N} X_i}{N} \tag{4.1}$$

converges to μ in the limit $N \to \infty$.

This result guarantees that the repeated measurement of the same unknown quantity will provide *asymptotically* an estimator of the true value.

An important application of the Large Number Law is the *Monte Carlo* method for numerical integration. Considering for simplicity the one-dimensional case, the integral of a function g over the interval $[a, b]$ can be estimated by the sample mean g evaluated at points u_i uniformly distributed in $[a, b]$. Indeed, the Law of Large Number guarantees that:

$$I_N = \frac{(b-a)}{N} \sum_i^N g(u_i) \to (b-a) \int_a^b du \, f(u) \, g(u) = \int_a^b du \, g(u). \tag{4.2}$$

Another benefit of the MC integration compared to other approximate analytical methods is that the error on the integral scales like $N^{-1/2}$ independently of the dimensionality.

Theorem 4.2 (Central Limit theorem) *Let $\{X_1, \ldots, X_N\}$ be a sequence of N independent random variables with mean μ_i and variances σ_i^2. Under suitable hypothesis (Lyapunov condition), the p.d.f. of the variable*

$$Z_N = \frac{\sum_{i=1}^{N}(X_i - \mu_i)}{\sqrt{\sum_i^N \sigma_i^2}} \tag{4.3}$$

converges to a normal distribution $\mathcal{N}(0, 1)$ in the limit $N \to \infty$.

The importance of the Central Limit theorem lies in its prediction that an experimental measurement, which is affected by a sufficiently large number of independent random fluctuations of *a priori* unknown distribution, but finite variance, will ultimately have a Gaussian distribution.

The likelihood function for point estimation

Let $\mathbf{X} = \{X_1, \ldots, X_N\}$ be a set of N independent observations of a random variable X (in general, X can a vector of random variables), and let $f(X|\boldsymbol{\theta})$ be the p.d.f. of X which depends on a set of parameters $\boldsymbol{\theta}$. The domain of X is further assumed to be independent of $\boldsymbol{\theta}$. The *likelihood function* of the data is defined by

$$L(\mathbf{X}|\boldsymbol{\theta}) = \prod_{i=1}^{N} f(X_i \,|\, \boldsymbol{\theta}). \tag{4.4}$$

The maximum likelihood estimator of $\boldsymbol{\theta}$, indicated by $\hat{\boldsymbol{\theta}}$, is a maximum of the likelihood function (4.4). For the case that L is a differentiable function of $\boldsymbol{\theta}$, the ML estimator is the root of the equation:

$$\left[\sum_{i=1}^{N} \frac{\partial}{\partial \boldsymbol{\theta}} \ln f(X_i \mid \boldsymbol{\theta}) \right]_{\hat{\theta}} = 0 \qquad (4.5)$$

In general, several roots may exist for finite N (local maxima), but the global maximum will always asymptotically converge to the true value. The ML estimator thus defined has the following asymptotic properties:

Theorem 4.3 (Asymptotic properties of the ML estimator) *The maximum likelihood estimator is asymptotically:*

- *a consistent and unbiased estimator (Law of Large Numbers), i.e. it gets arbitrarily "close" to the true parameter value;*
- *a sufficient estimator (Darmois theorem), i.e. it encodes all of the information on the parameters that the data contain;*
- *the most efficient estimator (Cramer–Rao inequality).*
- *The covariance matrix of the ML estimator is*

$$V(\hat{\boldsymbol{\theta}})_{ij} = -\mathrm{E}\left[\left(\frac{\partial^2 \ln L(\mathbf{X}|\boldsymbol{\theta})}{\partial \theta_i \, \partial \theta_j} \right)^{-1} \right] = N^{-1} \mathrm{E}\left[\left(\frac{\partial^2 \ln f(X|\boldsymbol{\theta})}{\partial \theta_i \, \partial \theta_j} \right)^{-1} \right] = (N \, I_{\boldsymbol{\theta}})^{-1},$$

$$(4.6)$$

where $I_{\boldsymbol{\theta}}$ is the information matrix. Thus, the quantity $\sqrt{N}(\hat{\boldsymbol{\theta}} - \boldsymbol{\theta})$ is distributed with p.d.f. $\mathcal{N}(0, I_{\boldsymbol{\theta}}^{-1})$.

- *The quantity at the left-hand side of Eq. (4.5) is also normally distributed with mean 0 and variance $N I_{\boldsymbol{\theta}}$.*

Owing to its attractive properties, the maximum likelihood method for point estimation is the most used technique for the cases where a likelihood model of the data is available, which is often the case in particle physics where the prior knowledge is usually provided by some theoretical model.

The likelihood function for hypothesis testing

Let $\boldsymbol{\theta}$ represent a point of the k-dimensional parameter space on which we wish to make inference from the data. Each point corresponds to a certain hypothesis, which can be *simple* if it consists of just one point, e.g. $\boldsymbol{\theta} = \boldsymbol{\theta}_0$, or *composite*, it consists of multiple simple hypotheses, e.g. $\boldsymbol{\xi}(\boldsymbol{\theta}) = 0$ with $\boldsymbol{\xi}$ is a vector of constraints of dimension $r < k$. Given two such hypotheses $\boldsymbol{\theta}_0$ and $\boldsymbol{\theta}_1$, the likelihood-ratio test statistic λ is defined as:

$$\lambda = \frac{L(\mathbf{X}|\boldsymbol{\theta}_0)}{L(\mathbf{X}|\boldsymbol{\theta}_1)}. \qquad (4.7)$$

The importance of the likelihood-ratio test statistic relies on its optimal properties to discriminate between two hypotheses, and on its known asymptotic behaviour.

Theorem 4.4 (Properties of the likelihood-ratio test statistic) *The likelihood-ratio test statistic is:*

- *the best test statistic for simple hypotheses testing (Neyman–Pearson lemma);*
- *asymptotically the most powerful test statistic to discriminate between two composite hypotheses (Wald's theorem [2]).*
- *If the maximum likelihood estimators of $\boldsymbol{\theta}$ are given, the asymptotic distribution of the likelihood ratio*

$$\lambda = \frac{L(\mathbf{X}|\hat{\hat{\boldsymbol{\theta}}})}{L(\mathbf{X}|\hat{\boldsymbol{\theta}})} \tag{4.8}$$

is known (Wilks' theorem [3]). Here, $\hat{\boldsymbol{\theta}}$ is the ML estimator obtained by maximising the likelihood with respect to all parameters, while $\hat{\hat{\boldsymbol{\theta}}}$ is the ML estimator obtained by maximising the likelihood in a subset of the parameter space defined by the r constraints $\boldsymbol{\xi}(\boldsymbol{\theta}) = 0$ (conditional likelihood). In particular, if the hypothesis under test is true, the variable $q = -2 \ln \lambda$ is distributed like a χ^2 with as many degrees of freedom as the number of fixed parameters (r).

The special case $\xi(\theta_1) = \theta_1 - \mu = 0$ gives:

$$q(\mu) = -2 \ln \lambda(\mu) = -2 \ln \frac{L(\mathbf{X}|\mu, \hat{\hat{\theta}}_2, \ldots, \hat{\hat{\theta}}_k)}{L(\mathbf{X}|\hat{\theta}_1, \ldots, \hat{\theta}_k)} = \frac{(\mu - \hat{\mu})^2}{\sigma_{\hat{\mu}}^2} + \mathcal{O}(N^{-\frac{1}{2}}), \tag{4.9}$$

where $\hat{\mu}$ is the ML estimator of μ, which is asymptotically Gaussian distributed around μ with standard deviation $\sigma_{\hat{\mu}}$. See Ref. [4] for more details. The 68.3 and 95.4% CL intervals on μ can be then determined as the points at which $q = 1$ and $q = 4$, respectively.

Problems

Problem 4.1 Let X be a random variable with p.d.f. $f(X|\theta)$, and $\mathbf{X}$ an array of independent observations of X. Prove that the ML estimator $\hat{\theta}$ provides a consistent estimator of θ.

Discussion

The maximum likelihood estimator is an instance of the so-called *implicitly defined estimators*, defined as the roots of the experimental observable:

$$\xi_N(\hat{\theta}) = \frac{1}{N} \sum_i g(X_i, \hat{\theta}) = 0, \tag{4.10}$$

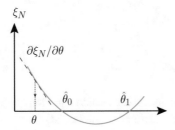

Fig. 4.1 Illustration of an implicitely defined estimator. For finite N, one or multiple roots of the equation $\xi_N = 0$ exist. One of such solutions will however converge to the true value θ for $N \to \infty$, provided that the derivative of ξ_N is asymptotically non-zero

where g satisfies the constraint $E[g(X, \theta)] = 0$. By the Law of Large Numbers, the estimator defined by Eq. (4.10) converges to the true value θ, provided that:

$$\left| \lim_{N \to \infty} E\left[\frac{\partial \xi_N(\theta)}{\partial \theta} \right] \right| > 0 \qquad (4.11)$$

This last condition guarantees that the first equation can be inverted to yield a solution, see Fig. 4.1.

Solution

Since f is a p.d.f., it must satisfy the normalisation condition

$$\int dX\, f(X|\theta) = 1 \qquad (4.12)$$

for any value of θ. Differentiating both sides of Eq. (4.12) with respect to θ, we thus have:

$$0 = \frac{\partial}{\partial \theta} \int dX\, f(X|\theta) = \int dX\, \frac{\partial f(X|\theta)}{\partial \theta} = \int \frac{1}{f} \frac{\partial f(X|\theta)}{\partial \theta} f\, dX =$$
$$= E\left[\frac{\partial \ln f(X|\theta)}{\partial \theta} \right] \qquad (4.13)$$

By differentiation twice Eq. (4.12) with respect to θ, we obtain:

$$0 = \frac{\partial}{\partial \theta} \int \frac{\partial \ln f(X|\theta)}{\partial \theta} f(X|\theta)\, dX = \int \left[\frac{\partial^2 \ln f(X|\theta)}{\partial \theta^2} + \left(\frac{\partial \ln f(X|\theta)}{\partial \theta} \right)^2 \right] f\, dX \qquad (4.14)$$

Hence, if we indicate by $\xi_N(\theta)$ the sum of Eq. (4.5) divided by N, the Law of Large Numbers states that

$$\lim_{N \to \infty} \xi_N(\theta) = \mathrm{E} \left[\frac{\partial \ln f(X|\theta)}{\partial \theta} \right] = 0 \tag{4.15}$$

$$\lim_{N \to \infty} \frac{\partial \xi_N}{\partial \theta} = \mathrm{E} \left[\frac{\partial^2 \ln f(X|\theta)}{\partial \theta^2} \right] = -\mathrm{E} \left[\left(\frac{\partial \ln f}{\partial \theta} \right)^2 \right] < 0 \tag{4.16}$$

The second condition guarantees that the first equation can be inverted to yield a solution. Thus, the ML estimator defined implicitely as the root of ξ_N will asymptotically converge to the true value θ.

Suggested readings

For more details on the proof, the reader can refer to Chap. 7 of Ref. [5].

Problem 4.2 An experiment measures $\tilde{n}$ events, with $\tilde{n} \gg 1$, in a channel where a signal of unknown yield s contribute. The channel is affected by a source of background for which an auxiliary measurement $\tilde{b} \pm \sigma_b$ is available ($\tilde{b}$ can be assumed to be Gaussian distributed with standard deviation σ_b). Verify explicitly Wilks' theorem by studying the behaviour of profiled likelihood ratio $q(s)$.

Solution

For $\tilde{n} \gg 1$, the number of measured events can be assumed to be normally distributed with standard deviation $\sqrt{\tilde{n}}$, see Problem 4.11. Under this assumption, the likelihood function of the data is given by:

$$L(\tilde{n}, \tilde{b} \,|\, s, b) = \frac{1}{\sqrt{2\pi}\sqrt{\tilde{n}}} \exp\left[-\frac{(\tilde{n} - s - b)^2}{2\tilde{n}} \right] \frac{1}{\sqrt{2\pi}\,\sigma_b} \exp\left[-\frac{(\tilde{b} - b)^2}{2\sigma_b} \right] \tag{4.17}$$

The ML estimators $\hat{n}$ and $\hat{b}$ of the signal and background yields are the solutions of the equation

$$0 = \partial_{s,b} \ln L(\tilde{n}, \tilde{b} \,|\, s, b) = \partial_{s,b} \left[\frac{(\tilde{n} - s - b)^2}{2\tilde{n}} + \frac{(\tilde{b} - b)^2}{2\sigma_b^2} \right] \quad \Rightarrow \quad \begin{cases} \hat{s} = \tilde{n} - \tilde{b} \\ \hat{b} = \tilde{b} \end{cases} \tag{4.18}$$

The constrained ML estimator $\hat{\hat{b}}$ is obtained by fixing s and maximising $-\ln L$ with respect to b, yielding:

$$0 = \partial_b \ln L(\tilde{n}, \tilde{b} \,|\, s, b) \quad \Rightarrow \quad \hat{\hat{b}} = \frac{\frac{\tilde{b}}{\sigma_b^2} + \frac{\tilde{n} - s}{\tilde{n}}}{\frac{1}{\sigma_b^2} + \frac{1}{\tilde{n}}}, \tag{4.19}$$

which coincides with the combined ML estimator from two independent measurements, see Eq. (4.83). The profile likelihood ratio $q(s)$ is then given by:

$$q(s) = -2 \ln \frac{L(s, \hat{\hat{b}})}{L(\hat{s}, \hat{b})} = -2 \left[-\frac{(\tilde{n} - s - \hat{\hat{b}})^2}{2\tilde{n}} - \frac{(\hat{\hat{b}} - \tilde{b})^2}{2\sigma_b^2} + \frac{(\tilde{n} - \hat{s} - \hat{b})^2}{2\tilde{n}} + \frac{(\hat{b} - \tilde{b})^2}{2\sigma_b^2} \right] =$$

$$= \frac{(\tilde{n} - s - \hat{\hat{b}})^2}{\tilde{n}} + \frac{(\hat{\hat{b}} - \tilde{b})^2}{\sigma_b^2} = \frac{\left(\hat{n} - s - \tilde{b}\right)^2}{\sigma_b^4 n \left(\frac{1}{\sigma_b^2} + \frac{1}{n}\right)} + \frac{\left(\hat{n} - s - \tilde{b}\right)^2}{\sigma_b^2 n^2 \left(\frac{1}{\sigma_b^2} + \frac{1}{n}\right)} =$$

$$= \frac{\left(s - \hat{n} + \tilde{b}\right)^2}{\sigma_b^2 + n} = \frac{(\hat{s} - s)^2}{\sigma_b^2 + n}. \tag{4.20}$$

Since $\hat{s}$ is Gaussian distributed (it is the linear combination of Gaussian distributed variables, see Problem 4.9), with variance $n + \sigma_b^2$, it follows that $q(s)$ is distributed like a χ^2 with one degree of freedom, in agreement with Wilks' theorem, see Eq. (4.9). Notice that the effect of the uncertainty σ_b on the background yield is to broaden the negative log-likelihood function compared to the expectation from the sole Poisson statistics.

Problem 4.3 An experiment measures the decay times of an unstable particle with lifetime τ, produced at rest in the laboratory frame. Denote the set of N measurements by $\mathbf{X} = \{t_1, \ldots, t_N\}$. Show that in the limit $N \to \infty$, the ML estimator $\hat{\tau}$ is distributed as $\mathcal{N}(\tau, (N\, I_\tau)^{-1})$, where I_τ is the information, whereas $\partial_\tau \ln L(\mathbf{X} \mid \tau)$ is distributed as $\mathcal{N}(0, N\, I_\tau)$.

Solution

The decay times t are distributed according to the exponential law

$$f(t \mid \tau) = \frac{1}{\tau} e^{-\frac{t}{\tau}} \quad \Rightarrow \quad \mathrm{E}[t] = \tau, \quad \mathrm{Var}[t] = \tau^2. \tag{4.21}$$

The information I_τ is given by:

$$I_\tau = \mathrm{E}\left[-\left(\frac{\partial^2 \ln f(t \mid \tau)}{\partial \tau^2} \right) \right] = -\int_0^\infty dt \left(-\frac{2t}{\tau^3} + \frac{1}{\tau^2} \right) \frac{1}{\tau} e^{-\frac{t}{\tau}} = \frac{1}{\tau^2}. \tag{4.22}$$

If the N measurements are all independent, the likelihood of the data $\mathbf{X}$ is given by:

$$L(\mathbf{X} \mid \tau) = \prod_{i=1}^N f(t_i \mid \tau) = \frac{1}{\tau^N} e^{-\frac{\sum_{i=1}^N t_i}{\tau}}, \qquad \ln L(\mathbf{X} \mid \tau) = -N \ln \tau - \frac{\sum_{i=1}^N t_i}{\tau}. \tag{4.23}$$

The ML estimator $\hat{\tau}$ is given by the roots of Eq. (4.5), i.e.

$$0 = \frac{\partial}{\partial \tau} \ln L(\mathbf{X} \mid \hat{\tau}) = \sum_{i=1}^{N} \frac{\partial}{\partial \tau} \ln f(t_i \mid \hat{\tau}) = -\frac{N}{\hat{\tau}} + \frac{\sum_{i=1}^{N} t_i}{\hat{\tau}^2},$$

$$\hat{\tau} = \frac{\sum_{i=1}^{N} t_i}{N} = \frac{\tau}{\sqrt{N}} \underbrace{\frac{\sum_{i=1}^{N} (t_i - \tau)}{\sqrt{N}\tau}}_{z_N} + \tau \tag{4.24}$$

In the last equation, we have isolated the variable z_N, which the Central Limit theorem predicts to be asymptotically distributed as $\mathcal{N}(0, 1)$. Since the ML estimator is a linear function of z_N, it will be also (asymptotically) normally distributed, with mean and variance:

$$\mathrm{E}[\hat{\tau}] = \tau, \qquad \mathrm{Var}[\hat{\tau}] = \frac{\tau^2}{N} = \frac{1}{N I_\tau}, \tag{4.25}$$

where we have made use of Eq. (4.22). Finally, from Eq. (4.23) we can see that

$$\frac{\partial}{\partial \tau} \ln L(\mathbf{X} \mid \tau) = -\frac{N}{\tau} + \frac{\sum_{i=1}^{N} t_i}{\tau^2} = \frac{\sum_{i=1}^{N} (t_1 - \tau)}{\tau^2} = \frac{\sqrt{N}}{\tau} z_N. \tag{4.26}$$

The left-hand side of Eq. (4.26) is asymptotically normally distributed with mean and variance given by:

$$\mathrm{E}\left[\frac{\partial}{\partial \tau} \ln L(\mathbf{X} \mid \tau) \right] = \frac{\sqrt{N}}{\tau} \mathrm{E}[z_N] = 0 \tag{4.27}$$

$$\mathrm{Var}\left[\frac{\partial}{\partial \tau} \ln L(\mathbf{X} \mid \tau) \right] = \frac{N}{\tau^2} \mathrm{Var}[z_N] = N I_\tau \tag{4.28}$$

Suggested readings

The reader is addressed to Chap. 5 of Ref. [5] for further details on the information and its relation with the ML estimator.

Problem 4.4 Show that for a given random variable X, the quantity

$$\mathrm{E}[f(X)] \cdot \sqrt{\mathrm{Var}[X]}, \tag{4.29}$$

where f is the p.d.f. of X, admits a lower bound. Estimate this bound.

Discussion

This exercise complements Problem 3.28 by proving that the luminosity of a bunched collider whose bunch densities are independent is inversely proportional to the bunch RMS in the transverse plane. Indeed, the luminosity in such a case is proportional to $\int dx\, n^2(x)$, where n is the bunch particle density in the transverse direction.

Solution

Without loss of generality, we can assume $E[X] = 0$. We then introduce the non-normalised function ρ, such that $f = \rho \left[\int dx\, \rho\right]^{-1}$. We define the functional F as:

$$
F[\rho] \equiv \underbrace{\left(\int dx\, \rho(x)\right)^{-5}}_{I} \underbrace{\left(\int dx\, \rho^2(x)\right)^{2}}_{K} \underbrace{\left(\int dx\, x^2\rho(x)\right)}_{J} \tag{4.30}
$$

where I, K, and J are assumed to be finite, and $\langle X^2 \rangle$ is the variance of X. We shall prove that F has a minimum for the family of parabolic distributions

$$
\rho_0(x) = \max\left[a\left(1 - \frac{16}{9}x^2\right), 0\right], \tag{4.31}
$$

with $a > 0$. Indeed, if we can consider the infinitesimal variation $\delta\rho$ at position x, Eq. (4.30) gives:

$$
\delta F = \left(\frac{4\rho_0}{K} + \frac{x^2}{J} - \frac{5}{J}\right) F\, \delta\rho\, dx. \tag{4.32}
$$

By requiring $\delta F = 0$ for any $\delta\rho$ and dx, we then have the condition:

$$
\frac{4\rho_0}{K} + \frac{x^2}{J} - \frac{5}{J} = 0 \quad \Rightarrow \quad \rho_0(z) = \begin{cases} \frac{K}{4}\left(\frac{5}{I} - \frac{z^2}{J}\right) & -\left(\frac{5J}{I}\right)^{\frac{1}{2}} \leq x \leq \left(\frac{5J}{I}\right)^{\frac{1}{2}} \\ 0 & \text{otherwise} \end{cases}
$$

$$\tag{4.33}$$

However, I, K, and J are not independent. In particular, the definition is consistent provided that:

$$
I = \int dx\, \rho_0(x) = \frac{K}{4}\left(\frac{5}{I}\int dx - \frac{1}{J}\int dx\, x^2\right) = \frac{K}{4}\left(\frac{5J}{I}\right)^{\frac{1}{2}}\left[\frac{10}{I} - \frac{10}{3I}\right] = \frac{K}{3}\frac{5^{\frac{3}{2}}J^{\frac{1}{2}}}{I^{\frac{3}{2}}},
$$

$$
J = \int dx\, x^2\rho_0 = \frac{K}{4}\left(\frac{5}{I}\int dx\, x^2 - \frac{1}{J}\int dx\, x^4\right) = \frac{K}{4}\left(\frac{5J}{I}\right)^{\frac{3}{2}}\left[\frac{10}{3I} - \frac{2}{I}\right] = \frac{K}{3}\frac{5^{\frac{3}{2}}J^{\frac{3}{2}}}{I^{\frac{5}{2}}},
$$

$$
K = \int dx\, \rho_0^2 = \frac{K}{4}\left(\frac{5}{I}\int dx\, \rho_0 - \frac{1}{J}\int dx\, x^2\rho_0\right) = K. \tag{4.34}
$$

Hence, for $\rho_0(x)$ to be consistent, we need to have:

$$I^{\frac{5}{2}} J^{-\frac{1}{2}} K^{-1} \equiv F[\rho_0]^{-\frac{1}{2}} = \frac{5^{\frac{3}{2}}}{3} \quad \Rightarrow \quad F[\rho_0] = \frac{3^2}{5^3} = 0.072. \tag{4.35}$$

A straightforward calculation shows that the solution (4.30) satisfies Eq. (4.35) for any $a > 0$. Therefore:

$$E[f(X)] \cdot \sqrt{\mathrm{Var}[X]} \geq \sqrt{F[\rho_0]} = 0.268. \tag{4.36}$$

However, the functional F is not bounded from above: for example, consider the Cauchy distribution $\mathscr{C}$; in this case, $E[\mathscr{C}]$ is finite, but the variance is not, so that the functional F is infinite.

Suggested readings

This exercise is based on Ref. [6].

Bando n. 13153/2009

Problem 4.5 Discuss a method to simulate a data sample distributed according to a given function $f(x)$.

Discussion

Consider n random variables $\mathbf{x}$ distributed with probability density function $f(\mathbf{x})$. Let $y = h(\mathbf{x})$ be a generic function of $\mathbf{x}$. The variable y is also a random variable. The p.d.f. g of y is given by

$$g(y) = \int d\mathbf{x}\, f(\mathbf{x})\, \delta(y - h(\mathbf{x})) = \sum_{\tilde{x}_1} \int dx_2 \ldots dx_n \frac{f(\tilde{x}_1, x_2, \ldots, x_n)}{\left| \frac{\partial h}{\partial x_1}(\tilde{x}_1, x_2, \ldots, x_n) \right|}, \tag{4.37}$$

where $\tilde{x}_1$ are the roots of the equation $y - h(\mathbf{x}) = 0$.

Solution

A sample of data points distributed according to a generic p.d.f. f can be obtained by applying an analytical transformation to a sample of randomly distributed points $x \in [0, 1]$, i.e. with p.d.f. $g(x) = I_{[0,1]}$. If the numbers are randomly distributed on a different interval, one can always redefine x such that it is always contained within the $[0, 1]$ interval. Let's define

$$F(t) \equiv \int_{-\infty}^{t} dt'\, f(t'), \qquad y \equiv F^{-1}(x). \tag{4.38}$$

By using Eq. (4.37), it easy to verify that the sample of y values thus generated is distributed according to the target p.d.f. f. Indeed:

$$\frac{g\,(F(y))}{\left|\frac{dF^{-1}}{dx}\right|} = \left|\frac{dF}{dy}\right| = f(y), \tag{4.39}$$

where we have used the fact that $g(x) = 1$ and the chain rule $(f^{-1})' = (f')^{-1}$.

For the special case that f is Gaussian function of arbitrary mean μ and standard deviation σ, a cheaper method to generate the desired distribution relies on the Central Limit Theorem 4.2 which guarantees that the variable

$$z_N = \frac{\sum_i^N x_i - N/2}{\sqrt{N/12}} \tag{4.40}$$

for uniformly distributed points x_i is asymptotically distributed like $\mathcal{N}(0, 1)$. By considering the variable $y = \sigma z + \mu$, we can then obtain the desired target distribution.

If $f \sim e^{-x/2} x^{n/2-1}$, i.e. a χ_n^2 distribution with n degrees of freedom, the above method can be used to generate n independent variables z_i, and then considering the sum $y = \sum_{i=1}^{n} z_i^2$.

Suggested readings

The method of generating events distributed according to a prior model by means of random numbers is called *Monte Carlo* generation, and represents one of the most used tool to simulate multi-dimensional problems. Examples relevant for HEP include the simulation of collision events and of particle interaction with the detector. For an overview on MC methods, see Ref. [5].

Bando n. 1N/R3/SUB/2005

Problem 4.6 If a particle with mean lifetime τ has not decayed after a time t, what is the probability that it decays within the following time interval Δt?

Solution

The exponential law has "no memory". Indeed, the probability that the particle decays at a time greater than t_1, provided that the particle has not yet decayed at the time t_0, can be computed by Bayes theorem:

$$\text{Prob}\,[t \geq t_1 \,|\, t \geq t_0] = \frac{\text{Prob}\,[t \geq t_1,\, t \geq t_0]}{\text{Prob}\,[t \geq t_0]} = \frac{e^{-t_1/\tau}}{e^{-t_0/\tau}} = e^{-(t_1 - t_0)/\tau}, \tag{4.41}$$

with $t_1 > t_0$. Hence, the probability of surviving over a time $\Delta t = t_1 - t_0$, given that the particle has survived up to a time t_0, depends only on Δt and not on the intermediate time t_0. We can therefore conclude that the probability of decaying in a time window Δt starting from any time at which the particle is still undecayed, is always given by $1 - e^{-\Delta t/\tau}$.

Problem 4.7 Determine the p.d.f. of the variable $t = t_1 + t_2$, where t_1 and t_2 are independent random variables with exponential distribution with mean τ_1 and τ_2, respectively.

Solution

Since t is a known function of the random variables t_1 and t_2, whose joint p.d.f. is also known, we can use Eq. (4.37) to determine the p.d.f. $g(t)$:

$$g(t) = \frac{1}{\tau_1 \tau_2} \int_0^\infty dt_1\, dt_2\, e^{-\frac{t_1}{\tau_1} - \frac{t_2}{\tau_2}} \delta(t - t_1 - t_2) = \frac{1}{\tau_1 \tau_2} e^{-t/\tau_2} \int_0^t dt_1\, e^{-t_1\left(\frac{1}{\tau_1} - \frac{1}{\tau_2}\right)} =$$

$$= \frac{1}{\tau_2 - \tau_1} e^{-t/\tau_2} \left[1 - e^{-t\frac{\tau_2 - \tau_1}{\tau_1 \tau_2}} \right] = \frac{e^{-t/\tau_2} - e^{-t/\tau_1}}{\tau_2 - \tau_1}, \tag{4.42}$$

with $t \geq 0$. As made clear by Eq. (4.42), the p.d.f. $g(t)$ is *not* of exponential form.

Discussion

The p.d.f. $g(t)$ describes the differential time distribution of a stable isotope C produced from the nuclear decay chain $A \rightarrow B \rightarrow C$, see Problem 4.8. If one of the two lifetimes is much larger than the other, the p.d.f. $g(t)$ reduces to an exponential law with lifetime given by the maximum between τ_1 and τ_2.

Bando n. 18211/2016

Problem 4.8 A nucleus A undergoes α-decay to a nucleus B with a lifetime $\tau_A = 2$ min. The nucleus B then decays to a nucleus C with lifetime $\tau_B = 5 \times 10^3$ s. At the beginning there are 2.7×10^7 nuclei of type A. Compute the activity of B after 1.2 s and after 5×10^3 s.

Discussion

The time evolution of the three populations N_A, N_B, and N_C can be obtained analytically by solving a joint system of ODE that describes the detailed input-output balance:

$$\begin{cases} \dot{N}_A = -\lambda_A N_A \\ \dot{N}_B = +\lambda_A N_A - \lambda_B N_B \\ \dot{N}_C = +\lambda_B N_B \end{cases} , \quad \text{with} \quad \begin{cases} N_A(0) = N_0 \\ N_B(0) = 0 \\ N_C(0) = 0 \end{cases} \tag{4.43}$$

with $\lambda_i = 1/\tau_i$. The first of Eq. (4.43) can be readily integrated yielding an exponential decay $N_A(t) = N_0\, e^{-\lambda_A t}$, which can be then inserted in the second equation giving a first-order polynomial ODE with a uniform term. The solution can be obtained by using the general algorithm for solving this type of ODE, or by trying an *ansatz* solution given by an arbitrary linear combinations of $e^{-\lambda_A t}$ and $e^{-\lambda_B t}$ functions. The equation and the boundary conditions then fix the unknown coeffcients.

Solution

Instead of solving Eq. (4.43), we take a "probabilistic" approach where the time of a nuclear decay is treated as a random variable: the p.d.f $g(t)$ derived in Eq. (4.42) thus describes how the arrival of the nuclei C is distributed in time. The population $N_C(t)$ is thus proportional to the cumulative function of g evaluated at time t:

$$N_C(t) = N_0 \int_0^t dt'\, g(t') = \frac{N_0}{\tau_B - \tau_A} \left[\tau_B \left(1 - e^{-t/\tau_B}\right) - \tau_A \left(1 - e^{-t/\tau_A}\right)\right] =$$

$$= N_0 \left[1 + \frac{\tau_A\, e^{-t/\tau_A} - \tau_B\, e^{-t/\tau_B}}{\tau_B - \tau_A}\right] \tag{4.44}$$

In order to find $N_B(t)$, we first notice that the total number of nuclei $N_A + N_B + N_C$ must be conserved in time, so that:

$$N_A(t) + N_B(t) + N_C(t) = N_0, \qquad N_B(t) = N_0 - N_A(t) - N_C(t) =$$

$$= N_0 \left[1 - e^{-t/\tau_A} - \left(1 + \frac{\tau_A e^{-t/\tau_A} - \tau_B e^{-t/\tau_B}}{\tau_B - \tau_A}\right)\right] = \frac{N_0\, \tau_B}{\tau_B - \tau_A} \left(e^{-t/\tau_B} - e^{-t/\tau_A}\right).$$

$$\tag{4.45}$$

The activity $\mathscr{A}_B$ of the nucleus B is defined as the rate of $B \to C$ decays in the sample, i.e.:

$$\mathscr{A}_B(t) = \lambda_B\, N_B(t) = \frac{N_0}{\tau_B - \tau_A} \left(e^{-t/\tau_B} - e^{-t/\tau_A}\right). \tag{4.46}$$

Given that $\tau_B \gg \tau_A$, we can approximate the activities at the time $t_1 = 1.2\,\text{s} \ll \tau_A$ and $t_2 = 5 \times 10^3\,\text{s} = \tau_B \gg \tau_A$ as:

$$\mathscr{A}_B(t_1) \approx \frac{N_0}{\tau_B} \left(1 - \frac{t_1}{\tau_B} - 1 + \frac{t_1}{\tau_A}\right) \approx \frac{N_0\, t_1}{\tau_B\, \tau_A} = 0.54 \times 10^2\,\text{Bq} \tag{4.47}$$

$$\mathscr{A}_B(t_2) \approx \frac{N_0}{\tau_B} \frac{1}{e} = 0.20 \times 10^4\,\text{Bq}, \tag{4.48}$$

where we have expressed the results in the standard units of $\text{Bq} = 1\,\text{Hz}$. This result is intuitive: after a time interval small compared to the lifetime of B, almost all the decayed nuclei A have been transmuted into B; the sample thus contains a population of B with size $\lambda_A\, t_1\, N_0$, and the activity of B is $\lambda_B\, \lambda_A\, t_1\, N_0$. After a time interval t_2 much larger than τ_A, the initial sample of nuclei A has fast transmuted into B: the activity of B is thus the same as if the sample at time $t = 0$ had been entirely made of nuclei B.

Suggested readings

For more details on the decay chain problem see e.g. Sect. 1.11 of Ref. [7].

Problem 4.9 Prove that the sum of an arbitrary number of Gaussian random variables is still a Gaussian variable.

Discussion

This can be more elegantly proved by going into the Fourier-transform space, or, in probability language, by using the *characteristic function* of a p.d.f.. Given a random variable X with p.d.f. $f(X)$, its characteristic $\phi_X(t)$ is defined as

$$\phi_X(t) = \mathrm{E}\left[e^{itX}\right] = \int_{-\infty}^{+\infty} dx\, e^{itx} f(x), \qquad f(x) = \frac{1}{2\pi} \int_{-\infty}^{+\infty} dt\, e^{-itx} \phi_X(t).$$

$$(4.49)$$

The second equation shows how to revert back to the x space. From its definition, it also follows that the characteristic of the sum of N independent variables is given by the product of the N characteristic:

$$\phi_{\sum X_i}(t) = \prod_{i=1}^{N} \phi_{X_i}(t).$$

$$(4.50)$$

The method of characteristics is often used in probability to compute the p.d.f. of composite variables, including discrete ones, see e.g. the classical proof of the Central Limit theorem.

Solution

Let's define $Z = \sum_i^N X_i$. We can apply Eq. (4.50) to give:

$$\phi_Z(t) = \prod_{i=1}^{N} \phi_{X_i}(t) = \prod_{i=1}^{N} \exp\left[i\,\mu_i\,t - \frac{1}{2}\sigma_i^2 t^2\right] = \exp\left[i\underbrace{\left(\sum_i^N \mu_i\right)}_{\mu} t - \frac{1}{2}\underbrace{\left(\sum_i^N \sigma_i^2\right)}_{\sigma^2} t^2\right]$$

$$g(Z) = \frac{1}{2\pi} \int_{-\infty}^{+\infty} dt\, e^{-i(Z-\mu)t - \frac{1}{2}\sigma^2 t^2} = \mathcal{N}(Z \mid \mu, \sigma).$$

$$(4.51)$$

Hence, $g(Z)$ is still a Gaussian p.d.f. Notice that the mean and variance of Z could have been guessed *a priori* as a consequence of the linear dependence with respect to the N variables.

Problem 4.10 Show that the sum of an arbitrary number of independent Poisson variables is still a Poisson variable.

Solution

It suffices to consider the sum of two variables: if their sum is still Poisson-distributed, by recursive sum of pairs of variables we can prove that also the sum of an arbitrary number of such variables is still Poisson-distributed. Let's denote these two variables by m and n, and their means by μ and ν, respectively. We want to prove that $k = m + n$ is Poisson-distributed with mean $\kappa = \mu + \nu$. Under the assumption that m and n are independent, the probability of observing k is given by

$$
\text{Prob}\,[k] = \sum_{m=0}^{k} \mathscr{P}(m \mid \mu)\, \mathscr{P}(k - m \mid \nu) = \sum_{m=0}^{k} \frac{e^{-(\mu+\nu)}\mu^m \nu^{k-m}}{m!\,(k-m)!} =
$$

$$
= \frac{e^{-(\mu+\nu)}(\mu+\nu)^k}{k!} \underbrace{\sum_{m=0}^{k} \frac{k!}{m!\,(k-m)!} \left(\frac{\mu}{\mu+\nu}\right)^m \left(1 - \frac{\mu}{\mu+\nu}\right)^{k-m}}_{=1} =
$$

$$
= \mathscr{P}(k \mid \mu + \nu). \tag{4.52}
$$

Hence, k is a Poisson variable with mean $\mu + \nu$.

Discussion

This fact has important implications in counting experiments where the channel under study receives contribution from several uncorrelated processes (e.g. a signal process plus number of independent sources of background): the number of events counted in the bin of the observable under study is still Poisson-distributed with mean given by the sum of means of all contributing processes.

Problem 4.11 Prove that a Poisson p.d.f. approaches a Gaussian p.d.f. with the same mean and variance in the limit $N \to \infty$.

Solution

We can use Stirling formula to approximate the asymptotic behaviour of $n!$:

$$
n! \approx \sqrt{2\pi n}\left(\frac{n}{e}\right)^n. \tag{4.53}
$$

By making use of Eq. (4.53), the Poisson p.d.f. becomes:

$$
\mathscr{P}(n \mid \mu) \approx \frac{1}{\sqrt{2\pi n}} e^{-\mu}\mu^n \left(\frac{e}{n}\right)^n = \frac{1}{\sqrt{2\pi n}} e^{-(n-\mu)+n\ln\frac{\mu}{n}} =
$$

$$
= \frac{1}{\sqrt{2\pi n}} e^{-(n-\mu)+n\left(-\frac{n-\mu}{n} - \frac{(n-\mu)^2}{2n^2} + \mathscr{O}(\frac{1}{n^2})\right)} = \mathscr{N}(n \mid \mu, \sqrt{\mu}) + \mathscr{O}(n^{-\frac{1}{2}}),
$$

$$
\tag{4.54}
$$

Fig. 4.2 Comparison between $\mathscr{P}(n \mid \mu)$ (solid line) and $\mathscr{N}(n \mid \mu, \sqrt{\mu})$ (dashed line) for different values of μ. For the Poisson p.d.f., the $n!$ term is replaced by its analytical continuation $\Gamma(x+1)$

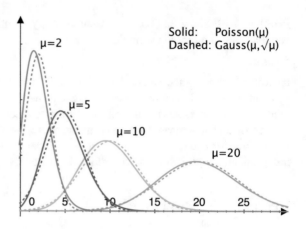

where we have Taylor-expanded the logarithm around $\mu/n \approx 1$. For illustration, Fig. 4.2 compares $\mathscr{P}(n \mid \mu)$ and $\mathscr{N}(n \mid \mu, \sqrt{\mu})$ for different values of μ.

Discussion

Thanks to this result, the bin contents n_i of a histograms, such that $n_i \gg 1$, can be treated as gaussian variables with standard deviation $\sigma_i = \sqrt{n_i}$. This proves very useful for statistical tests, like the Pearson χ^2 test, since the distribution of the test statistic becomes that of a χ^2 variable. Notice that we could have predicted the large sample behaviour of the Poisson p.d.f. by invoking the Central Limit theorem. Indeed, since any Poisson process with large mean can be seen as the sum of an arbitrarily large number of Poisson variables with finite variance, see Problem 4.10, the limiting distribution must be a Gaussian.

Bando n. 1N/R3/SUB/2005

Problem 4.12 An experiment selects signal events at a rate s and background events at a rate b. Determine the data taking time needed to observe the signal with a statistical significance of n standard deviations.

Solution

Let's denote the data taking time by T_{run}. The statistical significance for a counting experiment with large event counts is defined as the signal yields N_S in units of the standard deviations of the background, which for large event counts can be estimated by $\sqrt{N_B}$. Hence:

$$n \leq \frac{N_S}{\sqrt{N_B}} = \frac{s\, T_{\text{run}}}{\sqrt{b\, T_{\text{run}}}} \quad \Rightarrow \quad T_{\text{run}} \geq n^2 \frac{b}{s^2} \tag{4.55}$$

For a given significance, the data taking time scales with an increased signal rate more favourably than for the background: for a small relative increase $s \to (1 + \varepsilon_S)s$, the background is allowed to undergo a twofold relative increase without changing the significance for a given data taking time.

Problem 4.13 Prove that the number of countings in a fixed time window T of a random process characterised by a uniform probability of occurrence per unit time λ is distributed according to the Poisson p.d.f.

Solution

Let's divide the time window T into N equal subranges of duration $\tau = T/N$. For sufficiently large N, τ will become much smaller than λ^{-1}, so that the probability of single-event in a given subrange will be given by $\lambda\tau \ll 1$, while the probability of double-event $(\lambda\tau)^2$ can be neglected. The probability of observing n outcomes can be computed by using the Binomial law for n positive events (i.e. the number of single-events in any of the subranges) given N trials, each with success probability $\lambda\tau$:

$$\text{Prob}\,[n \mid N, \lambda, T] = \frac{N!}{(N-n)!\,n!} \left(\frac{\lambda T}{N}\right)^n \left(1 - \frac{\lambda T}{N}\right)^{N-n} \tag{4.56}$$

Taking the limit $N \to \infty$ such that λT is finite, we get:

$$\lim_{N \to \infty} \text{Prob}\,[n \mid N, \lambda, T] = \frac{(\lambda T)^n}{n!} e^{-\lambda T} \lim_{N \to \infty} \frac{N(N-1)\ldots(N-n+1)}{N^n} e^{+\lambda T \frac{n}{N}} =$$

$$= \frac{(\lambda T)^n\, e^{-\lambda T}}{n!} \equiv \frac{\mu^n\, e^{-\mu}}{n!} = \mathscr{P}(n \mid \mu), \tag{4.57}$$

with $\mu = \lambda T = \text{E}\,[n] = \text{Var}\,[n]$.

Bando n. 13153/2009

Problem 4.14 The surface of a given detector is irradiated by a unform flux. The counting rate on a single channel is ν. Let the sampling time be T. Determine the probability of having empty events and the probability of having single events.

Solution

The number of events integrated over the sampling time T is distributed with a Poisson p.d.f.. Indeed, if the flux is uniform, the interaction of each particle with the detector can be seen as a random process characterised by a uniform probability of occurrence per unit time, thus giving a Poisson distribution for the number of interactions in a fixed time window, see Problem 4.13. Hence:

$$\text{Prob}\,[n = 0] = \mathscr{P}(0 \mid \nu T) = e^{-\nu T}$$

$$\text{Prob}\,[n = 1] = \mathscr{P}(1 \mid \nu T) = e^{-\nu T}(\nu T) \tag{4.58}$$

By measuring the zero-event probability, the flux ν can be measured by simply inverting the first of Eq. (4.58). Also the single-event probability can be used to derive ν through the second of Eq. (4.58). The solution is however twofold, as we have already seen for the analogous problem of measuring the dead time of a paralyzable system, see Problem 3.38.

Bando n. 13153/2009

Problem 4.15 A Monte Carlo program simulates a scattering process. After running for some time, a sample of $N \gg 1$ events is generated, of which N_+ events have weight $w > 0$ while N_- of events with weight $-w$. Let the cross section of the simulated process be σ. Determine the equivalent integrated luminosity of this sample and the relative uncertainty on the simulated event yield.

Discussion

When a sample consists of N weighted events each with weight w_i, $i = 1, \ldots, N$, the event yield Y and its variance σ_Y^2 are given by:

$$\begin{cases} Y = \sum_{i=1}^{N} w_i \\ \sigma_Y^2 = \sum_{i=1}^{N} w_i^2 \end{cases} \quad \Rightarrow \quad \frac{\sigma_Y}{Y} = \frac{\sqrt{\sum_{i=1}^{N} w_i^2}}{\sum_{i=1}^{N} w_i}. \tag{4.59}$$

If $N \gg 1$, the Law of Large Numbers allows us to approximate the values of Y and σ_Y^2 by using the p.d.f. of the weights w, $f(w)$ (which we assume to be well defined, i.e. that $\int dw \, f(w) = 1$). We can therefore write:

$$\begin{cases} Y \approx N \, \mathrm{E}[w] = N \, \mu_w \\ \sigma_Y^2 \approx N \, \mathrm{E}[w^2] = N(V_w + \mu_w^2) \end{cases} \quad \Rightarrow \quad \frac{\sigma_Y}{Y} \approx \frac{1}{\sqrt{N}} \left(1 + \frac{V_w}{\mu_w^2} \right)^{\frac{1}{2}}, \tag{4.60}$$

where μ_w and V_w are the mean and variance of the weights w, respectively. As one can see, the relative uncertainty is always equal or larger than for unweighted events, the equality being realised when $w_i \equiv w$ for any i, i.e. $V_w = 0$.

The use of weighted MC events is pretty common in HEP analyses. For example, when the MC events consists of one or more correlated variables, the use of event weights allows to modify the conditional distributions compared to the model used for the generatorion. In some cases, like the simulation of scattering events at NLO accuracy, or statistical tools like *sPlot* [8], the events weights can be negative, although observables should be always associated with positive cross sections.

Solution

The total event yield and the equivalent integrated luminosity of the sample are given by

$$Y = \sum_i w_i = w\,(N_+ - N_-) = w\,N\,(1 - 2f),$$

$$\mathscr{L}_{int.} = \frac{Y}{\sigma} = \frac{w\,N\,(1 - 2f)}{\sigma}, \tag{4.61}$$

where we have introduced the fraction of negative events $f = N_-/N$. The mean and RMS of the weight distributions can be readily computed:

$$\mu_w = -w\,f + w\,(1 - f) = w\,(1 - 2f)$$
$$V_w = w^2\,f + w^2\,(1 - f) - w^2\,(1 - 2f)^2 = w^2\left[1 - (1 - 2f)^2\right]. \tag{4.62}$$

From Eq. (4.60), we can determine the relative uncertainty on the event yield:

$$\frac{\sigma_Y}{Y} = \frac{1}{\sqrt{N}}\left(\frac{1}{1 - 2f}\right). \tag{4.63}$$

Therefore, the presence of a fraction f of negative-weight events increases the relative uncertainty of the simulated sample by a factor of $(1 - 2f)^{-1}$ compared to the Poisson expectation from unweighted events.

_____

Bando n. 1N/R3/SUB/2005

Problem 4.16 Given an event containing four b quarks, what is the probability of observing at least one lepton if the branching ratio $b \to X\,\ell\,\nu$ is equal to 20%?

Solution

The b decays are independent from each other. The number n_ℓ of leptonic decays is therefore distributed according to a Binomial law with single-event probability $\varepsilon = 0.2$, hence:

$$\text{Prob}\,[n_\ell \geq 1] = 1 - \mathscr{B}(0 \mid N = 4, \varepsilon = 0.2) = 1 - (1 - \varepsilon)^4 = 59\%. \tag{4.64}$$

Discussion

Thanks to a lifetime of $\mathscr{O}(10^{-12}$ s), B hadrons generated from the hadronisation of b quarks of sufficiently large momentum give rise to distinctive signatures that can help discriminate them from jets initiated by lighter quarks or gluons. With the use of modern vertex detectors, rejection power against light jets as large as $(1 - \beta)^{-1} \sim 100$ can be obtained for efficiencies in excess of $1 - \alpha \sim 50\%$ (we have used the symbols α and β for the errors of first and second kind, respectively). See Problem 2.46 for more details. Another distinctive feature of b-initiated jets is the presence of a charged lepton. Besides the prompt lepton production $b \to X\,\ell\,\nu$ mentioned by the exercise, an additional source of charged leptons arises from the semileptonic decay of charmed mesons produced in the decay chain of the initial

B hadrons. Overall, the branching ratio for producing at least one charged lepton is large as 35% [9].

Suggested readings

For more details on recent b tagging techniques at hadron colliders, the reader is addressed to Refs. [10, 11]

Bando n. 13705/2010

Problem 4.17 A crystal contains two kinds of impurities A and B in equal amount. The impurity A absorbs photons without emitting electrons, while the impurity B absorbes photons by emitting electrons. The absorbtion cross section of A is 99 times larger than the cross section for B. Let the dimension of the crystal be such that all entering photons get absorbed, and let us assume that 200 photons impinge the crystal. What is the probability that at least three electrons are emitted?

Solution

Given an input photon, the probability that it gets absorbed by the impurity B is

$$\varepsilon = \frac{\sigma_B}{\sigma_A + \sigma_B} = 0.01 \tag{4.65}$$

The number of emitted electrons $n_{p.e}$ given $N = 200$ incident photons is distributed according to a Binomial law with single-event probability ε. Since $\varepsilon N = 2$ is finite, we can approximate this distribution with a Poissonian of mean εN. The probability of emitting at least three electrons is therefore given by:

$$\text{Prob}\left[n_{p.e.} \geq 3\right] = 1 - \sum_{i=0}^{2} \mathscr{P}(i \mid \varepsilon N) =$$
$$= 1 - (0.1353 + 0.2706 + 0.2707) \approx 32\%. \tag{4.66}$$

Notice that the result obtained by using a Binomial p.d.f. agrees with the Poissonian estimate to better than 10^{-5}.

Problem 4.18 An experiment measures events containing a pair of particles reconstructed as electrons. The collected data sample, denoted by PP, is affected by two sources of background: the first consists of one real electron and one pion mis-identified as an electron; the second consists of two mis-identified pions. The pion mis-identification probability $f_{1,2}^{i}$ for both reconstructed objects and for each event i is known from simulation. Two independent data samples, containing, respectively, exactly one (PF) and exactly zero (FF) reconstructed electrons are also measured. Assume a prefect identification probability for real electrons. Estimate the total background yield in the PP sample.

Solution

Under the assumption that the electron identification probability is perfect, the PF sample is contaminated by both one- and two-pion events, whereas the FF sample contains only two-pion events. The probabilities for all possible combinations of wrong and correct identification of a background events are:

$$
\begin{cases}
\text{Prob}\,[\text{FF} \mid \pi\,\pi] = (1 - f_1^i)(1 - f_2^i) \\
\text{Prob}\,[\text{PF} \mid \pi\,\pi] = f_1^i(1 - f_2^i) + (1 - f_1^i)f_2^i \\
\text{Prob}\,[\text{PP} \mid \pi\,\pi] = f_1^i f_2^i \\
\text{Prob}\,[\text{PF} \mid \pi\,e] = 1 - f_a^i \\
\text{Prob}\,[\text{PP} \mid \pi\,e] = f_a^i
\end{cases}
\tag{4.67}
$$

where the index $a = 1, 2$ refers to the particle not-reconstructed as an electron. We can express the number of background events in the PP region *per* event falling in the FF or PF region by taking the ratio between the respective probabilities. The number of two-pion events in the PP sample can be therefore estimated as

$$
N_{\text{PP} \mid \text{FF}}^{\pi\,\pi} = \sum_{i \in \text{FF}} \frac{f_1^i f_2^i}{(1 - f_1^i)(1 - f_2^i)}.
\tag{4.68}
$$

Since the PF sample contains both one- and two-pion events, the quantity

$$
N_{\text{PP} \mid \text{PF}}^{\pi\,e + \pi\,\pi} = \sum_{j \in \text{PF}} \frac{f_a^j}{1 - f_a^j},
\tag{4.69}
$$

where the sum runs over the PF sample, provides the correct background yield in the PP sample from one-pion events, but it underestimates the contribution from two-pion events. In particular, by summing together Eqs. (4.68) and (4.69), the contribution from two-pion events would be overestimated because some events get double-counted. The contribution of two-pion events to the sum in Eq. (4.69) can be however estimated from the FF sample to be

$$
N_{\text{PP} \mid \text{PF} \mid \text{FF}}^{\pi\,\pi} = \sum_{i \in \text{FF}} \underbrace{\frac{f_1^i}{1 - f_1^i}}_{\text{PF} \mid \text{FF}} \underbrace{\frac{f_2^i}{1 - f_2^i}}_{\text{PP} \mid \text{PF}} + \underbrace{\frac{f_2^i}{1 - f_2^i}}_{\text{PF} \mid \text{FF}} \underbrace{\frac{f_1^i}{1 - f_1^i}}_{\text{PP} \mid \text{PF}} = 2 \sum_{i \in \text{FF}} \frac{f_1^i}{1 - f_1^i} \frac{f_2^i}{1 - f_2^i}
\tag{4.70}
$$

The total background yield in the PP region is therefore given by:

$$
\underbrace{N_{\text{PP} \mid \text{FF}}^{\pi\,\pi}}_{\pi\,\pi} + \underbrace{N_{\text{PP} \mid \text{PF}}^{\pi\,e + \pi\,\pi} - N_{\text{PP} \mid \text{PF} \mid \text{FF}}^{\pi\,\pi}}_{\pi\,e} = \sum_{j \in \text{PF}} \frac{f_a^j}{1 - f_a^j} - \sum_{i \in \text{FF}} \frac{f_1^i}{1 - f_1^i} \frac{f_2^i}{1 - f_2^i}.
\tag{4.71}
$$

Notice that we made the assumption that the mis-identification probabilities factorise. This exercise illustrates an example of a data-driven technique for estimating the background contamination in events with mis-identified particles.

4.2 Error Propagation

Error propagation refers to the estimation of the variance of a random variable z which is a known funtion h of an array of random variables $\mathbf{x}$, whose covariance matrix V is assumed to be known. The linearised error propagation method is based on a first-order expansion of the function h around the experimental point $\bar{\mathbf{x}}$:

$$z = h(\mathbf{x}) = h(\bar{\mathbf{x}}) + \sum_i \partial_i h(\bar{\mathbf{x}}) \, (x_i - \bar{x}_i) + \dots \tag{4.72}$$

Given the linear approximation of Eq. (4.72), the variance of z can be determined as:

$$\mathrm{Var}\,[z] \approx \sum_{ij} \partial_i h(\bar{\mathbf{x}}) \, V_{ij} \, \partial_j h(\bar{\mathbf{x}}) \tag{4.73}$$

Since V is a symmetric positive-definite matrix, $\mathrm{Var}\,[z]$ from Eq.(4.73) is always positive.

Problems

Problem 4.19 Consider a random variable $z = h(x, y)$, where (x, y) is a pair of random variables with mean value (x_0, y_0) and covariance matrix V. Derive an expression for σ_z^2 by assuming that h can be expanded to first order around (x_0, y_0). Then, specialise the result to two special cases: $h = a\,x + b\,y$ and $h = x/y$.

Solution

For the case $h = a\,x + b\,y$, the linear expansion of Eq. (4.72) is exact. The gradient of h is constant and given by $\nabla h_0 = (a, b)$. Equation (4.73) then gives:

$$\sigma_z^2 = (a\ \ b) \begin{pmatrix} \sigma_x^2 & \rho\sigma_x\sigma_y \\ \rho\sigma_x\sigma_y & \sigma_y^2 \end{pmatrix} \begin{pmatrix} a \\ b \end{pmatrix} = a^2\sigma_x^2 + b^2\sigma_y^2 + 2\,a\,b\,\rho\,\sigma_x\,\sigma_y \tag{4.74}$$

For the special case $\rho = 0$, we get the result $\sigma_z = (a\,\sigma_x) \oplus (b\,\sigma_y)$.

For the case $h = x/y$, the gradient of the function at the mean value x_0, y_0 is given by $\nabla h_0 = y_0^{-1}(1, -x_0/y_0)$. Equation (4.73) then gives:

$$\sigma_z^2 = \frac{1}{y_0^2} (1 \quad - x_0/y_0) \begin{pmatrix} \sigma_x^2 & \rho\sigma_x\sigma_y \\ \rho\sigma_x\sigma_y & \sigma_y^2 \end{pmatrix} \begin{pmatrix} 1 \\ -x_0/y_0 \end{pmatrix} =$$

$$= \frac{1}{y_0^2} \left[\sigma_x^2 - 2\frac{x_0}{y_0}\rho\sigma_x\sigma_y + \sigma_y^2\frac{x_0^2}{y_0^2} \right]$$

$$= z_0^2 \left[\left(\frac{\sigma_x}{x_0}\right)^2 + \left(\frac{\sigma_y}{y_0}\right)^2 - 2\rho\left(\frac{\sigma_x}{x_0}\right)\left(\frac{\sigma_y}{y_0}\right) \right] \qquad (4.75)$$

Notice that this result could have been obtained directly from Eq. (4.75) by applying the propagation of error to $\ln z = \ln x - \ln y$ and noticing that $d \ln z = dz/z$. For the special case $\rho = 0$, we get the result $\sigma_z/z = (\sigma_x/x) \oplus (\sigma_y/y)$.

Suggested readings

There is a large number of textbooks dedicated to statistical methods for particle physics. For example, the reader is addressed to Refs. [5, 12] for a more detailed discussion.

Bando n. 1N/R3/SUB/2005

Problem 4.20 After having collected an integrated luminosity $\mathscr{L}_{int} = 10\,\text{fb}^{-1}$, an analysis of $B^0 \to J/\Psi\,K_S^0$, with $J/\Psi \to \ell^+\ell^-$ and $K_S \to \pi^+\pi^-$, selects $N = 100$ candidate events. By using $N_{MC} = 1000$ simulated events, the selection efficiency is estimated to be $\varepsilon = 37\%$. What are the measured BR, the related statistical uncertainty, and the systematic uncertainty due to having simulated too few events? Assume $\sigma_{B^0} \approx 1\,\text{nb}$.

Discussion

Given a data sample of N scattering events corresponding to an integrated luminosity $\mathscr{L}_{int}$, the measured cross section is

$$\sigma = \frac{N - N_B}{\varepsilon\,\mathscr{L}_{int}}, \qquad (4.76)$$

where N_B is the estimated background yield in the data sample and ε is the total efficiency (including the acceptance) of the experiment to the signal. The latter is usually estimated by using a Monte Carlo simulation of the signal process.

Solution

By using Eq. (4.76), the branching ratio for $B^0 \to J/\Psi\,K_S$ can be measured as:

$$\text{BR}_{J/\Psi\,K^0} = \frac{\sigma_{\ell^+\ell^-\pi^+\pi^-}}{\sigma_{B^0}\,\text{BR}_{\pi^+\pi^-}\,\text{BR}_{\ell^+\ell^-}} = \frac{N}{\varepsilon\,\mathscr{L}_{int}\,\sigma_{B^0}\,\text{BR}_{\pi^+\pi^-}\,\text{BR}_{\ell^+\ell^-}} =$$

$$= \frac{100}{0.37 \cdot 10\,\text{fb}^{-1} \cdot 1\,\text{nb} \cdot 0.69 \cdot (0.0596 \times 2)} = 3.3 \times 10^{-4}, \qquad (4.77)$$

where we have used the PDG values for $BR_{\pi^+\pi^-}$ and $BR_{\ell^+\ell^-}$ [9]. Under the assumption that the uncertainty on the integrated luminosity and on the inclusive cross section are negligible, the relative uncertainty on the branching ratio can be estimated from Eq. (4.75):

$$\frac{\delta BR_{J/\Psi K^0}}{BR_{J/\Psi K^0}} \approx \sqrt{\left(\frac{1}{\sqrt{N}}\right)^2 + \left(\frac{\delta\varepsilon}{\varepsilon}\right)^2}, \tag{4.78}$$

where the first term is the pure statistical uncertainty, while the second represent a systematic uncertainty. The latter be evaluated by using the expectation from a Binomial distribution:

$$\left(\frac{\delta\varepsilon}{\varepsilon}\right)^2 = \frac{N_{MC}\,\varepsilon\,(1-\varepsilon)}{N_{MC}^2\,\varepsilon^2} = \frac{(1-\varepsilon)}{N_{MC}\,\varepsilon}. \tag{4.79}$$

Putting everything together, we thus have:

$$\frac{\delta BR_{J/\Psi K^0}}{BR_{J/\Psi K^0}} \approx \sqrt{\frac{1}{100} + \frac{1-0.37}{1000\cdot 0.37}} \approx 11\%. \tag{4.80}$$

The extra uncertainty due to the limited size of MC simulated events is therefore $11\% - 1/\sqrt{100} \approx 1\%$.

Suggested readings

The decay channel $B^0 \to J/\Psi\, K_S^0$ (branching ratio 8.7×10^{-4} [9]) plays an important role in B physics as a tool to measure one of the angles of the unitarity triangle. More precisely, the propagation of a physical neutral B meson initially produced as a B_0 or $\bar{B}_0$ features time oscillations due to $B_0 - \bar{B}_0$ mixing. The amplitude of such oscillations as measured via the $J/\Psi\, K_S^0$ decay, a channel which can be produced by both neutral B states, provides a direct measurement of the CP-violating parameter $\sin 2\beta$. See Problem 5.43 for more details on this subject.

Bando n. 13705/2010

Problem 4.21 The ratio between neutral and charged current interactions from neutrinos on nucleus are measured by two independent experiments, CTF at Fermilab and CDHS at CERN, to be: CTF: 0.27 ± 0.02 and CDHS: 0.295 ± 0.01. What is the combined result? Are the two measurements consistent with each other?

Discussion

Point-estimation from a combination of measurements with covariance matrix V can be performed by using the *least squares method*, which coincides with the ML estimator for normally distributed variables. The least squares estimator of the mean of an ensemble of measurements is the solution of the equation

$$0 = \frac{\partial}{\partial \mu} \left[\frac{1}{2} (\mathbf{x} - A\mu)^\mathsf{T} V^{-1} (\mathbf{x} - A\mu) \right] \tag{4.81}$$

$$\Rightarrow \quad \begin{cases} \hat{\mu} = \left(A^\mathsf{T} V^{-1} A \right)^{-1} A^\mathsf{T} V^{-1} \mathbf{x} = \frac{\sum_{ij} (V^{-1})_{ij} x_i}{\sum_{ij} (V^{-1})_{ij}} \\ \sigma_{\hat{\mu}}^{-2} = A^\mathsf{T} V^{-1} A = \sum_{ij} (V^{-1})_{ij} \end{cases} \tag{4.82}$$

where we have defined $E[\mathbf{x}] = A\mu = (1, 1, \ldots, 1)^\mathsf{T} \mu$. For the special case of uncorrelated measurements, V^{-1} is diagonal and the formula simplifies to the well-known result:

$$\hat{\mu} = \frac{\sum_i \frac{x_i}{\sigma_i^2}}{\sum_i \frac{1}{\sigma_i^2}}, \qquad \sigma_{\hat{\mu}}^2 = \frac{1}{\sum_i \frac{1}{\sigma_i^2}}. \tag{4.83}$$

Solution

The combined measurement can be obtained by using the least square estimator of Eq. (4.83):

$$\hat{\mu} \pm \sigma_{\hat{\mu}} = 0.290 \pm 0.009. \tag{4.84}$$

The compatibility between the two measurements can be quantified by using the χ^2 test-statistic t defined as:

$$t = \sum_i \frac{(x_i - \hat{\mu})^2}{\sigma_i^2} = \frac{(0.27 - 0.290)^2}{(0.02)^2} + \frac{(0.295 - 0.290)^2}{(0.01)^2} = 1.25. \tag{4.85}$$

If the measurements are nornally distributed, the test statistic t is distributed like a χ^2 with $n = 1$ degree of freedom. From tabulated tables, one can find that the p-value for this experiment is 26%. The two measurements are therefore well compatible with each other.

Suggested readings

For a more formal treatment of methods for point-estimation, the reader is addressed to Chap. 7 of Ref. [5].

Bando n. 13153/2009

Problem 4.22 An invariant mass distribution shows a Gaussian peak centered at a mass M, it contains N events, and has a width at half maximum equal to D. Determine the precision on the mass measurement M.

Solution

For a Gaussian distributed variable, the relation between its standard deviation and FWHM is given by Eq. (2.103). The combination of N independent measurementes gives an uncertainty on the mean:

$$\sigma_M = \frac{\sigma}{\sqrt{N}} = \frac{D}{2.35\sqrt{N}}. \tag{4.86}$$

Bando n. 18211/2016

Problem 4.23 A sample of 2500 measurements of a quantity x are normally distributed. From this sample, a mean value of $\bar{x} = 34.00 \pm 0.06$ (68% CL) is obtained. What is the probability that a subsequent measurement results in a value $x \geq 37$?

Solution

The variable $\bar{x}$ is Gaussian distributed, see Problem 4.9, hence its two-sided confidence interval at CL $= 1 - \alpha = 68\%$ has a width of 2σ. The latter is related to the standard deviation of the single-measurement, σ_x, by

$$\sigma_{\bar{x}} = \frac{\sigma_x}{\sqrt{N}} \quad \Rightarrow \quad \sigma_x = 0.06 \cdot \sqrt{2500} = 3.0 \tag{4.87}$$

Since $\sigma_{\bar{x}}/\sigma_x \ll 1$, we can assume the mean value μ to be $\bar{x}$. The probability p that a Gaussian variable of mean $\mu = \bar{x}$ fluctuates more than $(37 - \bar{x})/\sigma_x = 1.0$ standard deviations from its mean is $p = \alpha/2 = 16\%$.

Bando n. 13705/2010

Problem 4.24 A radioactive source emits two kinds of uncorrelated radiation A and B. This radiation is observed through a counter which is able to distinguish A from B. In a given time interval, the countings are $N_A = 200$ and $N_B = 1000$. What is the statistical uncertainty on the ratio $R = N_A/N_B$? In case the observation time were reduced by a factor of 250, do you think that the new ratio $r = n_A/n_B$ would represent an unbiased estimator of the true ratio between the two activities? In case of negative answer, how would you correct for it?

Solution

The variable R thus defined coincides with the ML estimator for the parameter $\rho = \mu_A/\mu_B$. This can be easily verified. Indeed, the likelihood of the two observations can be parametrised in terms of ρ and one of the two mean counting rates, e.g. μ_B, as:

$$L(N_A, N_B; \rho, \mu_B) = \frac{e^{-\mu_B(1+\rho)} \rho^{N_A} \mu_B^{N_A+N_B}}{N_A! N_B!}. \tag{4.88}$$

Setting the derivatives of L with respect to both parameters to zero, one gets the expected result for the ML estimators:

$$\hat{R} = N_A/N_B, \qquad \hat{\mu}_B = N_B. \tag{4.89}$$

The asymptotic properties of the ML estimator [5] guarantee that $\hat{R}$ is an unbiased estimator, if a large number of observations are made. For just one experiment, this is in general not the case, see e.g. Chap. 7 of Ref. [5]. However, since the variance of the individual rates is known, it is possible to estimate the variance of R by standard error propagation, which gives the result:

$$\frac{\sigma_R}{R} = \sqrt{\left(\frac{\sigma_A}{N_A}\right)^2 + \left(\frac{\sigma_B}{N_B}\right)^2} = \sqrt{\frac{1}{N_A} + \frac{1}{N_B}} = 7.7\%, \tag{4.90}$$

see Eq. (4.73). This result assumes that the function $R(N_A, N_B)$ can be expanded around the mean values μ_A and μ_B, and that the relative variations of the countings of order σ_A/A and σ_B/B around these values are small. This assumption breaks up if at least one of the two countings is small. In particular, if the observation time gets reduced by a factor of 250, the mean counting rates for A and B will be of order 1 and 4, respectively, thus invalidating the estimate of Eq. (4.90). The variance of the estimator r will be larger.

A possible way to overcome this problem is to repeat the measurements $N \gg 1$ times, which reduces the uncertainty on the estimator of the mean number of decays n_A and n_B by a factor of $1/\sqrt{N}$: for sufficiently large values of N, the ratio $r = n_A/n_B$ becomes again well behaved.

4.3 Confidence Intervals

In the literature, one can find two approaches to the definition of confidence levels, that follow either the *frequentist* or *Bayesan* approach. Only the former is considered here. According to the frequentist paradigm, the parameter μ is said to be contained inside a confidence interval $I_\mu(\mathbf{X})$ at a confidence level CL $= 1 - \alpha$ if:

$$\text{Prob}\left[\mu \in I_\mu(\mathbf{X})\right] = \alpha, \tag{4.91}$$

irrespectively of the true (and possibly unknown) value of μ. In almost all the cases, this is only possible inference on the true value of the parameter μ that an experimental can provide. The quantity α, which is coventional and should be chosen *a priori*, quantifies the amount of error of the first kind associated with the measurement, i.e. the probability that the true value does not lie within the confidence interval. For a more refined discussion on the subject, the reader is addressed to dedicated textbooks like Refs. [5, 12].

Problems

Bando n. 13153/2009

Problem 4.25 What is the correct statistical approach to assess the efficiency of a detector and of its related uncertainty?

Solution

The efficiency of a detector is defined as the number of output signals M per input events N, i.e. $\hat{\varepsilon} = M/N$, where $\hat{\varepsilon}$ is the estimator of the unknown detector efficiency ε which is assumed to be independent of the time of arrival of the signals or their rate. The number of output signals for a fixed value of N is distributed according to a binomial law:

$$\mathscr{B}(M \mid N, \varepsilon) = \frac{N!}{M!\,(N-M)!}\varepsilon^M\,(1-\varepsilon)^{N-M}. \tag{4.92}$$

A frequentist confidence interval I_ε at the confidence level CL $= 1 - \alpha$ can be built by using the Neyman construction [13], i.e. as the interval of ε values for which the p-value corresponding to the measured value M is smaller than α (for one sided-intervals), or $\alpha/2$, for symmetric and double-sided intervals. An example of Neyman construction for a generic confidence interval is shown in Fig. 4.3.

Suggested readings

The concept of confidence interval is comprehensively illustrated in Ref. [14].

Problem 4.26 Determine the one-sided 95% CI on the efficiency ε of a detector that reconstructs each of the N pulses sent in, for the cases $N = 50$ and $N = 5000$.

Solution

Referring to Problem 4.25 with $M = N$, the best estimate for the efficiency is $\hat{\varepsilon} = 1$ and the one-sided confidence interval at a given confidence level CL $= 1 - \alpha$ is given by

$$I_\varepsilon = \left[\alpha^{\frac{1}{N}}, 1\right] = \begin{cases} [0.9418, 1.0] & N = 50 \\ [0.9994, 1.0] & N = 5000 \end{cases} \tag{4.93}$$

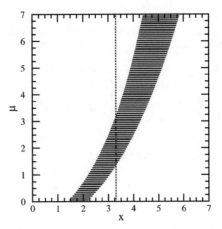

Fig. 4.3 A generic confidence interval construction. For each value of μ, one draws a horizontal acceptance interval $[x_1, x_2]$ such that $\text{Prob}\,[x \in [x_1, x_2]|\mu] = \alpha$. Upon performing an experiment to measure x and obtaining the value x_0, one draws the dashed vertical line through x_0. The confidence interval $[\mu_1, \mu_2]$ is the union of all values of μ for which the corresponding acceptance interval is intercepted by the vertical line (from Ref. [14])

By construction, such intervals contain the true value with a probability $\text{CL} = 95\%$ regardless of the true, unknown value of ε.

Problem 4.27 A counting experiment measures the number of outcomes of a process of unknown yield μ. Compare the 95% CL upper limit on μ as a function of the experimental outcome N obtained by using the Neyman construction with the confidence interval at the same confidence level obtained by using Wilks' approximation of the negative log-likelihood function.

Solution

The upper limit at $1 - \alpha = 95\%$ CL, denoted by μ_{ul}, can be obtained by using the Neyman construction of confidence levels, i.e.:

$$\alpha = \text{Prob}[n \leq N] = \sum_{n=0}^{N} \frac{\exp\left[-\mu_{\text{ul}}\right] \mu_{\text{ul}}^n}{n!}. \tag{4.94}$$

Under the assumption that only one process contributes, the likelihood of the data is given by

$$L(n \mid \mu) = \frac{\exp\left[-\mu\right] \mu^n}{n!}, \tag{4.95}$$

Table 4.1 The upper limits at 95% CL on the mean μ of a Poisson process as a function of the outcome N, obtained by using the classical Neyman construction (second column) and by using the asymptotic approximation of the negative log-likelihood ratio (third column)

N	Neyman	Wilks	Rel. diff. (%)
1	4.75	3.65	23
2	6.30	5.31	16
5	10.5	9.6	8
10	17.0	16.1	5
20	29.1	28.3	3
50	63.3	62.6	1
100	118.1	117.4	0.6

Fig. 4.4 The negative log-likelihood ratio $q(\mu)$ from a Poisson process with mean μ for different values of the outcome N. For illustration, the functions $q(\mu)$ are compared with their parabolic approximation

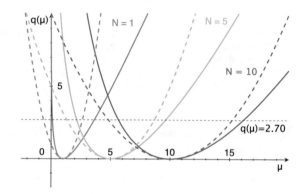

from which one obtains the intuitive result for the ML estimator: $\hat{\mu} = N$. Wilks' theorem states that the negative log-likelihood ratio q is asymptotically distributed like a χ^2. For the experiment under consideration, one has:

$$q(\mu) = -2 \ln \frac{L(N \mid \mu)}{L(N \mid \hat{\mu})} = -2 (N - \mu) - 2 N \log \frac{\mu}{N}. \qquad (4.96)$$

If $q(\mu)$ were distributed like a χ^2 with one d.o.f., the upper limit at 95% CL would be given by the root of the solution $q(\mu_{ul}) = 2.70$. Table 4.1 reports the upper limits at 95% CL obtained by using the two alternative approaches. The values have been obtained by using the program reported in Appendix 4.3. Already for $N = 10$, the asymptotic approximation is better than 5%. Indeed, for large values of N the Poisson distribution approaches a Gaussian, see Problem 4.11, which satisfies Wilks' theorem by construction. For illustration, Fig. 4.4 shows the function $q(\mu)$ for three particular values of N, and compares them to the respective parabolic approximation.

Problem 4.28 In a Cherenkov counter, on average $n_\gamma = 20$ photons are produced per incident particle. If the efficiency of conversion of each photon into photoelectrons is $\varepsilon_Q = 20\%$, given $N = 1000$ incoming particles, what is the mean number of particles (and relative uncertainty) that are undetected?

Solution

The number of p.e. produced per incident particle is distributed like a Poissonian with mean $\varepsilon_Q n_\gamma$. The probability of emitting at least one p.e., which is the condition for the particle to be detected, is therefore given by:

$$\varepsilon = \text{Prob}\left[n_{\text{p.e.}} > 0\right] = 1 - \mathscr{P}(0 \mid \varepsilon_Q n_\gamma) = 1 - e^{-\varepsilon_Q n_\gamma} = 0.982. \qquad (4.97)$$

Given N incoming particles, the number of detected particles is distributed according to a binomial p.d.f., see Eq. (4.92), with efficiency ε given by Eq. (4.97) and variance $N \varepsilon (1 - \varepsilon)$. For large values of N such that εN is finite, the binomial distribution approaches a Poissonian of mean εN, which approaches a Gaussian for large values of εN, see Problem 4.11. We can therefore use the interval with boundaries $\varepsilon N \pm \sqrt{N \varepsilon (1 - \varepsilon)}$ as an estimator of the 68% quantile for ε, i.e.

$$\bar{N} = N \varepsilon \pm \sqrt{N \varepsilon (1 - \varepsilon)} = 982 \pm 4 \quad \text{at} \quad 68\% \text{ CL.} \qquad (4.98)$$

Appendix 1

This computer program computes the frequentist 95% CL upper limit (Neyman construction) and the approximate 95% CL upper limit from the asymptotic properties of the negative log-likelihood ratio (Wilks' theorem), for a Poisson process of unknown mean value μ, see Problem 4.27. A numerical scan is performed in order to find the appropriate isocontour levels.

```
import math

# grid points for numerical scan
nmax=1000

# the\index{Poisson density} Poisson p.d.f.
def pois(n=1,mu=1):
    return math.exp(-mu)*math.pow(mu, n)/math.factorial(n)

def get_CL_neyman(n=1, CL=0.95):
    alpha = 1-CL
    muL = n
    muH = n+3*math.sqrt(n)+5 # a 'reasonable' upper limit
```

```
        step = (muH-muL)/nmax
    for mu_step in range(nmax):
        mu = muL+step*mu_step
        prob = 0.0
        for x in range(n+1):
            prob += pois(n=x, mu=mu)
        if prob<alpha:
            break
    return [mu, prob]


def get_CL_wilks(n=1., chi2=2.705):
    muL = n
    muH = n+3.*math.sqrt(n)+5.
    step = (muH-muL)/nmax
    for mu_step in range(nmax):
        mu = muL+step*mu_step
        nll = -2*(n-mu) - 2*n*math.log(mu/n)
        if nll > chi2:
            break
    return [mu,nll]


# run!
for n in [1,2, 5, 10, 20, 50, 100]:
    ney = get_CL_neyman(n=n, CL=0.95)
    wil = get_CL_wilks(n=n, chi2=2.705)
    print n, ney[0],  wil[0], (wil[0]-ney[0])/ney[0]
```

References

1. J. Jacod, P. Protter, *Probability Essentials*, 2nd edn. (Springer, Berlin, 2004)
2. A. Wald, Trans. Am. Math. Soc. **54**(3), 426–482 (1943). https://doi.org/10.2307/1990256
3. S.S. Wilks, Ann. Math. Stat. **9**, 60–62. https://doi.org/10.1214/aoms/1177732360
4. G. Cowan et al., Eur. Phys. J. C **71**, 1554 (2011). https://doi.org/10.1140/epjc/s10052-011-1554-0
5. F. James, *Statistical Methods in Experimental Physics*, 2nd edn. (World Scientific, Hackensack, 2006)
6. H.G. Hereward, CERN/MPS/DL 69–12 (1969)
7. W.R. Leo, *Techniques for Nuclear and Particle Physics Experiments*, 2nd edn. (Springer, Berlin, 1993)
8. M. Pivk, F.R. Le Diberder, Nucl. Instrum. Meth. A **555**, 356–369 (2005). https://doi.org/10.1016/j.nima.2005.08.106
9. C. Patrignani et al., (Particle Data Group). Chin. Phys. C **40**, 100001 (2016)
10. ATLAS Collaboration, JINST **11**, P04008 (2016). https://doi.org/10.1088/1748-0221/11/04/P04008
11. C.M.S. Collaboration, JINST **8**, P04013 (2013). https://doi.org/10.1088/1748-0221/8/04/P04013
12. G. Cowan, *Statistical Data Analysis* (Oxford University Press, Oxford, 1998)
13. J. Neyman, Phil. Trans. Royal Soc. Lond. Ser. A **236**, 333–380 (1937)
14. G.J. Feldman, R.D. Cousin, Phys. Rev. D **57**, 3873–3889 (1998). https://doi.org/10.1103/PhysRevD.57.3873

Chapter 5
Subnuclear Physics

Abstract The problems collected in the fifth chapter deal with particle physics from a more theoretical perspective. They are mostly concerned with the phenomenological predictions of Standard Model of particle interactions. After introducing the concept of symmetry, which is of cardinal importance in the formulation of a good theory, the proposed exercises will focus on the physics of electroweak and strong interactions. The two last sections are dedicated to flavour physics and to the Higgs boson.

5.1 Conservation Laws

The principles of quantum mechanics and the invariance of physics under transformations of the Poincaré group impose stringent constraints on the mathematical structure of a consistent theory. Nature preferentially realises a number of additional symmetries which, once introduced in the theory, provide further constraints. By Noether's theorem, the invariance of the Lagrangian $\mathscr{L}$ under a group of transformations gives rise to a conserved current, or, in quantum-mechanical language, to a conserved quantum number. Classifications of states in terms of their conserved quantum numbers proves of the greatest help to give a rationale to the experimental observations.

Problems

Problem 5.1 Show that rotation-invariance and Bose–Einstein symmetry implies that a spin-1 particle cannot decay into two photons.

Solution

The quantum-mechanical amplitude for the decay $X \to \gamma\,\gamma$, with $J_X = 1$ must be a Lorentz-scalar and linear in the polarisation vectors of the particle X and of the two photons. We denote these vectors in the centre-of-mass by $\boldsymbol{\eta}$, $\boldsymbol{\varepsilon}_1$, and $\boldsymbol{\varepsilon}_2$. The other independent vector with a defined $1 \leftrightarrow 2$ symmetry is the relative momentum $\mathbf{k} = \mathbf{k}_1 - \mathbf{k}_2$. Furthermore, for a massless spin-1 particle, the Klein–Gordan equation

© Springer International Publishing AG 2018
L. Bianchini, *Selected Exercises in Particle and Nuclear Physics*,
UNITEXT for Physics, https://doi.org/10.1007/978-3-319-70494-4_5

$\partial_\nu^2 A_\mu = 0$ implies $\boldsymbol{\varepsilon} \cdot \mathbf{k} = 0$. Out of these four vectors, the only combinations linear in the polarisation vectors and with a definite symmetry with respect to the two photon indices are:

$$(\boldsymbol{\varepsilon}_1 \cdot \boldsymbol{\varepsilon}_2)(\boldsymbol{\eta} \cdot \mathbf{k})$$

$$(\boldsymbol{\varepsilon}_1 \times \boldsymbol{\varepsilon}_2) \cdot \boldsymbol{\eta}, \qquad (\boldsymbol{\varepsilon}_1 \times \boldsymbol{\varepsilon}_2) \cdot (\boldsymbol{\eta} \times \mathbf{k}), \qquad (\boldsymbol{\varepsilon}_1 \times \boldsymbol{\varepsilon}_2)(\boldsymbol{\eta} \cdot \mathbf{k}) \qquad (5.1)$$

A term like $(\boldsymbol{\varepsilon}_1 \times \mathbf{k}) \cdot (\boldsymbol{\varepsilon}_2 \times \mathbf{k})(\boldsymbol{\eta} \cdot \mathbf{k})$ is equivalent to the first of Eq. (5.1) modulo a function of $\mathbf{k}^2$. The first, second, and fourth are odd under $1 \leftrightarrow 2$ exchange, whereas Bose–Einstein symmetry requires the amplitude to be symmetric under photon label exchange. The third combination is identically zero since:

$$(\boldsymbol{\varepsilon}_1 \times \boldsymbol{\varepsilon}_2) \cdot (\boldsymbol{\eta} \times \mathbf{k}) = \varepsilon_{ijk} \, \varepsilon_1^j \varepsilon_2^k \, \varepsilon^{ilm} \, \eta_l \, k_m = (\delta_j^l \delta_k^m - \delta_k^l \delta_j^m) \varepsilon_1^j \, \varepsilon_2^k \, \eta_l \, k_m =$$

$$= (\boldsymbol{\varepsilon}_1 \cdot \boldsymbol{\eta})(\boldsymbol{\varepsilon}_2 \cdot \mathbf{k}) - (\boldsymbol{\varepsilon}_2 \cdot \boldsymbol{\eta})(\boldsymbol{\varepsilon}_1 \cdot \mathbf{k}) = 0. \qquad (5.2)$$

Since no amplitude consistent with the selection rules can be constructed, the decay must be forbidden.

Discussion

This result, known as *Yang's theorem*, has been advocated at the time when the first evidence of the 125 GeV Higgs boson was being gathered in the diphoton channel $H \to \gamma \gamma$ as a proof that the particle had to be a boson, since it decays to a pair of spin-1 particles, and furthermore $J \neq 1$. This theorem also implies that the decay of the neutral pseudo-vector mesons, like $\phi^0, \omega^0, \rho^0, J/\Psi, \Upsilon$, to a pair of photons is forbidden, as well as the radiative decay $Z^0 \to \gamma \gamma$.

Suggested Readings

This result was first formalised by Yang in 1950, see his original paper [1].

Bando n. 18211/2016

Problem 5.2 Which of the following electromagnetic transitions:

$$\frac{1^+}{2} \to \frac{1^-}{2} \qquad 0^+ \to 0^- \qquad \frac{3^+}{2} \to \frac{1^-}{2} \qquad 2^+ \to 1^+ \qquad 1^+ \to 0^+, \qquad (5.3)$$

are allowed in electric dipole (E1) or magnetic dipole (B1) approximation?

Discussion

The electric and magetic dipole approximations consist in assuming an interaction Hamiltonian of the form

$$\mathscr{H}_{\text{E1}} = e\,\mathbf{r} \cdot \mathbf{E}, \qquad \mathscr{H}_{\text{B1}} = -\frac{e}{2m}\,(\mathbf{L} + g\mathbf{S}) \cdot \mathbf{B}. \tag{5.4}$$

Under rotations, both $\mathbf{r}$ and $\mathbf{L}$ transform as vectors. According to the *Wigner–Eckart theorem*, the matrix elements of a tensor operator T_q^k between states of definite total momentum take the form:

$$\langle J\,m|T_q^k|J'\,m'\rangle = C_{m'\ q\ m}^{J'\ k\ J}\,\underbrace{\langle J||T^k||J'\rangle}_{\text{reduced element}}, \tag{5.5}$$

where $C_{m_1\ m_2\ M}^{j_1\ j_2\ J}$ are the *Clebsh–Gordan coefficients*, defined by the relation:

$$|j_1\,m_1\rangle|j_2\,m_2\rangle = \sum_J C_{m_1\ m_2\ m_1+m_2}^{j_1\ j_2\ J}|j_1\,j_2\mid J\,m_1+m_2\rangle. \tag{5.6}$$

For the case $j_1 = 0$, $j_2 = 1$, the Clebsh–Gordan coefficients vanish because the tensor product of two states with total angular momentum zero does not overlap with states of angular momentum one, and indeed:

$$C_{m_1\ m_2\ 0}^{j_1\ j_2\ 0} = (-1)^{j_1-m_1}\frac{1}{\sqrt{2j_1+1}}\delta_{j_1\,j_2}\delta_{m_1\,m_2}. \tag{5.7}$$

Hence, for vector operators ($k = 1$), the matrix elements between states with $J = 0$ vanish: the transition $0 \to 0$ cannot occur through dipole operators. The other selection rule for E1 and B1 transitions between states of definite total angular momentum is:

$$\Delta J = 0, \pm 1, \qquad \Delta m_J = 0, \pm 1. \tag{5.8}$$

While $\mathbf{r}$ is a proper vector, $\mathbf{L}$ and $\mathbf{S}$ are pseudovectors. This implies an important selection rule: E1 transitions change the parity of the state, whereas B1 transitions do not.

Solution

The first and third transitions have $\Delta J = 0, 1$ and involve a change in parity: they are therefore compatible only with a E1 transition. The fourth and fifth have $\Delta J = 1$ and do not change parity: they must be B1 transitions. The second one is forbidden for both transitions: it has to proceed through scalar or tensor operators.

Suggested Readings

The Wigner–Eckart theorem is discussed in several textbooks of classical Quantum Mechanics. See Appendix B of Ref. [2] for a compendium of formulas.

Problem 5.3 Why is the decay $n \to p e^-$ forbidded even without postulating lepton-number conservation?

Solution

A putative $n \to p e^-$ decay would violate angular momentum conservation, i.e. invariance of the Hamiltonian under rotations. Indeed, a state made of two spin-$1/2$ particles can only have integer total angular momentum, while the initial state has $J = 1/2$.

Discussion

Lepton-number is an accidental symmetry of the SM Lagrangian, see also Problem 5.47, bringing to stringent selection rules on the decays of hadrons. The most stringent limits on lepton-number violating decays of neutrons into an antilepton (e^+, μ^+) and a charged meson (π^-, ρ^-, K^-) have been provided by the water-Cherenkov IMB-3 experiment at the Fairport Salt Mine [3], giving lower bounds on the lifetimes of order 10^{32} years at 90% CL. The most stringent limits on lepton-violating proton decays come from the Super-Kamiokande experiment, see Problem 2.29, ranging between 10^{33} and 10^{34} years at 90% CL, depending on the channel under study [4]. In several BSM extensions, like GUT theories where the quark and lepton doublets are embedded into multiplets of higher dimensions, the lepton number L and baryon number B are not individually conserved.

Another reaction which would bring to an evidence of lepton-number violation is the neutrinoless-double β decay ($0\nu\beta\beta$). In this case, the lepton number symmetry would be broken by the Majorana mass term.

Suggested Readings

For more details on the limits on the nucleon lifetimes, the reader is addressed to Refs. [3, 4].

Problem 5.4 What is the total wavefunction of deuterium? The binding energy of deuterium is $\mathscr{B}_D = -2.22$ MeV. Argue why this implies that the mass of the neutron has to be fine-tuned compared to the proton and electron mass at the 10^{-3} level to allow for star nucleosynthesis.

Solution

The determination of the total wavefunction of deuterium (D) relies on symmetry arguments and experimental observations. First, one notices that a proton-neutron state can either have $I = 0$ ($\Psi \propto |pn\rangle - |np\rangle$) or $I = 1$ ($\Psi \propto |pn\rangle + |np\rangle$). The first state is antisymmetric under exchange $p \leftrightarrow n$, the second is symmetric. Deuterium is a mass eigenstate of the strong Hamiltonian, hence it must have a definite value of

I. Let us assume $I = 1$. In this case, one should expect two more degenerate states, $|pp\rangle$ and $|nn\rangle$, which are not observed neither as stable particles, nor as resonances in nucleon-nucleon scattering. Hence, $I = 0$. Since the deuterium wavefunction needs to be overall antisymmetric under $p \leftrightarrow n$ (Fermi–Dirac statistics), the rest of the wavefunction needs to be symmetric. For a pair of spin-1/2 particles, the symmetry is $(-1)^L(-1)^{S+1}$. The observation that $J(D) = 1$ implies three possible values for the angular momentum L:

1. $L = 0$, $S = 1$: the symmetry is $+1$;
2. $L = 1$, $S = 0, 1$: the symmetry is $+1$ for $S = 0$, and -1 for $S = 1$; hence only the former is acceptable;
3. $L = 2$, $S = 1$: the symmetry is $+1$;

Therefore, deuterium can be either a superimposition of $^3S_1$ and $^3D_1$, or a pure $^1P_1$ state: mixing between the three would result in a state of undefined parity, since $P = (-1)^L$, which is not possible since the strong dynamics conserves parity. A pure $L = 1$ state is however discarded by the observation that the nuclear magnetic moment of deuterium agrees within $\approx 2\%$ with the sum of the proton and nuclear magnetic moments, which is the expectation from a $S = 1$ state (aligned spins). Finally, the observation of a finite quadrupole moment requires the presence of a small fraction of $^3D_1$, hence:

$$\Psi_D = \frac{|pn\rangle - |np\rangle}{\sqrt{2}} \times \left(\alpha|^3S_1\rangle + \sqrt{1 - \alpha^2}|^3D_1\rangle \right).$$

Coming to the fine-tune question, we notice that deuterium plays a key role in the nuclear synthesis occurring in stars, since the first exothermic reaction towards helium synthesis is:

$$p\,p \rightarrow D\,e^+\,\nu_e.$$

This reaction can take place provided that:

$$2\,m_p > m_D + m_e = m_p + m_n - \mathscr{B}_D + m_e \quad \Leftrightarrow \quad (m_n - m_p + m_e) < 2.22\ \text{MeV}.$$

This inequality is indeed satisfied, since $m_n - m_p + m_e \approx (1.3 + 0.5)$ MeV $=$ 1.8 MeV. However, if only the neutron mass had been only 0.4 MeV heavier than what it is, the fusion reaction would be endothermic. One can also notice that a neutron mass smaller than 0.8 MeV compared to its nominal value, would make the decay $n \rightarrow p\,e^-\,\bar{\nu}_e$ not possible, with dramatic implications for nucleosynthesis. We can therefore conclude that the neutron mass is fine-tuned for nuclear synthesis to better than $\mathscr{O}(1\ \text{MeV})/m_n \approx 0.1\%$.

Problem 5.5 Consider a heavy nucleus $^A_Z X$ decaying to a lighter nucleus $^{A'}_{Z'} Y$ plus additional particles. The reaction actually involves a neutral atom. Write down the reaction in terms of the initial nucleus and of its decay products for the following decays, together with the equation for the Q-value:

- β^- decay
- β^+ decay
- α decay
- electron capture (EC)

Solution

Let's denote the atomic mass of the neutral atom by $\mathcal{M}$, and the binding energy of the atomic electrons by $\mathcal{B}_e$. After a nuclear decay, the atom is in an excited or incomplete electronic configuration, $\mathcal{M}^*$, since the electron orbitals must also adapt to the new nuclear configuration. Unless the electron belongs to a low-lying level (K-shell), this binding energy is negligible compared to the other mass scales, see e.g. Eq. (2.24). We then have:

$$\beta^- \text{ decay}: \quad {}^A_Z X \quad \rightarrow \quad {}^A_{Z+1}Y \, e^- \, \bar{\nu}_e \tag{5.9}$$
$$\mathcal{M}({}^A_Z X) > \mathcal{M}^*({}^A_{Z+1}Y) + m_e = \mathcal{M}({}^A_{Z+1}Y) + \mathcal{B}_e$$

$$\beta^+ \text{ decay}: \quad {}^A_Z X \quad \rightarrow \quad {}^A_{Z-1}Y \, e^+ \, \nu_e \tag{5.10}$$
$$\mathcal{M}({}^A_Z X) > \mathcal{M}^*({}^A_{Z-1}Y) + m_e = \mathcal{M}({}^A_{Z-1}Y) - \mathcal{B}_e + 2\,m_e$$

$$\alpha \text{ decay}: \quad {}^A_Z X \quad \rightarrow \quad {}^{A-4}_{Z-2}Y \, \alpha \tag{5.11}$$
$$\mathcal{M}({}^A_Z X) > \mathcal{M}({}^{A-4}_{Z-2}Y) + \mathcal{M}({}^4_2 \text{He})$$

$$\text{EC}: \quad {}^A_Z X \quad \rightarrow \quad {}^A_{Z-1}Y \, \nu_e \tag{5.12}$$
$$\mathcal{M}({}^A_Z X) > \mathcal{M}^*({}^A_{Z-1}Y) = \mathcal{M}({}^A_{Z-1}Y) + \mathcal{B}^K_e$$

Discussion

The EC is energetically favoured compared to the β^+ decay, in the sense that if β^+ is allowed, then also EC must be possible with larger Q-value. This can be easily seen from the equations above, since $\mathcal{B}^K_e \approx 10 \div 100$ KeV $\ll 2\,m_e \approx 1$ MeV.

Suggested Readings

For an introduction to nuclear decays, the reader is addressed to Chap. 1 of Ref. [5].

Bando n. 1N/R3/SUB/2005

Problem 5.6 A π^- beam on a polarised target produces Λ baryons, which are analysed in momentum $\mathbf{p}$ and polarisation $\boldsymbol{\varepsilon}$. Explain how this measurement allows one to test if the dynamics conserves parity.

Discussion

A standard method to test the parity-conservation of strong interactions is through the detection of an asymmetry between the inclusive cross sections when the polarisations of the beam or of the target is flipped. When strange baryons are produced, like Λ or Σ, their polarisation can be measured from the distribution of the decay products in the centre-of-mass. Since the Λ and Σ baryons decay by the parity-violating electroweak interaction, the centre-of-mass angle between the decay products and the baryon momentum, taken as quantisation axis, is distributed as

$$\frac{1}{\Gamma}\frac{d\Gamma}{d\Omega^*} = 1 + \alpha\, P_L\, \cos\theta^*, \tag{5.13}$$

where α is determined by the $V - A$ interaction, while P_L is the longitudinal polarisation of the baryon. When the polarity of the outgoing hadrons cannot be analysed through their decay products, as it is the case for protons and neutrons, the measurement of a *left-right asymmetry* with respect to a transverse polarisation, which is instead compatible with invariance under parity, can be used to measure the level of longitudinal polarisation: the latter can be turned into a transverse polarisation by means of resonant magnetic fields.

Solution

The polarisation of the Λ baryons can be extracted from the distribution of the $N\,\pi$ decay angle in the centre-of-mass according to Eq. (5.13). If parity is conserved at the production level, the Λ baryons should not be longitudinally polarised. Only transverse polarisation is allowed, if parity needs to be conserved. Indeed, a polarisation orthogonal to the scattering plane,

$$P_T = \langle S \cdot \hat{\mathbf{n}} \rangle, \quad \text{with} \quad \hat{\mathbf{n}} = \frac{\mathbf{p} \times \mathbf{p}'}{|\mathbf{p} \times \mathbf{p}'|}, \tag{5.14}$$

is unaffected by the parity operation, since $\hat{\mathbf{n}} \to \hat{\mathbf{n}}$. On the contrary, a longitudinal polarisation P_L, i.e. a polarisation onto the plane spanned by $\mathbf{p}$ and $\mathbf{p}'$ changes sign under parity. Hence, and indication of longitudinal polarisation would indicate that parity is violated by the production dynamics.

Suggested Readings

The use of Λ and $\bar{\Lambda}$ particles as polarimeters has been explored by the ALEPH Collaboration in $e^+ e^-$ events at the Z^0 peak [6]. Since the neutral-weak interactions are parity violating, the produced Λ ($\bar{\Lambda}$) developed a negative (positive) longitudinal polarisation, $|P_L| \approx 0.32$, a leftover of the preferred helicity state of the left-handed

s ($\bar{s}$) quark. See Ref. [7] for a broader discussion on the experimental tests of parity violation.

Problem 5.7 Determine the C, P, and $C P$ eigenvalues for a particle-antiparticle bound state for both spin-0 and spin-1/2 particles.

Solution

We can use Landau's rule, which relates the properties of transformation of particle-antiparticle bound states in a given configuration of spin S and orbital angular momentum L to the spin-statistics properties of the constitutents. For a spin-0 particle-antiparticle state, we have:

$$\begin{cases} P = \underbrace{(-1)^L}_{\mathbf{p}\leftrightarrow\mathbf{p'}} \\ \underbrace{+1}_{(1)\leftrightarrow(2)} = \underbrace{(-1)^L}_{\mathbf{p}\leftrightarrow\mathbf{p'}} \underbrace{C}_{b^\dagger\leftrightarrow d^\dagger}, \quad C = (-1)^L \end{cases} \quad \Rightarrow \quad C P = +1. \tag{5.15}$$

For a spin-1/2 particles-antiparticle state, we have instead:

$$\begin{cases} P = \underbrace{(-1)}_{\eta_P=-\bar{\eta}_P} \underbrace{(-1)^L}_{\mathbf{p}\leftrightarrow\mathbf{p'}} = (-1)^{L+1} \\ \underbrace{-1}_{(1)\leftrightarrow(2)} = \underbrace{(-1)^L}_{\mathbf{p}\leftrightarrow\mathbf{p'}} \underbrace{C}_{\chi\leftrightarrow\bar{\chi}} \underbrace{(-1)^{S+1}}_{s\leftrightarrow s'}, \quad C = (-1)^{L+S} \end{cases} \quad \Rightarrow \quad C P = (-1)^{S+1}.$$

$$\tag{5.16}$$

In Eq. (5.16), we have used the fact that, for Dirac fermions, particle and antiparticles are assigned opposite parity.

Discussion

Among the others, these properties imply selection rules on the allowed decay of mesons, which are $q\bar{q}$ bound states. Equation (5.16) applied to the positronium, motivates the existence of two positronium states of different lifetimes. Indeed, for $S = 0$ one has $C = +1$ for the ground state $L = 0$. Since the photon carries $C = -1$, the *parapositronium* ("para" = antiparallel spins), can only decay to two photons, with a width:

$$\Gamma(\text{p-Ps} \rightarrow \gamma\gamma) = \frac{m_e \alpha^5}{2} = 8.0 \times 10^9 \text{ s}^{-1}. \tag{5.17}$$

For $S = 1$, the *ortopositronium* ("orto" = parallel spin) has $C = -1$, hence it can only decay to 3γ with a smaller width

$$\Gamma(\text{o-Ps} \to \gamma\,\gamma\,\gamma) = \frac{2(\pi^2 - 9)}{9\pi} \frac{m_e\,\alpha^6}{2} = 7.2 \times 10^6 \text{ s}^{-1}. \qquad (5.18)$$

The fact that the o-Ps $\to \gamma\,\gamma$ is forbidden is also a consequence of Yang's theorem, see Problem 5.1.

Suggested Readings

The reader is addressed to Ref. [8] for more application of these selection rules and for the derivation of the parapositronium decay width from the solution of the Schroedinger equation of the hydrogen atom.

Problem 5.8 Analyze the decay of the $\eta^0, \rho^0, \omega^0, \phi^0$ mesons into two and three pions in terms of the P, C, I, and G symmetry.

Discussion

The strong interaction conserves parity, charge conjugation, isospin, and, hence, G-parity. The latter is defined as

$$G = C\,e^{i\,\pi\,I_2}, \qquad (5.19)$$

i.e. as the charge conjugation operation followed by a rotation of $180°$ by one of the two generators of isospin that are not diagonal (here taken to be I_2). Remember that $SU(3)$ has rank $N - 1 = 2$, hence it admits two diagonal generators, which are chosen to be I_3 and Y. The electromagnetic interaction conserves parity, charge conjugation, but violates isospin I, although it preserves the third component, I_3, see Problem 5.9. As a consequence, G-parity is also violated. The charged-current interaction does not conserve either of these symmetries. In particular, it maximally violates P and C, but conserves $C\,P$, at least at leading-order. As for isospin, it behaves as a combination of $\Delta I = 1/2$ and $I = 3/2$ operators.

For a fermion-antifermion bound state with orbital momentum L, spin S, and isospin I, we have:

$$P = (-1)^{L+1}, \qquad C = (-1)^{L+S}, \qquad G = (-1)^{L+S+I} \qquad (5.20)$$

see Problem 5.7, which gives to the assignments $J^{CPG}(\eta^0) = 0^{+-+}$, $J^{CPG}(\rho^0) = 1^{--+}$, and $J^{CPG}(\omega_0, \phi_0) = 1^{---}$. A final state with orbital momentum L has orbital parity $(-1)^L$. For $n = 3$ pions, one has to consider the angular momentum of each pair of pions in their centre-of-mass frame, L_{12}, and the angular momentum of the pair with respect to the third pion, L_3. A particle with $J = 1$ decaying to three pions can thus have both values of parity. A final state containing n pions has $G = (-1)^n$. By using these rules, and remembering that $J^{CP}(\pi) = 0^{+-}$, we can determine the P, C, G, and I numbers of the initial and final states.

Table 5.1 P, C, G, and I eigenvalues of the initial and final states for the decay of the neutral light mesons into two and three pions

Decay	P	C	G	I
$\eta^0 \to \pi^0 \pi^0$	$(-1) \neq (-1)^2(-1)^0$	$(+1) = (-1)^0$	$(+1) \neq (-1)^2$	$0 = 0, 2$
$\eta^0 \to \pi^+ \pi^-$	$(-1) \neq (-1)^2(-1)^0$	$(+1) = (-1)^0$	$(+1) = (-1)^2$	$0 = 0, 1, 2$
$\eta^0 \to \pi^0 \pi^0 \pi^0$	$(-1) = (-1)^3(-1)^0$	$(+1) = (-1)^0$	$(+1) \neq (-1)^3$	$0 = 0, 1, 2, 3$
$\eta^0 \to \pi^+ \pi^- \pi^0$	$(-1) = (-1)^3(-1)^0$	$(+1) = (-1)^0$	$(+1) \neq (-1)^3$	$0 = 0, 1, 2, 3$
$\rho^0 \to \pi^0 \pi^0$ a	$(-1) = (-1)^2(-1)^1$	$(-1) \neq (-1)^0$	$(+1) = (-1)^2$	$1 \neq 0, 2$
$\rho^0 \to \pi^+ \pi^-$	$(-1) = (-1)^2(-1)^1$	$(-1) = (-1)^1$	$(+1) = (-1)^2$	$1 = 0, 1, 2$
$\rho^0 \to \pi^0 \pi^0 \pi^0$	$(-1) = \pm 1$	$(-1) \neq (-1)^0$	$(+1) \neq (-1)^3$	$1 = 1, 3$
$\rho^0 \to \pi^+ \pi^- \pi^0$	$(-1) = \pm 1$	$(-1) = (-1)^1$	$(+1) \neq (-1)^3$	$1 = 0, 1, 2, 3$
$\omega^0, \phi^0 \to \pi^0 \pi^0$ a	$(-1) = (-1)^2(-1)^1$	$(-1) \neq (-1)^0$	$(-1) \neq (-1)^2$	$0 = 0, 2$
$\omega^0, \phi^0 \to \pi^+ \pi^-$	$(-1) = (-1)^2(-1)^1$	$(-1) = (-1)^1$	$(-1) \neq (-1)^2$	$0 = 0, 1, 2$
$\omega^0, \phi^0 \to \pi^0 \pi^0 \pi^0$	$(-1) = \pm 1$	$(-1) \neq (-1)^0$	$(-1) = (-1)^3$	$0 \neq 1, 3$
$\omega^0, \phi^0 \to \pi^+ \pi^- \pi^0$	$(-1) = \pm 1$	$(-1) = (-1)^1$	$(-1) = (-1)^3$	$0 = 0, 1, 2, 3$

[a] Forbidden by Bose–Einstein symmetry

Solution

The results for the decay of the neutral pseudo-scalar and vector mesons into pions are reported in Table 5.1. The main features to be noted are:

- the η^0 decay to two pions violates P, hence it cannot occur through neither the strong nor the electromagnetic interaction. The decay into three charged pions, instead, conserves P but not G, hence it has to be an electromagnetic interaction, which explains the narrow width of the η_0 meson compared to the ρ^0;
- a spin-1 meson cannot decay to $\pi^0 \pi^0$, since the final state would be antisymmetric for exchange of the two particles, since the symmetry of the orbital wavefuction is $(-1)^L = -1$, thus violating Bose–Einstein symmetry;
- the decay of a vector meson with $C = -1$ into $\pi^0 \pi^0 \pi^0$ violates C, so it cannot be neither a strong nor an electromagnetic process. For ω^0 and ϕ^0, this decay would also violate isospin. Strong decays into $\pi^+ \pi^- \pi^0$ are instead possible for ω^0 and ϕ^0, but not for ρ^0 because of G-parity. In the latter case, the decay is electromagnetic, hence it is suppressed compared to the more likely $\rho^0 \to \pi^+ \pi^-$ decay.

Suggested Readings

For this kind of problems, a valid reference is the dedicated summary tables of the PDG [9]. Besides the measured decays, the PDG tables report the results of searches for decays that would violate some of the conservation laws expected from the SM.

Problem 5.9 Explain which selection rules forbid the electromagnetic decays $\pi^0 \rightarrow 3\gamma$ and $\Sigma^{*-} \rightarrow \Sigma^- \gamma$.

Discussion

The internal symmetries of the Hamiltonian determine selection rules on the decay of particles. The electromagnetic interaction conserves parity P, charge conjugation C, but violates isospin I, although it preserves the third component, I_3 (remember the relation $Q = I_3 + Y/2$). On the other hand, V-spin is conserved, since it is an internal symmetry of the (d, s) quarks, which have the same electric charge.

Solution

Since $\pi^0 \rightarrow 3\gamma$ would be an electromagnetic interaction, it should conserve C. However, the neutral pion has $C = 1$, while the final state has $C = (-1)^3 = -1$, thus forbidding such a decay. The $\Sigma^{*-} \rightarrow \Sigma^- \gamma$ decay would also be an electromagnetic process, but it should violate V-spin, since Σ^{*-} belongs to the $V = 3/2$ quadruplet $(\Delta^-, \Sigma^{*-}, \Xi^{*-}, \Omega^-)$, while Σ^- belongs to the $V = 1/2$ doublet (Σ^-, Ξ^-).

Bando n. 18211/2016

Problem 5.10 The J/Ψ meson has a mass of 3096 MeV and a width of 100 keV. Why is it so narrow?

Solution

The J/Ψ is a $c\bar{c}$ bound state. Strong decays into charmed mesons like $J/\Psi \rightarrow D\bar{D}$ are kinematically forbidden. Strong decays into lighter mesons, like pions and kaons, cannot be described, at leading order, by single-gluon exchange diagrams because gluons are charged under $SU(3)_c$. Two-gluon exchange is not possible either because the J/Ψ has $C = (-1)^{L+S} = -1$, while a two-gluon state has $C = +1$. Three-gluon exchange is possible, but the decay width is suppressed by α_s^3. Overall, the hadronic width turns out to be a factor of about 10 larger than the electromagnetic width responsible for $J/\Psi \rightarrow \ell\bar{\ell}$.

Discussion

The smallness of the J/Ψ decay width, which can be explained within the QCD quark model as discussed above, is an instance of the so-called *Okubo-Zweig-Iizuka rule* (OZI), which states that the decay width of a hadron whose Feynman diagram consists of unconnected lines is suppressed. Here, a $q\bar{q}$ pair produced from the partonic sea is considered as unconnected.

Suggested Readings

A good introduction to this topic can be found in Chap. 4 of Ref. [8].

Bando n. 1N/R3/SUB/2005

Problem 5.11 Motivate why the ϕ meson decays into three pions only 15% of the times, while the remaining decays consist mostly of $\phi \to K\,K$. For latter case, prove that:

1. $\phi \to K_S^0\,K_L^0$ is the only allowed decay into neutral kaons.
2. The width of ϕ decays into neutral kaons over the total width into kaons of any charge is about 0.4.

Discussion

Since the $SU(3)$ flavour symmetry is broken by the much larger mass of the strange quark, see Table 5.3, the $I = 0$ octet and singlet pseudo-vector states mix together, yielding the physical ω^0 and ϕ^0 states. The latter is almost entirely a $s\,\bar{s}$ bound state.

Solution

Thanks to its large mass, infact, larger than for the ω^0 and ρ^0, the decays $\phi \to K^+\,K^-$ and $\phi \to K^0\,\bar{K}^0$ are both kinematicall allowed, although the Q-value is very small (about 30 MeV). The ϕ meson decay to a pair of neutral pions is forbidden by Bose–Einstein symmetry, while the decay to charged pions is forbidden by G-parity conservation, see Problem 5.8, hence the ϕ meson can only decay strongly to three charged pions. The OZI rule, however, suppresses this decay channel since it involves unconnected quark lines. Overall, the decay to kaons dominates with a branching ratio of about 84%.

Since the two kaons are in p-wave, the symmetry of the orbital wavefunction is $(-1)^L = -1$. The flavour part of the wavefunction needs to be antisymmetric as required by the Bose–Einstein statistics. This leaves the only combination:

$$\frac{|K^0\,\bar{K}^0\rangle - |\bar{K}^0\,K^0\rangle}{\sqrt{2}}, \tag{5.21}$$

which, in terms of the weak eigenstates, is a pure $K_S^0\,K_L^0$ state. This can be also proved by noticing that CP is an almost perfect symmetry of the strong and weak Hamiltonian, and since the ϕ meson has $CP = (-1)^{S+1} = +1$, see Eq. (5.16), the final state needs to have also $CP = +1$. Since the orbital parity is -1, the only combination is $K_S^0\,K_L^0$, since $CP|K_S^0\rangle = +|K_S^0\rangle$ and $CP|K_L^0\rangle = -|K_L^0\rangle$. The antisymmetric $K^+\,K^-$ and $K^0\,\bar{K}^0$ states have $I = 0$, as required by the isospin invariance of the strong interaction. The ratio:

$$\alpha = \frac{\Gamma(\phi \to K_S^0 K_L^0)}{\Gamma(\phi \to K^+ K^-) + \Gamma(\phi \to K_S^0 K_L^0)}, \tag{5.22}$$

would be 0.5 from isospin invariance. However, the decay amplitude in the centre-of-mass frame must be a Lorentz-scalar and linear in the polarisation $\boldsymbol{\varepsilon}$ of the decaying meson, which gives the only combination $\mathscr{A} \propto \boldsymbol{\varepsilon} \cdot \mathbf{p}^*$, where $\mathbf{p}^*$ is the momentum in the centre-of-mass frame. Hence, the ratio α should be modified to:

$$\alpha = \frac{|\mathbf{p}_0^*|^2}{|\mathbf{p}_\pm^*|^2 + |\mathbf{p}_0^*|^2} = 0.43, \tag{5.23}$$

where $|\mathbf{p}_0^*| = 107$ MeV and $|\mathbf{p}_\pm^*| = 125$ MeV. The Coulomb barrier effect, present in the $K^+ K^-$ decay only, accounts for another 4%, thus giving $\alpha = 0.39$ [10].

Suggested Readings

See the original paper [10] on the measurement of the ϕ properties and the more extended discussion present in Chap. 5 of Ref. [11].

Bando n. 13153/2009

Problem 5.12 The Ξ^{*-} baryon can decay through the reaction (a) $\Xi^{*-} \to K^- \Sigma^0$, followed by (b) $K^- \to \pi^- \pi^0$, (c) $\Sigma^0 \to \Lambda^0 \gamma$, (d) $\pi^- \to \mu^- \bar{\nu}_\mu$, (e) $\mu^- \to e^- \nu_\mu \bar{\nu}_e$, (f) $\pi^0 \to \gamma \gamma$, and (g) $\Lambda^0 \to p e^- \bar{\nu}_e$. Classify the seven decays by interaction type by knowing that the quark contents of the hadrons taking place to the reaction chain are: $\Xi^{*-}(ssd)$, $\Sigma^0(sdu)$, $\Lambda^0(sdu)$, $p(uud)$, $K^-(s\bar{u})$, $\pi^-(d\bar{u})$, $\pi^0(u\bar{u} - d\bar{d})$.

Solution

The seven decay reactions can be classified by using the selection rules imposed by the properties of invariance of the strong, electroweak (CC), and electromagnetic (EM) interaction. The results are reported in Table 5.2. In particular, the strong interaction satisfies $\Delta S = \Delta I = \Delta I_3 = 0$, the electromagnetic interaction satisfies $\Delta S = \Delta I_3 = 0$, $\Delta I \neq 0$. Decays where $\Delta S \neq 0$ or $\Delta I_3 \neq 0$ are mediated by the charged-current interaction.

Suggested Readings

This topic is well discussed by almost all textbooks on particle physics, like Refs. [8, 12].

Problem 5.13 Determine the relation between the cross sections $\pi^+ p \to \pi^+ p$, $\pi^- p \to \pi^0 n$ and $\pi^- p \to \pi^- p$ at the Δ peak.

Table 5.2 Classification of the Ξ^{*-} decay chain proposed by the exercise

Decay	ΔS	ΔI	ΔI_3	Interaction
$\Xi^{*-} \to K^- \Sigma^0$	0	0	0	Strong
$K^- \to \pi^- \pi^0$	-1	$+3/2$	$-1/2$	CC
$\pi^- \to \mu^- \bar{\nu}_\mu$	0	-1	-1	CC
$\mu^- \to e^- \nu_\mu \bar{\nu}_e$	0	0	0	CC
$\pi^0 \to \gamma \gamma$	0	-1	0	EM
$\Sigma^0 \to \Lambda^0 \gamma$	0	-1	0	EM
$\Lambda^0 \to p e^- \bar{\nu}_e$	$+1$	$+1/2$	$+1/2$	CC

Discussion

The strong interactions are invariant under isospin transformations. Indeed, the $SU(3)$ flavour symmetry, which is at the basis of the observed pattern of hadrons, contains isospin I as a sub-symmetry, together with U-spin and V-spin.

Solution

Isospin invariance implies that the transition amplitude for a pion-nucleon scattering can be decomposed into a sum of amplitudes for different total isospin I. At the Δ peak, only the $I_{\pi N} = 3/2$ amplitude contributes. Using the Clebsch–Gordan coefficients, see e.g. Sect. 44 of Ref. [9], we can write the tensor-product of a $I_N = 1/2$ and $I_\pi = 1$ state as a linear sum of states of total isospin $I_{\pi N} = 1/2, 3/2$, namely:

$$|\pi^+ p\rangle = |\frac{3}{2}, +\frac{3}{2}\rangle$$

$$|\pi^- p\rangle = \sqrt{\frac{1}{3}}|\frac{3}{2}, -\frac{1}{2}\rangle - \sqrt{\frac{2}{3}}|\frac{1}{2}, -\frac{1}{2}\rangle$$

$$|\pi^0 n\rangle = \sqrt{\frac{2}{3}}|\frac{3}{2}, -\frac{1}{2}\rangle - \sqrt{\frac{1}{3}}|\frac{1}{2}, -\frac{1}{2}\rangle$$

$$(5.24)$$

from which we can deduce the proportion:

$$\sigma(\pi^+ p \to \pi^+ p) : \sigma(\pi^- p \to \pi^0 n) : \sigma(\pi^- p \to \pi^- p) =$$

$$= |\mathscr{A}_{\frac{3}{2}}|^2 : |\frac{\sqrt{2}}{3}\mathscr{A}_{\frac{3}{2}}|^2 : |\frac{1}{3}\mathscr{A}_{\frac{3}{2}}|^2 = 1 : \frac{2}{9} : \frac{1}{9} = 9 : 2 : 1. \qquad (5.25)$$

Suggested Readings

The first evidence for the Δ resonances in pion-hydrogen scattering was established by E. Fermi and collaborators at the Chicago cyclotron in 1952. The reader is encouraged to read their original paper [13]. The pattern of measured cross sections enforced the interpretation of the resonance as a new $I = 3/2$ particle. The angular distribution of the nucleon-pion allowed to determine its spin, see Problem 1.7.

Problem 5.14 Express the isospin wavefunction of the charged and neutral pions in terms of the fundamental (u, d) and $(\bar{u}, \bar{d})$ representation, using the fact that (π^-, π^0, π^+) form an isospin triplet.

Solution

Refering to Eq. (5.132) of Problem 5.36, we see that the isospin triplet states made of a quark-antiquark pair are given by:

$$|\pi^a\rangle = \frac{\sigma_{ij}^a}{\sqrt{2}}|\bar{q}_i \, q_j\rangle = \begin{cases} \frac{|\bar{d}\,u\rangle + |\bar{u}\,d\rangle}{\sqrt{2}} \\ \frac{-i|\bar{d}\,u\rangle + i|\bar{u}\,d\rangle}{\sqrt{2}} \\ \frac{|\bar{u}\,u\rangle - |\bar{d}\,d\rangle}{\sqrt{2}} \end{cases} \tag{5.26}$$

The three states are however not eigenstates of I_3. The latter can be obtained by a simple rotation of the first two, giving the three pion states:

$$\left(|\pi^-\rangle, |\pi^0\rangle, |\pi^+\rangle\right) = \left(|\bar{u}\,d\rangle, \frac{|\bar{u}\,u\rangle - |\bar{d}\,d\rangle}{\sqrt{2}}, |\bar{d}\,u\rangle\right) \tag{5.27}$$

Notice the sign "$-$" in the π^0 state vector, which differs from the more familiar composition of two spin-1/2 particles into a $S = 1$ state. In this case, the two states transform under the *same* representation of the $SU(2)$ group.

Bando n. 18211/2016

Problem 5.15 Why does the neutrinoless double-beta decay imply that the neutrinos are massive Majorana particles?

Discussion

A massive Majorana neutrino χ, assumed to be of left-handed chirality, is described by a free Lagrangian:

$$\mathscr{L}^{\text{M}}_{\text{free}} = i\chi^\dagger \bar{\sigma}^\mu \partial_\mu \chi + \frac{1}{2} \left[m\, \chi^T (i\sigma_2)\chi - m\, \chi^{*T}(i\sigma_2)\chi^* \right], \qquad (5.28)$$

where $\bar{\sigma}_\mu = (1, -\boldsymbol{\sigma})$. The Lagrangian does not admit a conserved neutrino number, since under an arbitrary $U(1)$ transformation, the mass term is not invariant under $\chi \to e^{i\phi}\chi$. When *two* left-handed Weil spinors $\chi_{1,2}$ are available, one can construct a generic free-Lagrangian as

$$\mathscr{L}^{\text{M}}_{\text{free}} = \sum_i i\chi_i^\dagger \bar{\sigma}^\mu \partial_\mu \chi_i + \frac{1}{2} \left[\chi_i^T M_{ij}(i\sigma_2)\chi_j + \text{h.c.} \right], \qquad (5.29)$$

where now M is 2×2 symmetric matrix. The special case $M_{11} = M_{22} = 0$, $M_{12} \equiv m > 0$ gives rise to a conserved lepton-number associated with the transformation $\chi_1 \to e^{i\phi}\chi_1$ and $\chi_2 \to e^{-i\phi}\chi_2$. In terms of a Dirac spinor $\chi^D \equiv \chi_1 + (i\sigma_2)\chi_2^*$, where now, $(i\sigma_2)\chi_2^*$ transforms like a right-handed spinor, the Lagrangian takes the familiar form:

$$\mathscr{L}^{\text{D}}_{\text{free}} = i\bar{\chi}^D \slashed{\partial} \chi^D - m\, \bar{\chi}^D \chi^D, \qquad (5.30)$$

which is the Lagrangian for a massive Dirac neutrino and admits a conserved lepton-number associated with the transformation $\chi^D \to e^{i\phi}\chi^D$. The outstanding difference between Eqs. (5.28) and (5.30) is that a new degree of freedom has been introduced to have Dirac-type neutrinos.

If no right-handed neutrinos are introduced, the left-handed neutrinos are massless and lepton number is conserved. If the three right-handed neutrinos are introduced, as demanded by the observation of neutrino oscillations, one should distinguish between three cases depending on the value of the right-handed neutrinos mass term $\frac{1}{2}M^R_{ij} N_i N_j$:

- The six-by-six mass matrix M, analogous to the one in Eq. (5.29), is a symmetric off-diagonal matrix, i.e. $M^R = 0$: the neutrinos are Dirac particles in full analogy with the charged leptons and quarks; lepton number is conserved.
- The elements of the matrix M^R are much larger than the Yukawa mass term: there are three Majorana-type, light neutrinos, mostly made of the weakly interacting ν_i, and three Majorana-type, heavy and approximately decoupled right-handed neutrinos. Lepton number is maximally violated.
- Any other combination between these two extreme cases gives rise to up to six distinct eigenvalues of the full neutrino matrix, with lepton number violation.

Since the second option has the virtue of explaining the lightness of the neutrinos by assuming a large value for M^R rather by postulating an unnaturally small value for the Yukawa coupling, it is also the more frequently considered. See Ref. [14] for more details.

Solution

The SM Lagrangian with Dirac-type neutrinos admits a conserved lepton number. The $0\nu\beta\beta$ decay of unstable nuclei would violate lepton number by two units, hence this decay should occur. If instead the neutrinos are of Majorana type, there is no conserved lepton number and this decay is allowed. Since a Majorana neutrino with $m_\nu = 0$ would still result in a conserved lepton number, we should expect the amplitude to be proportional to m_ν. Indeed, the $0\nu\beta\beta$ amplitude must contain the propagator of a Majorana neutrino. The latter is given by:

$$\int d^4x\, e^{ipx} \langle \chi(x)\, \chi(0) \rangle = -\frac{m_\nu\, \sigma_2}{p^2 - m_\nu^2}. \tag{5.31}$$

Since the weak eigenstates do not coincide with the mass eigenstates, as shown by neutrino oscillations, the mass term in Eq. (5.31) should be replaced by a combination of masses of the three neutrino eigenstates. In terms of the PMNS matrix, we have:

$$m_\nu = \sum_i U_{ei}^2\, m_i \quad \text{and} \quad \Gamma_{0\nu\beta\beta} = |m_\nu|^2 |\mathcal{M}|^2 \Phi, \tag{5.32}$$

where Φ is a known phase space factor and $\mathcal{M}$ is the nuclear matrix element.

Suggested Readings

Ref. [14] provides a concise overview on this subject. See Ref. [15] for a more formal treatment of Majorana particles.

Bando n. 18211/2016

Problem 5.16 Is the helicity of a free neutrino produced in a weak decay conserved?

Solution

As shown in Problem 1.6, spinors of a given chirality have a definite helicity in the limit $\beta \to 1$. If the particle is massive and is produced with a given chirality, as is it the case for the neutrinos, the spinor will contain, in general, a mixture of opposite helicity states. Assuming $m_\nu = \mathcal{O}(\text{eV})$, the boost factor for neutrinos produced by standard reactions is so large that all (anti)neutrinos can be assumed to have a constant helicity $h = -1\, (+1)$.

Problem 5.17 The ρ^+ and K^+ mesons can both decay into $\pi^+\pi^0$. What is the total isospin of the $\pi^+\pi^0$ state for the two decays?

Solution

The composition of two isospin-1 particles can give $I = 0, 1, 2$. However, the state $I = 0$ cannot contribute since $I_3 = +1$ for a $\pi^+\pi^0$ state. For the ρ decay, conservation of isospin by the strong interaction implies $I = 1$. The $K^+ \to \pi^+\pi^0$ is instead an electroweak transition, since $\Delta S = 1$, which does not conserve isospin. However, the final state is made of two bosons, hence its wave function must be symmetric under exchange of the two particle indices. Since the symmetry of the spatial component of the wavefunction is $(-1)^L = +1$, the isospin component has to be symmetric under $\pi^+ \leftrightarrow \pi^0$ exchange, hence $I = 2$.

Discussion

The $I = 3/2$ transition responsible for the $K^+ \to \pi^+\pi^0$ decay is suppressed, explaining the smallness of the $K^+ \to \pi^+\pi^0$ decay probability compared to e.g. $K_S^0 \to (\pi\pi)_{I=0}$.

Suggested Readings

For more details on the $\Delta I = 1/2$ rule for charged-current interactions, the reader is addressed to Chap. 7 of Ref. [11].

Bando n. 13153/2009

Problem 5.18 The Δ^0 and Λ^0 particles decay into $\pi^- p$ with a mean lifetime of about 10^{-23} s and 10^{-10} s, respectively. Motivate this large difference.

Solution

The Δ resonance is a non-strange baryon which decays strongly to $\pi^- p$. The Λ^0 baryon has strangeness $S = -1$. Its decay is mediated by the electroweak interaction, which explains the smaller transition probability, and, hence, the much larger lifetime. While the Δ is usually referred to as a *resonance* rather than a particle, the Λ^0 is a so-called *V-particle* because its lifetime is long enough to allow its production and decay vertex, made of two opposite-sign tracks in the form of a "V", to be experimentally separated already at moderate energies since $c\tau \approx 7.9$ cm.

Suggested Readings

Check the PDG reviews [9] for the decay modes and properties of the Δ and Λ baryons.

Problem 5.19 Prove that, within the quark parton model, the requirement that the difference between the number of quarks and antiquarks bound in a hadron should be a multiple of three implies that hadrons must have an integer electric charge, and *vice versa*.

Solution

Let's denote the multiplicity of $u, d, \bar{u}$, and $\bar{d}$ quarks by $n_u, n_d, n_{\bar{u}}$, and $n_{\bar{d}}$, respectively. Let $n_q = n_u + n_d$ and $n_{\bar{q}} = n_{\bar{u}} + n_{\bar{d}}$ be the quark and antiquark multiplicities. Then, the hadron charge Q is given by:

$$Q = \frac{2}{3}n_u - \frac{1}{3}n_d - \frac{2}{3}n_{\bar{u}} + \frac{1}{3}n_{\bar{d}},$$

$$3Q = (2n_u - n_d) - (2n_{\bar{u}} - n_{\bar{d}}) = 2(n_q - n_{\bar{q}}) + 3(n_{\bar{d}} - n_d). \tag{5.33}$$

Since $(n_{\bar{d}} - n_d) \in \mathbb{Z}$, then $Q \in \mathbb{Z}$ implies that $(n_q - n_{\bar{q}})$ is also a multiple of three. *Vice versa*, if $(n_q - n_{\bar{q}})$ is a multiple of three, then $Q \in \mathbb{Z}$. Indeed, all hadrons observed so far are composed of quarks and antiquarks in numbers such that $(n_q - n_{\bar{q}})$ is a multiple of three: mesons ($q\,\bar{q}$), baryons ($q\,q\,q$), anti-baryons ($\bar{q}\,\bar{q}\,\bar{q}$), tetraquarks ($q\,\bar{q}\,q\,\bar{q}$), and pentaquarks ($q\,q\,q\,q\,\bar{q}$).

Discussion

A number of putative tetraquark states ($X, Z, Y, \ldots$) manifesting in the two-body invariant mass of meson pairs, of which one is a heavy meson, has been reported by several experiments in recent years (Belle, BESIII, CDF, D0, LHCb). The observation of pentaquarks has been established in 2015 by the LHCb Collaboration [16]: an amplitude analysis of the decay $\Lambda_b^0 \to J/\Psi\, K^-\, p$ has been performed, finding a structure in the $J/\Psi\, p$ kinematics consistent with an intermediate $u\,u\,d\,c\,\bar{c}$ resonance: a charmonium-pentaquark.

Suggested Readings

For the experimental evidence of tetraquark-like states, see e.g. the D0 evidence for a particle $X(5586)$ decaying to $B_s^0\, \pi^{\pm}$, published in Ref. [17]. References to other searches can be found in the same article. The observations of an exotic structure in the $J/\Psi\, p$ channel, consistent with two pentaquark resonances, is documented in Ref. [16]. The description of hadronic properties in terms of the quark model is well summarised in Sect. 15 of Ref. [9].

Problem 5.20 Within the quark parton model, each quark is described by a parton density function (PDF) q such that $q(x)\, dx$ gives the amount of quarks of type q sharing a fraction $x' \in [x, x + dx]$ of the parent hadron mass. Assume the proton to be made up of six different quarks: $u, \bar{u}, d, \bar{d}, s, \bar{s}$. Make use the proton quantum numbers (a) electric charge $Q_p = +1$, (b) baryon number $B_p = +1$, (c) strangeness $S_p = 0$, to predict the values of:

$$\int_0^1 u_V(x')\, dx', \qquad \int_0^1 d_V(x')\, dx', \qquad \int_0^1 s_V(x')\, dx', \qquad (5.34)$$

where $q_V \equiv q - \bar{q}$.

Solution

From the constraints on the electric charge, baryon number, and strangeness, we get the following system of equations:

$$\begin{cases} Q_p = \int_0^1 dx' \left[\frac{2}{3}u(x') - \frac{1}{3}d(x') - \frac{1}{3}s(x') - \frac{2}{3}\bar{u}(x') + \frac{1}{3}\bar{d}(x') + \frac{2}{3}\bar{s}(x') \right] = 1 \\ B_p = \int_0^1 dx' \left[\frac{1}{3}u(x') + \frac{1}{3}d(x') + \frac{1}{3}s(x') - \frac{1}{3}\bar{u}(x') - \frac{1}{3}\bar{d}(x') - \frac{1}{3}\bar{s}(x') \right] = 1 \\ S_p = \int_0^1 dx' \left[-s(x') + \bar{s}(x') \right] = 0 \end{cases}$$

from which we can derive the relations:

$$\begin{cases} \int_0^1 \left[u(x') - \bar{u}(x') \right] dx' = \int_0^1 u_V(x')\, dx' = 2 \\ \int_0^1 \left[d(x') - \bar{d}(x') \right] dx' = \int_0^1 d_V(x')\, dx' = 1 \\ \int_0^1 \left[s(x') - \bar{s}(x') \right] dx' = \int_0^1 s_V(x')\, dx' = 0 \end{cases} \qquad (5.35)$$

Notice that this result agrees with the expectation from the flavour $SU(3)$ model of hadrons, which assigns the $u\,u\,d$ flavour content to the proton.

Discussion

Relations between the moments of the PDF for different types of partons, like Eq. (5.35), are known as *sum rules*, and prove useful in constraining parton-related observables that cannot be directed measured. A paradigmatic example is provided by the average gluon momentum in hadrons, $\int dx\, x\, g(x) \approx 0.5$, which can be inferred from lepton DIS data even in the absence of tree-level gluon interactions with leptons.

Suggested Readings

For a broader overview on sum rules in QCD, the reader is addressed to dedicated monographs like Refs. [18, 19].

5.2 Electroweak and Strong Interactions

The standard model of electroweak and strong interactions (SM) is a relativistic quantum field theory invariant under the gauge group:

$$G_{SM} = SU(3)_C \times SU(2)_L \times U(1)_Y, \tag{5.36}$$

where $SU(3)_C$ is the symmetry group associated with the strong force (*colour*), and $SU(2)_L \times U(1)_Y$ is the symmetry related to the electroweak interactions (*isospin* and *hypercharge*).

In a gauge theory, each generator of the group is associated with a spin-1 gauge boson, which can be seen as the force mediator between the matter particles of spin-1/2. In the SM, there are $8 + 3 + 1$ such gauge bosons. Following the standard notation, see e.g. Ref. [20], we denote the gauge fields associated to the three sub-groups of Eq. (5.37) as $G_\mu^a, a = 1, \ldots, 8$, $W_\mu^a, a = 1, 2, 3$ and B_μ, respectively. The Lagrangian density of the gauge fields can be expressed in a compact form in terms of the field strengths:

$$
\begin{aligned}
G_{\mu\nu}^a &= \partial_\mu G_\nu^a - \partial_\nu G_\mu^a + g_s \, f^{abc} \, G_\mu^b \, G_\nu^c \\
W_{\mu\nu}^a &= \partial_\mu W_\nu^a - \partial_\nu W_\mu^a + g_2 \, \varepsilon^{abc} \, W_\mu^b \, W_\nu^c \\
B_{\mu\nu} &= \partial_\mu B_\nu - \partial_\nu B_\mu,
\end{aligned}
\tag{5.37}
$$

where f^{abc} and ε^{abc} are the structure constants of $SU(3)_C$ and $SU(2)_L$, and two of the three adimensional coupling constants of the theory (g_s, g_2) have been introduced. The third coupling (g_1) is associated with the abelian sub-group $U(1)_Y$, and therefore it enters only through the interaction with matter fields.

In addition to the gauge bosons, the spectrum of the theory accommodates all the elementary matter particles observed in experiments: *quarks* and *leptons*. These come in three replicas, or *families*. One family is composed by fifteen independent Weil spinors:

- up and down left–handed quarks (Q) of three different colors;
- up and down right–handed quarks (u_R, d_R) of three different colors;
- up and down left–handed leptons (L) ;
- down right–handed lepton (e_R).

They can be conveniently arranged into a multiplet of fields:

$$\{Q(\mathbf{3}, \mathbf{2})_{1/3}, \, L(\mathbf{1}, \mathbf{2})_{-1}, \, u_R(\mathbf{3}, \mathbf{1})_{4/3}, \, d_R(\mathbf{3}, \mathbf{1})_{-2/3}, \, e_R(\mathbf{1}, \mathbf{1})_{-2}\}, \tag{5.38}$$

where the charges under G_{SM} have been explicitated: the first (second) number in parenthesis indicates under which representation of the color (isospin) group the fields transform, while the subscript corresponds to the hypercharge. One should possibly add a right-handed neutrino $N_R(\mathbf{1}, \mathbf{1})_0$ to accommodate massive neutrinos. The right–handed neutrino is however sterile because it does not couple with the gauge fields (it is a singlet under transformations of the group in Eq. (5.36)). Notice that all fields are charged under $U(1)_Y$, but only left-handed particles are charged under $SU(2)_L$ and only the quarks are charged under $SU(3)_C$. The representation in the multiplet basis of the group in Eq. (5.36) is therefore reducible to blocks.

The gauge symmetry dictates the structure of the Lagrangian: to achieve invariance under an arbitrary local transformation of Eq. (5.36), only a few combinations of the fields are allowed. A Lagrangian density of the gauge and of the matter fields of one family, renormalizable and invariant under an arbitrary local transformation of (5.37), is provided by:

$$
\begin{aligned}
\mathscr{L}_{\mathrm{SM}} = &-\tfrac{1}{4}G^a_{\mu\nu}G^{\mu\nu}_a - \tfrac{1}{4}W^a_{\mu\nu}W^{\mu\nu}_a - \tfrac{1}{4}B_{\mu\nu}B^{\mu\nu}+ \\
&i\,\bar{L}\,\slashed{D}\,L + i\,\bar{e}_R\,\slashed{D}\,e_R + i\,\bar{Q}\,\slashed{D}\,Q + i\,\bar{u}_R\,\slashed{D}\,u_R + i\,\bar{d}_R\,\slashed{D}\,d_R,
\end{aligned}
\tag{5.39}
$$

with the covariant derivative D_μ defined as

$$
D_\mu \psi \equiv \left(\partial_\mu - i\,g_s\,\boldsymbol{T}\cdot\boldsymbol{G}_\mu - i\,g_2\,\frac{\boldsymbol{\sigma}\cdot\boldsymbol{W}_\mu}{2} - i\,g_1\,\frac{Y}{2}B_\mu\right)\psi.
\tag{5.40}
$$

The matrices T^a and σ^a are a particular representation of the generators of $SU(3)_C$ and $SU(2)_L$, respectively. In the basis where the left-handed neutrino and electron are the first and second element of the isospin doublet, σ^a are chosen to be the three Pauli matrices. The λ_a matrices are taken to be the eight Gell-Mann matrices. The Y matrices are diagonal in the multiplet space. The first row of Eq. (5.39) describes the dynamic of the gauge fields, with the kinetic, triple and quartic self-interaction terms (the latter two are present only for the non-abelian groups); the second and third rows contain the kinetic lagrangian of the fermions plus the interactions with the gauge fields.

In the attempt to write all possible gauge-invariant and renormalizable combination of matter and gauge fields, mass terms are excluded from the right-hand side of (5.39) for both the vector bosons and the fermions, so that the spectrum of the theory would consist of only massless particles. However, the addition of a new $SU(2)_L$-doublet of complex scalar fields, called the *Higgs doublet*, with the correct quantum numbers allows for additional gauge-invariant terms. Indeed, with the inclusion of the Higgs doublet:

$$
\Phi = \begin{pmatrix} \phi^+ \\ \phi^0 \end{pmatrix} = \begin{pmatrix} \phi_1 + i\phi_2 \\ \phi_3 + i\phi_4 \end{pmatrix},
\tag{5.41}
$$

with $Y_\Phi = 1$, three more terms pop up:

$$
\mathscr{L}_Y = -\lambda_e\,\bar{L}\,\Phi\,e_R - \lambda_u\,\bar{Q}\,i\sigma_2\Phi^*\,u_R - \lambda_d\,\bar{Q}\,\Phi\,d_R + \text{h.c.}
\tag{5.42}
$$

With a suitable unitary transformation in flavour space, the λ_e matrix can be made diagonal. Conversely, by three additional rotations of the Q, u_R, and d_R fields, only one between λ_u and λ_d can be diagonalised. The other will remain non-diagonal, so that an additional rotation is needed to go into the canonical mass basis after EWSB, see Sect. 5.4. The rotation will mix the $SU(2)_L$ quark doublets of the three generations, thus changing the charged-current Lagrangian into:

$$\mathscr{L}_{CC} = \frac{g}{\sqrt{2}} V_{ij} \, \bar{u}_i \, \gamma_\mu \frac{(1 - \gamma_5)}{2} d_j \, W^{+\mu} + \text{h.c.} \tag{5.43}$$

where the unitary matrix V_{ij} is called *Cabibbo-Kobayashi-Maskawa* matrix (CKM) and will be discussed in more detail in Sect. 5.3.

Problems

Bando n. 13707/2010

Problem 5.21 The neutron, muon, and tau decay via the weak interaction with mean lifetimes of 886 s, 2.2 μs, and 0.29 ps, respectively: how can you explain the huge difference between their lifetimes?

Discussion

The three decays are mediated by the charged-current interaction, which can be described by the effective Fermi lagrangian of Eq. (2.82). Hence, the decay widths must be proportional to G_F^2. However, the width must have the dimensions of an energy, hence Γ must be proportional to another mass scale Δ to the fifth power. The hierarchy between the decay times reflects the hierarchy in the mass scale Δ.

Solution

For the neutron, one can use Sargent's rule, see Eq. (1.214) with proper modifications to account for the finite electron mass, which is comparable to the neutron-proton mass difference. Using the effective Fermi Lagrangian of Eq. (2.82), where the ratio between V and A hadron currents is parametrised by a parameter α of $\mathscr{O}(1)$, such that $\alpha = 1$ gives a pure $V - A$ interaction, the neutron width can be calculated at LO:

$$\Gamma_n = \frac{G_F^2 \, Q^5}{60\pi^3} \cos^2 \theta_C (1 + 3\alpha^2) \Phi, \tag{5.44}$$

where $\Phi = 0.47$ is a numerical factor (that would be unity if the electron mass were neglected relative to Q) and θ_C is the *Cabibbo angle*. Hence, the neutron width is proportional to Q^5, with $Q = m_n - m_p = 1.29$ MeV. For the muon decay, the width can be computed exactly using the Fermi Lagrangian, which provides the well-known result:

$$\Gamma_\mu = \frac{G_F^2 \, m_\mu^5}{192\pi^3}. \tag{5.45}$$

This time, the width is proportional to m_μ^5, with $m_\mu = 106$ MeV. For the tau decay, we can use the results from Problem 5.28, giving:

$$\Gamma_\tau \approx 5\,\Gamma(\tau \to e\,\nu_e\,\nu_\tau) = \frac{5\,G_F^2\,m_\tau^5}{192\pi^3}. \tag{5.46}$$

Modulo phase-space factors, the expression are pretty similar. If we want to relate all widths to the purely leptonic width of Eq. (5.45), we see that the numerical coefficient in Eq. (5.44) is about 6. Hence:

$$\tau_n : \tau_\mu : \tau_\tau = 1 : \frac{\Gamma_n}{\Gamma_\mu} : \frac{\Gamma_n}{\Gamma_\tau} \approx 1 : 6\left(\frac{Q}{m_\mu}\right)^5 : \frac{6}{5}\left(\frac{Q}{m_\mu}\right)^5 =$$

$$1 : 1.7 \times 10^{-9} : 2.6 \times 10^{-16}, \tag{5.47}$$

in fair agreement with the experimental result: $1 : 2.5 \times 10^{-9} : 3.3 \times 10^{-16}$.

Suggested Readings

The calculation of the muon decay with can be found in several textbooks on particle physics, see e.g. Ref. [8, 12]. An instructive dissertation about the history of the charged-current interaction can be found in Chap. 6 of Ref. [11]. The reader is addressed to Ref. [14] for more details on the neutron decay.

Bando n. 1N/R3/SUB/2005

Problem 5.22 Weak interactions are responsible for the decay of muons ($\tau \sim 10^{-6}$ s), of B mesons ($\tau \sim 10^{-12}$ s), and of neutrons ($\tau \sim 10^3$ s). Which are the dominant factors responsible for the large differences in lifetime?

Solution

The difference between the neutron and muon decay widths is discussed in Problem 5.21. The B mesons are fairly more long-lived particles compared to what one would expect with respect to charmed mesons. For example, from a pure Q^5 scaling, one would expect the B^+ lifetime to be of order 10^{-15} s, while it is found to be roughly three orders of magnitude larger. Consider for example the decay $B^+ \to D^0 e^+ \nu_e$ and its light-flavour counterpart $\pi^+ \to \pi^0 e^+ \nu_e$. The former has a branching ratio of 10^{-1}, while the latter occurs with a branching ratio of about 10^{-8}. The naive ratio between the two decay widths would then be:

$$\frac{\Gamma(B^+ \to D^0 e^+ \nu_e)}{\Gamma(\pi^+ \to \pi^0 e^+ \nu_e)} \approx \left(\frac{m_{B^+} - m_{D^0}}{m_{\pi^+} - m_{\pi^0}}\right)^5 \approx 5 \times 10^{14}, \tag{5.48}$$

while the observed ratio is:

$$\frac{\Gamma(B^+ \to D^0 e^+ \nu_e)}{\Gamma(\pi^+ \to \pi^0 e^+ \nu_e)} = \frac{c\,\tau_{\pi^+} \cdot 10^{-1}}{c\,\tau_{B^+} \cdot 10^{-8}} = 10^7 \cdot \frac{7.8\text{ m}}{5 \times 10^{-4}\text{ m}} \approx 2 \times 10^{11}. \quad (5.49)$$

The difference is due to the magnitude of the CKM matrix element $|V_{cb}| \approx 0.04$, thus giving a factor $|V_{cb}|^2 \sim 10^{-3}$ smaller width for B decays compared to the CKM-favoured $d \to u$ transition.

Suggested Readings

An instructive introduction to the physics of B hadrons can be found in Chap. 11 of Ref. [11].

Problem 5.23 Provide a quantitative explanation of why the lifetime of charged pions ($\tau_{\pi^+} = 2.6 \times 10^8$ s) is much larger than the lifetime of neutral pions ($\tau_{\pi^0} = 8.5 \times 10^{-17}$ s). How would you measure the latter?

Solution

The decay of the charged pion occurs through the charged-current interaction, which involves the production of an off-shell W boson. The amplitude is therefore suppressed by the small Fermi constant $G_F = 1.17 \times 10^{-5}$ GeV^{-2}. More specifically, the Fermi Lagrangian of Eq. (2.82), provides a decay width [21]

$$\Gamma(\pi^+ \to \mu^+ \nu_\mu) = \cos^2 \theta_C \frac{G_F^2}{8\pi} f_\pi^2 m_\mu^2 m_\pi \left(1 - \frac{m_\mu^2}{m_\pi^2}\right)^2 = 2.5 \times 10^{-8} \text{ eV}, \quad (5.50)$$

where $f_\pi \approx 130$ MeV is the pion decay constant defined by $\langle 0| J_A^\mu |\pi(p)\rangle = i p^\mu f_\pi$. From this value, one gets $\tau_{\pi^+} = 25$ ns, where we have used Eq. (1.8) to convert the result in SI.

The π^0 decay is instead an electromagnetic process which is not suppressed by the electroweak mass scale. The amplitude for this decay can be related to the three-point axial-vector-vector amplitude, which should be zero because of conservation of the vector and axial current in QCD (the symmetry would be exact in the limit of massless quarks). The fact that this decay occurs is a manifestation of the so-called *chiral anomaly*, i.e. the breaking of a classical symmetry by quantum effects, in this case by quantum fluctuations of the quark fields in the presence of a gauge field. The relation between the conservation of the axial current and the decay amplitude $\pi^0 \to \gamma\,\gamma$ can be traced to the fact that the pion fields are related to the divergence of the axial current by the relation:

$$\partial_\mu J_A^{a\,\mu} = f_\pi m_\pi^2 \phi_\pi^a(x), \quad \text{with} \quad J_A^{a\,\mu}(x) = \bar{\psi}_i(x)\frac{\sigma_{ij}^a}{2}\gamma^\mu \gamma^5 \psi_j(x). \quad (5.51)$$

The chiral anomaly provides a leading-order contribution to the decay width that can be parametrised via an effective Lagrangian

$$\mathscr{L}_{\text{chiral}} = \frac{\alpha}{8\pi} \frac{1}{f_\pi} \varepsilon_{\mu\nu\rho\sigma} F^{\mu\nu} F_{\rho\sigma} \phi_{\pi^0}. \tag{5.52}$$

Making the quark content of the pion explicit, the decay width becomes [22]:

$$\Gamma(\pi^0 \to \gamma\,\gamma) = N_C^2 \left(e_u^2 - e_d^2\right)^2 \frac{\alpha^2 m_{\pi^0}^3}{64\pi^3 f_\pi^2} = 7.76\,\text{eV}, \tag{5.53}$$

where N_C is the number of coulors and $e_{u,d}$ are the electric charges of the up- and down-type quarks, corresponding to $\tau_{\pi^0} = 0.8 \times 10^{-16}$ s.

There have been four different experimental methods which have been utilised to measure the π^0 lifetime. The first method consists in a direct measurement of the size of the π^0 decay region by relating the decay length d to the π^0 lifetime by the relation $d = (|\mathbf{p}|/m_{\pi^0})\,c\tau_{\pi^0}$, assuming the pion momentum to be known by other means. The early measurements of this kind studied the production and decay of neutral pions in emulsions, which are limited by the granularity of the medium. A better sensitivity is provided by taking into account the Dalitz decay $\pi^0 \to e^+ e^- \gamma$ (BR $\approx$ 1.2%), where the π^0 can be produced by either nuclear reactions (*stars*) or by the decay of stopped kaons via $K^+ \to \pi^+ \pi^0$. In order to increase the mean flight distance, the neutral pions need to be produced at large boosts. The latest and most sensitive of such measurements has been carried out by the NA-30 experiment at CERN [23]. In this experiment, neutral pions were produced from the interaction of 450 GeV protons against a target consisting of two tungsten folis of variable distance. The produced π^0 had an average momentum of 235 GeV, corresponding to an average flight distance $d \approx 50$ μm. By increasing the distance between the two foils in a range from 5 to 250 μm, the rate of e^+e^- conversion was changed because of the larger fraction of π^0 decay upstream of the second foil: the lifetime could be then measured from the rate of photon conversions as a function of the foil distance.

The second experimental technique makes use of the *Primakoff effect*, i.e. the photo-production of π^0's in the electromagnetic field of a heavy nucleus, see Fig. 5.1. The cross section for the reaction $\gamma\,A \to \pi^0\,X$, with $\pi^0 \to \gamma\,\gamma$ and A is a nucleus of atomic number Z, can be related to the π^0 width, since

$$\sigma(\gamma\,A \to \pi^0\,X) \propto Z^2\,\Gamma(\pi^0 \to \gamma\,\gamma) \cdot \text{BR}(\pi^0 \to \gamma\,\gamma), \tag{5.54}$$

with $\text{BR}(\pi^0 \to \gamma\,\gamma) = 0.988$ [9]. Indeed, an explicit calculation from Eq. (5.51) yields a cross section:

$$\sigma \approx \frac{Z^2\alpha^3}{3\pi^2 f_\pi^2} \left(1 - \frac{m_\pi^2}{E_\gamma^2}\right)^{\frac{3}{2}}, \tag{5.55}$$

Fig. 5.1 Leading-order diagram contributing to the reaction $\gamma\, A \to \pi^0\, X$, where A is a nucleus with atomic number Z

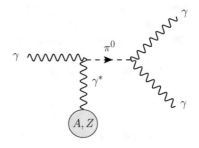

valid in the limit $E_\gamma \gtrsim m_\pi$, see e.g. Problem 84 of Ref. [2], to be compared with Eq. (5.53). This method was pioneered in 1965 by Bellettini et al. at the Frascati Laboratories and later exploited by several other experiments. See Ref [22] for a comprehensive review on this topic.

The third method exploited in experiments is based on measuring the cross section for $e^+ e^- \to e^+ e^- \pi^0$, where the π^0 is produced via photon-fusion, and which can be again related to $\Gamma(\pi^0 \to \gamma\gamma)$.

The fourth method is based on a measurement of the branching ratio of the radiative decay $\pi^+ \to e^+ \nu_e \gamma$, as carried out in the PIBETA experiment, which can be related to $\pi^0 \to \gamma\gamma$ by isospin invariance.

Suggested Readings

For a comprehensive review on this subject, the reader is addressed to Ref. [22].

Bando n. 13153/2009

Problem 5.24 Explain why the decay of the charged pion into an electron is disfavoured compared to the decay into a charged muon. Estimate the ratio between the two decay widths.

Solution

The leptonic decay of a charged pseudo-scalar meson can be described using the Fermi model of the charged-current interaction, giving an amplitude:

$$\mathscr{A}_{fi} = \frac{G_F}{\sqrt{2}} \langle 0| J_{A\,\alpha}(0) |\pi^+ \rangle \langle \mu^+ \, \nu_\mu | J_V^{\alpha\,\dagger}(0) - J_A^{\alpha\,\dagger}(0)|0\rangle, \tag{5.56}$$

where $J_{V,A}^\alpha$ are the vector and axial currents. The unknown hadronic matrix element is parametrised by the meson decay constant f_π. The leptonic matrix element selects leptons of left-handed chirality. Since the pion has $J = 0$, the two leptons must have the same helicity in the centre-of-mass frame. However, left-handed leptons with $|\mathbf{p}^*| > 0$ are polarised, see Problem 1.6, and have a unique helicity in the limit $\beta^* \to 1$, which would be opposite for particle and anti-particle. This decay is therefore helicity-suppressed. This is reflected by the fact that the decay amplitude is proportional to m_μ, as one can readily prove by noticing that:

$$\mathscr{A}_{fi} \propto \bar{u}_\nu \not{p}(1 - \gamma_5)v_\mu = -m_\mu \bar{u}_\nu(1 + \gamma_5)v_\mu, \tag{5.57}$$

where u_ν and v_μ are the Dirac spinors for particles and antiparticles, which must obey the Dirac equations $\not{p}_\nu u_\nu = 0$ and $(\not{p}_\mu + m_\mu)v_\mu = 0$, respectively. In Eq. (5.57), we have used the anti-commutation relation $\{\gamma_\mu, \gamma_5\} = 0$. Since the amplitude is between states of the same chirality, it must also vanish in the limit $\beta^* \to 0$. Indeed, the low-energy matrix element expansion of operators like $\bar{\psi}\gamma_\mu\psi$ and $\bar{\psi}\gamma_\mu\gamma_5\psi$ between two spin-$1/2$ particles are:

$$\begin{aligned} \bar{u}_1\gamma_\mu u_2 &\to (\chi_1^\dagger \chi_2, \mathbf{0}) \\ \bar{u}_1\gamma_\mu\gamma_5 u_2 &\to (0, \chi_1^\dagger \boldsymbol{\sigma} \chi_2) \end{aligned} \tag{5.58}$$

see e.g. Chap. 6 of Ref. [11]. For a state with $J = 0$, the first of these terms is zero since χ_1 and χ_2 must be orthogonal. The second must be contracted with $\mathbf{p}^*$, which is the only available vector. Hence, we should expect $|\mathscr{A}|^2$ to be proportional to the centre-of-mass momentum squared, $|\mathbf{p}^*|^2 \propto m_\pi^2 - m_\mu^2$. Finally, the decay width must be proportional to the two-body phase-space, which is again proportional to $|\mathbf{p}^*|^2$, see Eq. (1.189). The decay width can be easily computed by using the standard covariant formalism, see e.g. Ref. [21], yielding Eq. (5.50). The ratio between the decay width to muons and electrons is thus given by:

$$\frac{\Gamma(\pi^+ \to e^+ \nu_e)}{\Gamma(\pi^+ \to \mu^+ \nu_\mu)} = \frac{m_e^2 (m_\pi^2 - m_e^2)^2}{m_\mu^2 (m_\pi^2 - m_\mu^2)^2} = 1.28. \times 10^{-4}. \tag{5.59}$$

Once a radiative correction of about -3.9% is included [24], the theoretical result agrees with the observed ratio of 1.23×10^{-4} [9].

Suggested Readings

See Ref. [21] for a formal treatment of the charged pion decay. This topic is usually well discussed by classical textbooks like e.g. Ref. [8].

Bando n. 1N/R3/SUB/2005

Problem 5.25 What is the LO expectation for the ratio $R = \sigma_{\text{had}}/\sigma_{\mu\mu}$ for an e^+e^- machine operating at a centre-of-mass energy of 6 GeV? What is the the LO expectation for the BR of the W boson into a lepton-neutrino pair?

Discussion

In the history of QCD, the experimental observable $R = \sigma_{\text{had}}/\sigma_{\mu\mu}$ had a key role in providing direct evidence in favour of the quark model with $N_C = 3$ and fractional electric charge.

Solution

At LO, the production of a $\mu^+\mu^-$ pair as well as of quark-antiquark pair at $\sqrt{s} = 6$ GeV proceeds through a time-like photon in the s-channel. The cross section for $e^+e^- \to q\bar{q}$ must be proportional to the number of coulors N_C and to the charge squared of the quark, e_q. If n_f quark flavours are kinematically accessible, i.e. $m_q \leq \sqrt{s}/2$, and neglecting the mass of the produced quarks compared to their energy, so that the phase-space volume of Eq. (1.189) is the same for muons and for any of the quarks, then the LO ratio R is given by:

$$R_{\text{LO}} = N_C \sum_{q=u,d,s,c} e_q^2 = N_C \left[2 \cdot \left(\frac{2}{3}\right)^2 + 2 \cdot \left(-\frac{1}{3}\right)^2 \right] = \frac{10}{9} N_C \approx 3.3 \quad (5.60)$$

Given the experimental value for the quark masses, see Table 5.3, the assumptions above are valid at $\sqrt{s} = 6$ GeV. The experimental outcome is $R \approx 4$ [9], which is in fair agreement with the LO estimate. The NLO corrections to the ratio R gives a K factor [19]:

$$R_{\text{NLO}} = R_{\text{LO}} \left(1 + \frac{\alpha_s(s)}{\pi} \right) \approx 3.5, \quad (5.61)$$

which agrees even better with the experimental value. In obtaining the numerical value in Eq. (5.61), we have used the value $\alpha_s(s = (6\,\text{GeV})^2) \approx 0.2$, see e.g. Fig. 9.2 of Ref. [9].

Table 5.3 Mass and charge of the six quarks. From Ref. [18]

Flavour	Mass (MeV)	Charge (e)
d	$3 \div 9$	$-1/3$
u	$1 \div 5$	$+2/3$
s	$75 \div 170$	$-1/3$
c	$1150 \div 1350$	$+2/3$
b	$4000 \div 4400$	$-1/3$
t	1.74×10^5	$+2/3$

Suggested Readings

A formal treatment of R in perturbative QCD can be found in dedicated textbooks, like Chap. 3 of Ref. [19].

Bando n. 1N/R3/SUB/2005

Problem 5.26 What is the the LO expectation for the BR of the W boson into a lepton-neutrino pair?

Discussion

The coupling of the W boson to the leptons, here collected into three $SU(2)_L$ multiplets, is given by

$$\mathscr{L}_{CC}^{lep} = \frac{g}{\sqrt{2}} \bar{L} \gamma^\mu T^- P_L L W_\mu^+ + \text{h.c.} = \sum_{\ell=e,\mu,\tau} \frac{g}{2\sqrt{2}} \bar{\ell} \gamma^\mu (1 - \gamma_5) \nu_\ell W_\mu^+ + \text{h.c.},$$

$$(5.62)$$

where $T^\pm$ are the ladder operators of $SU(2)_L$ and P_L is the left-handed helicity projector. For the quark doublets Q, again seen as a multiplet in quark-flavour space, the interaction is similar to Eq. (5.62) modulo the insertion of the CKM matrix V:

$$\mathscr{L}_{CC}^{had} = \frac{g}{\sqrt{2}} \bar{Q} V \gamma^\mu T^- P_L Q W_\mu^+ + \text{h.c.} =$$

$$= \sum_{a,b=1}^{N_C} \sum_{i=u,c,t} \sum_{j=d,s,b} \frac{g}{2\sqrt{2}} \delta^{ab} \bar{u}_i^a V_{ij} \gamma^\mu (1 - \gamma_5) d_j^b W_\mu^+ + \text{h.c.} \qquad (5.63)$$

where δ^{ab} makes evident that the charged-current interactions are diagonal in colour space.

Solution

With the only exception of the CKM matrix, the charged-current coupling of fermions is universal. If we denote the LO width for the two-body decay by $\Gamma(W \to \ell\,\nu)$, the leptonic and hadronic widths are given by:

$$\Gamma_{lep} = \sum_{\ell=e,\mu,\tau} \Gamma(W \to \ell\,\nu) = 3\,\Gamma(W \to \ell\,\nu)$$

$$\Gamma_{had} = N_C \sum_{i=u,c} \sum_{j=d,s,b} |V_{ij}|^2 \Gamma(W \to \ell\,\nu) = 2\,N_C\,\Gamma(W \to \ell\,\nu)$$

$$(5.64)$$

Differences in phase-space are here neglected since they give rise to small corrections. In the second of Eq. (5.64), we have exploited the fact that the decay into final states with top quarks is kinematically forbidden, see Table 5.3, and made use of the unitarity of the CKM matrix, $V^{-1} = V^\dagger$. Hence, LO prediction is:

$$\Gamma_W = \Gamma_{\text{lep}} + \Gamma_{\text{had}} = (2N_C + 3)\Gamma(W \to \ell\,\nu) = 9\,\Gamma(W \to \ell\,\nu)$$

$$\text{BR}(W \to \ell\,\nu_\ell) = \frac{\Gamma_{\text{lep}}}{\Gamma_{\text{lep}} + \Gamma_{\text{had}}} = \frac{3}{2N_C + 3} = \frac{1}{3} \qquad (5.65)$$

Hence, the branching ratio to any lepton is roughly 10%.

Problem 5.27 Determine the width of the W boson at LO.

Solution

We can use the charged-current Lagrangian of Eqs. (5.62) and (5.63) to derive the amplitude $\mathcal{M}$ for the process $W^+ \to \ell^+\,\nu$, where ℓ and ν denote the up and down states of a generic $SU(2)_L$ doublet. Using Eq. (1.186), the decay width $\Gamma(W^+ \to \ell^+\,\nu)$ is given by

$$\Gamma(W^+ \to \ell^+\,\nu) = \frac{1}{2\,m_W} \int \frac{\sum |\mathcal{M}|^2}{2J + 1}\, d\Phi_2, \qquad (5.66)$$

where the sum runs over the polarisations and spin states of the fermions. The amplitude squared can be readily obtained by using the Feynman rules corresponding to the Lagrangian of Eq. (5.62), giving:

$$\frac{1}{3}\sum |\mathcal{M}|^2 = \left(\frac{g}{2\sqrt{2}}\right)^2 \left(\frac{1}{3}\sum_r \varepsilon_\mu(r)\varepsilon_\nu^*(r)\right) \text{Tr}\left[\ell\gamma_\mu(1 - \gamma_5)\slashed{\nu}\gamma_\nu(1 - \gamma_5)\right] =$$

$$= \left(\frac{g}{2\sqrt{2}}\right)^2 \frac{1}{3}\left(-g_{\mu\nu} + \frac{q_\mu q_\nu}{m_W^2}\right) 2\,\text{Tr}\left[\ell\gamma_\mu\slashed{\nu}\gamma_\nu(1 - \gamma_5)\right] =$$

$$= \left(\frac{g}{2\sqrt{2}}\right)^2 \frac{8}{3}\left[\left(-g_{\mu\nu} + \frac{q_\mu q_\nu}{m_W^2}\right)(\ell^\mu\nu^\nu + \nu^\mu\ell^\nu - (\ell\nu)g^{\mu\nu})\right] \quad (5.67)$$

In the last equality, we have used the fact that $P_L\ell = \ell P_R$ and $P_L^2 = 1$, and carried out the sum over the polarisation by using the completion relation. The trace containing γ_5 gives a vanishing contribution because it is totally antisymmetric, while the polarisation tensor is symmetric. By using the fact that $\ell\nu = m_W^2/2$, the contraction between the two tensors can be done trivially giving a factor of m_W^2. Hence:

$$\Gamma(W^+ \to \ell^+\,\nu) = \frac{1}{2\,m_W} \frac{g^2 m_W^2}{3} \int \frac{d\Omega^*}{32\pi^2} = \frac{g^2 m_W^2}{48\pi} = \frac{G_F m_W^3}{6\sqrt{2}\pi}, \qquad (5.68)$$

where we have used the relation

$$\frac{G_F}{\sqrt{2}} = \frac{g^2}{8\,m_W^2} \tag{5.69}$$

to conveniently express the result in terms of the Fermi constant. Finally, we can use Eq. (5.65) and get the result:

$$\Gamma_W = 9\,\Gamma(W^+ \to \ell^+\,v) = \frac{3G_F\,m_W^3}{2\sqrt{2}\pi} = 2.05\ \text{GeV}, \tag{5.70}$$

to be compared with the measured value of $\Gamma_W = (2.085 \pm 0.042)$ GeV [9].

Problem 5.28 Show that the branching ratios of the decay of τ leptons into muons, electrons, and hadrons are in the approximate ratio $1:1:3$. What fraction of the hadronic decays do you expect to contain at least one kaon?

Solution

The ratio between the leptonic and hadronic widths of the τ lepton can be predicted by using the results for the decays of the on-shell W boson, discussed in Problem 5.26. Indeed, the electroweak τ decay can be treated as the two-stage reaction $\tau \to W^*\,v_\tau$, $W^* \to$ anything. However, the much smaller centre-of-mass energy m_τ allows only two lepton flavours to contribute to the leptonic width and only the combination $u\,\bar{d}$ and $u\,\bar{s}$ to the hadronic width, so that:

$$\frac{\Gamma_{\text{lep}}}{\Gamma_{\text{had}}} \approx \frac{2}{N_C\,(|V_{ud}|^2 + |V_{us}|^2)} = \frac{2}{3}, \tag{5.71}$$

where we have used the fact that $|V_{ud}|^2 \approx 1 - |V_{us}|^2 = \cos^2\theta_C$. Hence, we get the relation:

$$\Gamma(\tau \to e\,v_e\,v_\tau) : \Gamma(\tau \to \mu\,v_\mu\,v_\tau) : \Gamma(\tau \to \text{hadrons}\,v_\tau) = 1:1:3. \tag{5.72}$$

This expectation is in fair agreement with the observed result $\text{BR}(\tau \to \ell\,v_\ell\,v_\tau) \approx 17\%$. QCD corrections decrease the ratio in Eq. (5.71) by roughly 20%. From Eq. (5.71), it is also evident that the decay width to strange mesons is Cabibbo-suppressed, with a typical ratio:

$$\frac{\Gamma(\tau \to K\,v_\tau)}{\Gamma(\tau \to \pi\,v_\tau)} \approx \frac{\sin^2\theta_C}{\cos^2\theta_C} = \left(\frac{0.22}{0.97}\right)^2 \approx 5\%. \tag{5.73}$$

Only one hadronic decay out of twenty is expected to contain a kaon.

Suggested Readings

For a modern review of τ lepton physics, the reader is addressed to Ref. [25].

Bando n. 18211/2016

Problem 5.29 Indicate at least one production process of the top quark at Tevatron and at LHC. How was it possible to determine the top quark mass at LEP?

Solution

The most abundant production mechanism of top quarks at hadron colliders is through the strong interaction $q\bar{q}$, $g\,g \to t\bar{t}$. Production of a single top quark via t-, tW-, and s-channel is suppressed since it envolves an electroweak coupling and depends on the b quark PDF (for t- and tW-channel).

The maximum LEP-2 energy of 209 GeV (see Problem 3.5) did not allow to produce $t\bar{t}$ pairs from s-channel γ^*/Z^0 production. The single-top channel $e^+ e^- \to W^- t\bar{b}$ is kinematically allowed, but its cross section is small and no such events were observed. The sensitivity to m_t came from electroweak precision observables [26] that could be measured to per mill accuracy [27], in particular the corrections to the ρ-parameter and to the $Z^0 \to b\bar{b}$ hadronic width:

$$\rho = \frac{m_W^2}{m_Z^2 \cos^2 \theta_W} = 1 + \Delta\rho = 1 + \frac{3\,G_F\,m_t^2}{8\pi^2\sqrt{2}} + \cdots$$

$$\Gamma_b = \Gamma_d \left(1 - \frac{20}{13}\Delta\rho + \cdots\right) \tag{5.74}$$

which are almost 1% corrections which could be measured at LEP.

Suggested Readings

See Ref. [28] for a comprehensive overview on the electroweak precision tests (EWP).

Bando n. 18211/2016

Problem 5.30 Estimate the minimum electron beam energy needed to study a system of linear dimension $d = 1$ fm.

Solution

By using the uncertainty principle, we can estimate the momentum of the exchanged virtual photon γ^* necessary to resolve a distance $d \sim 1$ fm to be

$$|\mathbf{q}| \gtrsim \frac{\hbar}{2}\frac{1}{d} \approx 2 \text{ GeV}/c, \tag{5.75}$$

where one can use the relation $\hbar c \approx 200$ MeV fm to convert a length into a momentum. Using the DIS notation, the four-momentum transfer $q = \ell - \ell'$ from the electron probe to the probed system of mass M is given by

$$q = (E - E', \boldsymbol{\ell} - \boldsymbol{\ell}') \equiv (v, \mathbf{q}) \tag{5.76}$$

If we denote the initial and final four-momentum of the system by P and P', conservation of energy and momentum implies

$$P + q = P' \quad \Rightarrow \quad \frac{-q^2}{2Mv} = 1 - \frac{W^2 - M^2}{2Mv} \equiv x, \tag{5.77}$$

where the *Bjorken variable* $x \in [0, 1]$ quantifies the degree of inelasticity of the scattering ($x = 1$ corresponds to a purely elastic scattering). From Eq. (5.76) and Eq. (5.77) we then have:

$$|\mathbf{q}|^2 = v^2 - q^2 = v^2 + 2v\,x\,M \quad \Rightarrow \quad |\mathbf{q}| = \sqrt{v\,(v + 2\,x\,M)}. \tag{5.78}$$

If both the energy and polar angle θ of the scattered electron can be measured, the momentum transfer is simply given by geometry:

$$|\mathbf{q}| = \sqrt{E^2 + E'^2 - 2EE'\cos\theta}, \tag{5.79}$$

where the small electron mass can be usually neglected. Depending on the energy/angular acceptance of the experiment, Eq. (5.78) or Eq. (5.79) together with Eq. (5.75) determine the necessary beam energy.

Notice that Eq. (5.75) should be taken as an order-of-magnitude estimate rather than a lower bound: if the probed system has a smooth charge distribution, the interaction with the virtual photon is suppressed at high energy by the lack of high-frequency modes. For example, if the system is a spin-1/2 particle of linear dimension d with no other intrinsic length scales, the generic quantum-mechanical current J_μ for an elastic scattering in momentum-space can be parametrised as

$$J_\mu(p, p') = \bar{u}(p')\left[F_1(q^2)\gamma_\mu + i\kappa\,\frac{q^\nu \sigma_{\mu\nu}}{2M}F_2(q^2)\right]u(p), \tag{5.80}$$

where $F_{1,2}(q^2)$ are the Lorentz-invariant form factors and paramerise the electric and magnetic dipole interaction with the photon, while κ is the anomalous magnetic moment ($\kappa = -1$ for electrons and $\kappa = 1.79$ for protons). Such a current gives rise to the differential cross section:

$$\frac{d\sigma}{d\Omega} = \frac{\alpha^2 \cos^2 \frac{\theta}{2}}{4E^2 \sin^4 \frac{\theta}{2}} \frac{E'}{E} \left[\left(F_1^2 - q^2 \frac{\kappa^2 F_2^2}{4M^2} \right) - \frac{q^2}{2M^2} (F_1 + \kappa F_2)^2 \tan^2 \frac{\theta}{2} \right]. \quad (5.81)$$

The form factors are proportional to the Fourier transform of the charge distribution of the hadron evaluated at the momentum transfer (in the centre-of-mass frame, $q^2 = -|\mathbf{q}|^2$). For smooth charge distributions characterised by a typical length scale d, the form factors are concentrated at values $|\mathbf{q}| \sim \hbar/d$[1] and such that they vanish in the limit of large momentum transfers, thus suppressing the cross section as shown by Eq. (5.81). Therefore, for systems of dimension d, the momentum transfer should not exceed a few times $\hbar/d$.

Bando n. 1N/R3/SUB/2005

Problem 5.31 A centre-of-mass energy of $270 + 270$ GeV as obtained at the proton-antiproton SPS collider at CERN was sufficient for producing W and Z^0 bosons, albeit with a reduced margin. Why?

Discussion

When two hadrons h_a and h_b undergo a large-momentum transfer scattering in a symmetric circular collider at a centre-of-mass energy $\sqrt{s} \gg m_h$, the final state produced by the binary parton interaction is characterised by zero transverse momentum (at LO) but a finite longitudinal momentum P_z and energy E. With the convention that h_a moves along the $+z$ direction, the total energy P_z and E are related to the parton fractions x_a and x_b by the relation:

$$\begin{cases} (x_a + x_b)\frac{\sqrt{s}}{2} = E \\ (x_a - x_b)\frac{\sqrt{s}}{2} = P_z \end{cases} \Rightarrow \quad x_{a,b} = \frac{E \pm P_z}{\sqrt{s}}. \quad (5.82)$$

The centre-of-mass energy $\sqrt{\hat{s}}$ of the partonic interaction is given by:

$$\sqrt{\hat{s}} = \sqrt{2\hat{p}_a \hat{p}_b} = \sqrt{x_a x_b s}. \quad (5.83)$$

The centre-of-mass energy available for the partonic collision is therefore a factor of $\sqrt{x_a x_b}$ smaller than the collider energy.

Solution

The dominant $W^\pm$ production channel at a proton-antiproton collider is via the Drell-Yan s-channel production:

---

[1] Consider for example the case of a linear "box" of half-size d: the Fourier transform is proportional to $\frac{\sin(d|\mathbf{q}|/\hbar)}{|\mathbf{q}|}$, which is concentrated in the region $|\mathbf{q}| \lesssim \hbar/d$.

$$u\,\bar{d} \rightarrow W^+, \qquad d\,\bar{u} \rightarrow W^- \qquad (5.84)$$

In both cases, the scattering can involve two valence quarks. The average momentum fraction of the incoming proton (antiproton) taken by u_V ($\bar{u}_V$) and d_V ($\bar{d}_V$) is given by:

$$\langle x_u \rangle_p = \langle x_{\bar{u}} \rangle_{\bar{p}} = \int_0^1 dx\, x\, u_V(x), \qquad \langle x_d \rangle_p = \langle x_{\bar{d}} \rangle_{\bar{p}} = \int_0^1 dx\, x\, d_V(x). \qquad (5.85)$$

Referring to Fig. 5.2 (right), we can estimate the integrals of Eq. (5.85) from the area below the corresponding function. Approximating these functions by triangles, we can estimate:

$$\langle x_u \rangle_p \approx \frac{1}{2} \cdot 0.5 \qquad \langle x_d \rangle_p \approx \frac{1}{2} \cdot 0.25, \qquad (5.86)$$

thus giving an average partonic centre-of-mass energy of:

$$\langle \sqrt{\hat{s}} \rangle \approx \sqrt{\langle x_u \rangle_p \langle x_d \rangle_p} \sqrt{s} = \sqrt{0.25 \cdot 0.12} \cdot 540\,\text{GeV} \approx 95\,\text{GeV}, \qquad (5.87)$$

which is just enough to produce both $W^\pm$ and Z^0 bosons.

Suggested Readings

The first direct production of $W^\pm$ bosons at a collider has been achieved at the proton-antiproton SPS synchrotron at CERN. The reader is encouraged to go through the discovery paper published by the UA1 Collaboration [29].

Bando n. 1N/R3/SUB/2005

Problem 5.32 Sketch the structure functions $x F_3(x)$, $F_2(x)$, and $g(x)$ at $Q^2 = 10$ GeV^2. How do the structure functions evolve as a function of the Q^2 of the interaction?

Discussion

The structure functions F_i are adimensional functions that enter the theoretical expressions for DIS cross sections. They are different for electromagnetic (F_i^γ) and charged-current interactions (F_i^ν, $F_i^{\bar{\nu}}$). To a more formal level, they are defined in terms of the hadronic tensor $H^{\mu\nu}(q, p)$, where $q = \ell - \ell'$ is the four-momentum transfer and p is the initial hadron four-momentum, by the relation:

$$H^{\mu\nu}(p, q) = -F_1\, g^{\mu\nu} + F_2\, p^\mu p^\nu + i\, F_3\, \varepsilon^{\mu\nu}{}_{\sigma\tau}\, p^\sigma q^\tau. \qquad (5.88)$$

Given that these functions are adimensional, they must depend on the ratio between q^2, pq, and of any other intrinsic mass scale M^2. In the LO parton-model, they are functions of $x = -q^2/2pq$ and are proportional to particular combinations of parton density functions evaluated at the Bjorken fraction x. At NLO, they receive Q^2-dependent corrections of $\mathcal{O}(\alpha_s)$, which make them evolve with energy as predicted by the *Dokshitzer-Gribov-Lipatov-Altarelli-Parisi* (DGLAP) evolution equations.

The LO prediction for the structure functions is:

$$
\begin{aligned}
F_1^\gamma(x) &= \frac{1}{2}\left[\sum_{d,u} \frac{1}{9}(d(x) + \bar{d}(x)) + \frac{4}{9}(u(x) + \bar{u}(x))\right] \\
F_1^\nu(x) &= d(x) + \bar{u}(x), \qquad F_1^{\bar{\nu}}(x) = u(x) + \bar{d}(x) \\
F_2^{\gamma,\nu,\bar{\nu}}(x) &= 2x\, F_1^{\gamma,\nu,\bar{\nu}}(x) \\
F_3^\gamma(x) &= 0
\end{aligned}
\tag{5.89}
$$

$$
F_3^\nu(x) = 2\left[\sum_{d,u} d(x) - \bar{u}(x)\right], \qquad F_3^{\bar{\nu}}(x) = 2\left[\sum_{d,u} u(x) - \bar{d}(x)\right]
$$

Here, u and d are the PDF for up- and down-type quarks in the proton. The middle equation is known as *Callan-Gross* relation and is valid for spin-1/2 quarks.

Solution

The function $x F_3^{\nu+\bar{\nu}}(x)$, averaged for neutrino and antineutrino data, is proportional to the difference between the quark and antiquark PDF:

$$
x F_3^{\nu+\bar{\nu}} = 2x \sum (u - \bar{u} + d - \bar{d}) \propto x\, q_V.
\tag{5.90}
$$

The structure function $F_2^\nu(x)$ is instead proportional to the sum of the quark PDF's. The difference between $x F_3$ and F_2 is therefore proportional to the antiquark density $\bar{q}$.

The structure functions evolve with the Q^2 of the interaction according to the DGLAP evolution equations, see Problem 5.38. The relative change of the structure function F_i with respect to Q^2 can be estimated to be:

$$
\frac{dF_i}{F_i} \sim \alpha_s \frac{dQ^2}{Q^2}.
\tag{5.91}
$$

At larger momentum transfer, valence quarks q of large momentum fraction x are resolved more and more into collinear $q\,g$ pairs, thus suppressing the q density at large x-values compared to low x-value, see Problem 5.38. In turn, the gluon and sea quark densities are enhanced at higher Q^2. Figure 5.2 shows the functions $xf(x,\ Q^2)$ for different parton flavours and for two values of Q^2 [9].

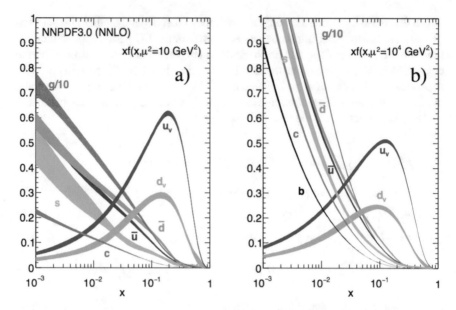

Fig. 5.2 The bands are x times the unpolarized parton distributions $f(x)$ (where $f = u_v, d_v, u, d, s \approx \bar{s}, c = \bar{c}, b = \bar{b}, g$) obtained in NNLO NNPDF3.0 global analysis at scales $Q^2 = 10$ GeV^2 (left) and $Q^2 = 10^4$ GeV^2 (right), with $\alpha_s(M_Z^2) = 0.118$ (from Ref. [9])

Suggested Readings

The DGLAP evolution equations are discussed in detail in dedicated textbooks like Ref. [18, 19].

Bando n. 1N/R3/SUB/2005

Problem 5.33 How do the electromagnetic and strong coupling constants evolve as a function of Q^2?

Discussion

The running of the coupling constants as a function of the renormalisation scale μ^2 is described by a *Renormalisation Group Equations* (RGE). In the SM, there are three independent gauge couplings $g_1 \equiv \sqrt{5/3}\, g_Y,^2\ g_2$, and g_3 associated with the $U(1)_Y$, $SU(2)_L$, and $SU(3)_C$ groups. At one loop, the three RGE's are given by:

---

[2]The normalisation factor of $\sqrt{5/3}$ is conventional and is inspired by GUT theories.

$$\frac{d\alpha_i}{d\ln\mu^2} = -\beta_0\,\alpha_i^2, \quad \text{with} \quad \beta_0 = \begin{cases} -\frac{41}{10\times4\pi} & g_1 \\ +\frac{19}{6\times4\pi} \cdot g_2 \\ +\frac{7}{4\pi} & g_3 \end{cases} \tag{5.92}$$

where $\alpha_i = g_i^2/4\pi$. See e.g. Ref. [30] for the full expression up to four-loops.

Solution

An ODE like Eq. (5.92) gives rise to a solution of the form:

$$\alpha(\mu^2) = \frac{\alpha(\mu_0^2)}{1 + \beta_0\,\alpha(\mu_0^2)\ln\left(\frac{\mu^2}{\mu_0^2}\right)}. \tag{5.93}$$

If $\beta_0 > 0$, the solution gives rise to a monotonically decreasing coupling constant and to asymptotic freedom (a "free" theory at high energy). This is the case for the non-abelian groups like $SU(3)_C$, see Problem 5.34. For $U(1)_Y$, the solution is instead monotonically increasing and develops a *Landau pole* at a scale:

$$\Lambda = \mu_0\exp\left[-\frac{1}{2\beta_0\alpha(\mu_0^2)}\right]. \tag{5.94}$$

The electromagnetic coupling is given by

$$e = g_2\sin\theta_W = \frac{g_Y}{\sqrt{1 + g_Y^2/g_2^2}}. \tag{5.95}$$

with $(g_Y/g_2)^2 = \tan^2\theta_W \approx 0.30$ at $\mu^2 = m_Z^2$. Since e is mostly given by g_Y, it also evolves like g_1, and thus develops a Landau pole at an very large energy scale, indeed far above the Plank scale $M_P = (\hbar c/G_N)^{1/2} = 1.2 \times 10^{19}$ GeV. At low energy, where only the quark and charged lepton loops contribute, the one-loop RGE for α is given by:

$$\frac{d\alpha}{d\ln\mu^2} = \frac{\alpha^2}{3\pi}\sum_i q_i^2\,N_i\,\Theta(\mu - m_i), \tag{5.96}$$

where N_i is the number of coulours for particle i. This equation can be solved to yield:

$$\alpha(\mu^2) \approx \alpha(\mu_0^2)\left[1 - \frac{\alpha(\mu_0^2)}{3\pi}\sum_i q_i^2\,N_i\ln\frac{\mu^2}{m_i^2}\right]^{-1}. \tag{5.97}$$

For example, when running from the electron mass up to the Z^0 mass, α increases by about 6%, moving from $1/137.0$ to about $1/128.9$.

Suggested Readings

A recent review of the RGE's in the SM at the light of the recently discovered Higgs boson can be found in Ref. [30].

Problem 5.34 The evolution equation of the running strong coupling α_s is given by:

$$\frac{d\alpha_s}{d \ln Q^2} = \beta(\alpha_s) = -\alpha_s \left[\alpha_s \beta_0 + \alpha_s^2 \beta_1 + \cdots \right] \tag{5.98}$$

where the coefficients β_0 and β_1 are defined by:

$$\beta_0 = \frac{11 \, C_A - 4 \, T_R \, n_f}{12\pi}, \qquad \beta_1 = \frac{11 \, C_A^2 - (6 \, C_F + 10 \, C_A) T_R \, n_f}{24\pi^2}, \tag{5.99}$$

with T_R, C_F, and C_A the Dinkin index and the Casimir of the fundamental and adjoint representation, respectively, while n_f indicates the number of active quark flavours.

1. Solve the evolution equation at one-loop level with the boundary condition $\alpha_s(\Lambda_{QCD}^{(0)}) = \infty$. What is the value of $\Lambda_{QCD}^{(0)}$ given that $\alpha_s(M_Z^2) = 0.118$, $n_f = 5$, and $M_Z = 91.187$ GeV?
2. Solve the evolution equation at two-loop level with the boundary condition $\alpha_s(\Lambda_{QCD}^{(1)}) = \infty$.

Discussion

The introduction of a running coupling constant $\alpha_s(Q^2)$ in the calculation of physical quantities removes the dependence on the unphysical renormalisation scale μ introduced by dimensional regularisation: the explicit μ-dependence of the physical amplitude is re-absorbed by defining a running coupling $\alpha_s(Q^2)$ evaluated at the *physical* scale Q^2: by using the functional form $\alpha_s(Q^2)$ and a measurement at some reference scale Q_0^2 through a physical process, the running strong coupling can be predicted at any other scale Q^2.

Solution

At the one loop-level, only the β_0 term is retained in Eq. (5.98), giving

$$Q^2 \frac{d\alpha_s}{dQ^2} = -\beta_0 \, \alpha_s^2, \qquad \frac{d\alpha_s}{\alpha_s^2} = -\beta_0 \frac{dQ^2}{Q^2}, \qquad d\left(-\frac{1}{\alpha_s}\right) = -\beta_0 \, d \ln Q^2$$

$$\frac{1}{\alpha_s(Q^2)} - \frac{1}{\alpha_s(Q_0^2)} = \beta_0 \ln \frac{Q^2}{Q_0^2} \tag{5.100}$$

If we choose $Q_0 \equiv \Lambda_{\mathrm{QCD}}^{(0)}$ such that $\alpha_s(\Lambda_{\mathrm{QCD}}^{(0)}) = \infty$, then:

$$\alpha_s(Q^2) = \frac{\beta_0}{\ln(Q^2/\Lambda_{\mathrm{QCD}}^{(0)\,2})}. \tag{5.101}$$

By inverting the formula above evaluated at some reference scale Q_0:

$$\Lambda_{\mathrm{QCD}}^{(0)} = Q_0 \exp\left[-\frac{1}{2\beta_0}\frac{1}{\alpha_s(Q_0)}\right]. \tag{5.102}$$

By using the measured value $\alpha_s(M_Z^2) = 0.118$ together with the group-theoretical value

$$\beta_0 = \frac{11\,C_A - 4\,T_R\,n_f}{12\pi} = \frac{11 \times 3 - 4 \times \frac{1}{2} \times 5}{12\pi} = 0.610, \tag{5.103}$$

one obtains:

$$\Lambda_{\mathrm{QCD}}^{(0)} = 91.2 \exp\left(-\frac{1}{2 \times 0.610 \times 0.118}\right) \text{ GeV} \approx 88 \text{ MeV}. \tag{5.104}$$

Another popular way of writing Eq. (5.101) is:

$$\alpha_s(Q^2) = \frac{\alpha_s(Q_0^2)}{1 + \beta_0\,\alpha_s(Q_0^2)\ln\frac{Q^2}{Q_0^2}}, \tag{5.105}$$

where an arbitrary scale Q_0^2 is introduced in place of Λ_{QCD}.

At the two loop-level:

$$Q^2\frac{d\alpha_s}{dQ^2} = -\beta_0\,\alpha_s^2 - \beta_1\,\alpha_s^3, \qquad \left(\frac{1}{\beta_0}\right)\frac{d\alpha_s}{-\alpha_s^2\left(1 + \frac{\beta_1}{\beta_0}\alpha_s\right)} = d\ln Q^2 \tag{5.106}$$

Next, we break the left-hand side of this equation into the sum of functions with simple primitive:

$$\left(-\frac{1}{\beta_0}\right)\frac{1}{\alpha_s^2\left(1 + \frac{\beta_1}{\beta_0}\alpha_s\right)} = \left(-\frac{1}{\beta_0}\right)\left[\frac{A}{\alpha_s^2} + \frac{B}{\alpha_s} + \frac{C}{1 + \frac{\beta_1}{\beta_0}\alpha_s}\right], \tag{5.107}$$

giving the solution:

$$A = 1, \qquad B = -\frac{\beta_1}{\beta_0}, \qquad C = +\left(\frac{\beta_1}{\beta_0}\right)^2. \tag{5.108}$$

Replacing the constants with the values of Eq. (5.108), we get:

$$\left(\frac{1}{\beta_0}\right)\left[-\frac{d\alpha_s}{\alpha_s^2}+\frac{\beta_1}{\beta_0}\frac{d\alpha_s}{\alpha_s}-\left(\frac{\beta_1}{\beta_0}\right)^2\frac{d\alpha_s}{1+\frac{\beta_1}{\beta_0}\alpha_s}\right]=d\ln Q^2$$

$$\left(\frac{1}{\beta_0}\right)\left[d\left(\frac{1}{\alpha_s}\right)+\frac{\beta_1}{\beta_0}d\left(\ln\frac{\alpha_s}{1+\frac{\beta_1}{\beta_0}\alpha_s}\right)\right]=d\ln Q^2,$$

$$\left(\frac{1}{\beta_0}\right)\left[\frac{1}{\alpha_s}|_{Q_0^2}^{Q^2}+\frac{\beta_1}{\beta_0}\ln\frac{\alpha_s}{1+\frac{\beta_1}{\beta_0}\alpha_s}|_{Q_0^2}^{Q^2}\right]=\ln\frac{Q^2}{Q_0^2} \qquad (5.109)$$

We now choose $Q_0 \equiv \Lambda_{QCD}^{(1)}$ such that $\alpha_s(\Lambda_{QCD}^{(1)})=\infty$, then:

$$\frac{1}{\alpha_s(Q^2)}+\frac{\beta_1}{\beta_0}\ln\frac{\beta_1\alpha_s}{\beta_0+\beta_1\alpha_s(Q^2)}=\beta_0\ln\frac{Q^2}{\Lambda_{QCD}^{(1)\,2}}. \qquad (5.110)$$

It is convenient to introduce the variable $t\equiv\left[\ln Q^2/\Lambda_{QCD}^{(1)\,2}\right]^{-1}$, and work in the assumption that t is small (indeed, modulo the factor $\beta_0\approx 0.6$, t coincides with α_s at one-loop, so it is of order 10^{-1} at the EW scale). In this case, we can make the approximation:

$$\frac{1}{\alpha_s(Q^2)}\approx\frac{\beta_0}{t}-\frac{\beta_1}{\beta_0}\ln\frac{t}{\beta_0^2/\beta_1+t}, \qquad \alpha_s(Q^2)=\frac{1}{\beta_0}t\frac{1}{1+\frac{\beta_1}{\beta_0^2}t\ln\frac{\beta_0^2/\beta_1+t}{t}}, \qquad (5.111)$$

where we also use the fact that $x\ln x\to 0$ for $x\to 0$, so that the second term at the left-hand side of Eq. (5.111) can be neglected. By expanding at first order:

$$\left[1+C_1t\ln\left(C_2\frac{1}{t}+C_3\right)\right]^{-1}=1-C_1t\ln\frac{1}{t}+\mathcal{O}(t), \qquad (5.112)$$

we have:

$$\alpha_s(Q^2)=\frac{1}{\beta_0\log\frac{Q^2}{\Lambda_{QCD}^{(1)\,2}}}\left[1-\frac{\beta_1}{\beta_0^2}\frac{\log\left(\log\frac{Q^2}{\Lambda_{QCD}^{(1)\,2}}\right)}{\log\frac{Q^2}{\Lambda_{QCD}^{(1)\,2}}}\right]. \qquad (5.113)$$

Figure 5.3 shows a summary of measurements of α_s as a function of the energy scale Q.

Fig. 5.3 Summary of measurements of α_s as a function of the energy scale Q. The respective degree of QCD perturbation theory used in the extraction of α_s is indicated in brackets (from Ref. [9])

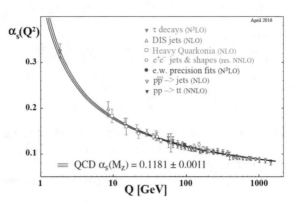

Suggested Readings

The reader is addressed to Chap. 4.3 of Ref. [18] for a formal treatment of renormalisation and for the derivation of the β-function in QCD.

Problem 5.35 Find the relation between the colour factors T_R, C_F, and C_A and the branching probabilities for $q \to q\,g$, $g \to g\,g$, and $g \to q\,\bar{q}$.

Solution

Consider the emission of a gluon g_a, with $a = 1, \ldots, 8$, from a quark q_i, with $i = 1, 2, 3$, which then transforms into a quark q_j. The colour part of the amplitude is proportional to

$$\mathscr{A}(q_i \to q_j\, g_a) \propto T_{ji}^a. \tag{5.114}$$

Summing over the final colours and averaging over the initial ones:

$$\overline{\sum} \mathscr{A}\mathscr{A}^* = \frac{1}{N_C} \sum_a \sum_{ij} T_{ji}^a T_{ji}^{a*} = \frac{1}{N_C} \text{Tr}\left[\sum_a T^a T^a\right] = \frac{N_C\, C_F}{N_C} = C_F \tag{5.115}$$

For the $g_a \to g_b\, g_c$ splitting, the colour part is proportional to:

$$\mathscr{A}(g_a \to g_b\, g_c) \propto f_{abc} \tag{5.116}$$

Summing over the final colours and averaging over the initial ones:

$$\overline{\sum \mathscr{A} \mathscr{A}^*} = \frac{1}{N_A} \sum_{abc} f_{abc} f^*_{abc} = \frac{1}{N_A} \sum_{abc} (-i f_{abc})(-i f_{acb}) = \frac{1}{N_A} \sum_{abc} \mathscr{T}^a_{bc} \mathscr{T}^a_{cb}$$

$$= \frac{1}{N_A} \mathrm{Tr} \left[\sum_a \mathscr{T}^a \mathscr{T}^a \right] = C_A. \tag{5.117}$$

Here, we have used the antisymmetry of the structure constants f_{abc} and the fact that $\mathscr{T}^a_{bc} = -i f_{abc}$ are the generators of the adjoint representation of $SU(3)$, with Casimir $C_A \mathbb{1}_{N_A}$.

For the $g_a \to q_i \bar{q}_j$ splitting, the colour part is proportional to:

$$\mathscr{A}(g_a \to q_i \bar{q}_j) \propto T^a_{ji}. \tag{5.118}$$

Summing over the final colours and averaging over the initial ones we get:

$$\overline{\sum \mathscr{A} \mathscr{A}^*} = \frac{1}{N_A} \sum_a \sum_{ij} T^a_{ij} T^{a*}_{ij} = \frac{1}{N_A} \mathrm{Tr} \left[\sum_a T^a T^a \right] = \frac{N_C C_F}{N_A} = T_R. \tag{5.119}$$

The last equality has been obtained by using the relations:

$$\sum_{ij} T^a_{ij} T^a_{ji} = T_R \delta^{aa}, \qquad \sum_{aj} T^a_{ij} T^a_{ji} = C_F \delta^{ii} \tag{5.120}$$

The left-hand side of the two equations are identical upon summation over all indices, which therefore implies:

$$T_R \mathrm{Tr} \left[\mathbb{1}_{N_A} \right] = C_F \mathrm{Tr} \left[\mathbb{1}_{N_C} \right] \quad \Rightarrow \quad T_R N_A = C_F N_C. \tag{5.121}$$

Discussion

For $SU(3)$, the following group identities hold:

$$\frac{T_R}{C_F} = \frac{N_C}{N_C^2 - 1} = \frac{3}{8}, \quad \frac{C_A}{C_F} = \frac{2 N_C^2}{N_C^2 - 1} = \frac{9}{4}, \quad N_A = N_C^2 - 1 = 8. \tag{5.122}$$

With the standard normalisation $T_R = 1/2$, we then have $C_F = 4/3$ and $C_A = 3$. When doing more involved calculations, tricks can be used to speed up the computation of colour factors. In particular, the antisymmetric tensor f_{abc} can be always exchanged for traces of T^a matrices using the relation:

$$T^a T^b - T^b T^a = i f_{abc} T^c \quad \Rightarrow \quad f_{abc} = -\frac{i}{T_R} \mathrm{Tr} \left[T^a T^b T^c - T^b T^a T^c \right] \tag{5.123}$$

Suggested Readings

The calculation of the colour factors in the $q\bar{q} \to g g$ amplitude provides an instructive example of coulour algebra. A guided calculation can be found in Sect. 3.3.4 of Ref. [18]

Problem 5.36 At the lowest order in QED, the interaction between an electron and a positron in a bound state is represented by a diagram where a photon is exchanged between a quark and an antiquark. Similarly, we can think of the QCD potential that binds quarks inside hadrons as being mediated by the exchange of a gluon. The potential is given by:

$$V_{q\bar{q}}^{QCD} = f\frac{\alpha_s}{r}, \tag{5.124}$$

where f is a colour factor which depends on the colour of the quark and antiquark states. Calculate the colour factor f for:

1. colour singlet quark-antiquark bound state $|h\rangle \propto \sum_{ij} \delta_{ij}|q_i\bar{q}_j\rangle$.
2. colour octet quark-antiquark bound state $|h; a\rangle \propto \sum_{ij} \lambda_{ij}^a|q_i\bar{q}_j\rangle$.
3. totally antisymmetric state builded with three quarks $|h\rangle \propto \sum_{ijk} \varepsilon_{ijk}|q_iq_jq_k\rangle$.

Discussion

The potential $V(\mathbf{x})$ can be calculated from the Fourier-transform of the non-relativistic $q q' \to q q'$ amplitude corresponding to a single-gluon exchange. The result from QED is the well-known Coulomb potential

$$V^{QED}(\mathbf{r}) = \pm\frac{\alpha}{|\mathbf{r}|}, \tag{5.125}$$

where the sign "+" applies to the $q q \to q q$ case, while the sign "−" applies to $q\bar{q} \to q\bar{q}$, see e.g. the discussion in Ref. [15]. Hence, the potential is predicted to be repulsive (attractive) for a quark-quark (quark-antiquark) interaction. This result depends on the Lorentz structure ($\bar{\psi}\gamma^\mu\psi$) of the fermionic current in QED, and is identical for the case of QCD. The only difference between the two theories arises from the colour structure of the quark-quark-gluon vertex and from the colour wave-function of the bound state, giving rise to the replacement

$$\alpha \to f\alpha_s, \tag{5.126}$$

where f_C is the colour factor that needs to be calculated.

We also prove that the three bound states above transform as singlets or octets, as anticipated. First, we recall that under the same infinitesimal transformation $U(\theta) = \mathbb{1} + i\theta_a T^a$, $\theta_a \ll 1$, quarks and antiquarks undergo the following transformation:

$$q_i \rightarrow q_i' = (\delta_{ij} + i\theta_a T_{ij}^a) q_j \tag{5.127}$$

$$\bar{q}_i \rightarrow \bar{q}_i' = (\delta_{ij} - i\theta_a T_{ij}^{a*}) \bar{q}_j = (\delta_{ij} - i\theta_a T_{ji}^a) \bar{q}_j \tag{5.128}$$

This allows us to prove that for the meson singlet:

$$\begin{aligned}
\delta_{ji}\, q_i\, \bar{q}_j &\rightarrow \delta_{ji}\, (\delta_{ik} + i\theta_a T_{ik}^a)(\delta_{lj} - i\theta_a T_{lj}^a)\, q_k\, \bar{q}_l \\
&\approx \delta_{ji}\, q_i\, \bar{q}_j + i\theta_a(\delta_{ji}\delta_{jl} T_{ik}^a - \delta_{ij}\delta_{ik} T_{lj}^a)\, q_k\, \bar{q}_l \\
&= \delta_{ji}\, q_i\, \bar{q}_j + i\theta_a(T_{lk}^a - T_{lk}^a) = \delta_{ji}\, q_i\, \bar{q}_j .
\end{aligned} \tag{5.129}$$

It is easy to show that this holds true also for a finite transformation U, since:

$$\delta_{ji}\, q_i\, \bar{q}_j \rightarrow \delta_{ji}\, U_{il}\, U_{mj}^*\, q_l\, \bar{q}_m = (U\, U^\dagger)_{lm}\, q_l\, \bar{q}_m = \delta_{ji}\, q_i\, \bar{q}_j . \tag{5.130}$$

For the meson octet:

$$\begin{aligned}
T_{ji}^a\, q_i\, \bar{q}_j &\rightarrow T_{ji}^a\, (\delta_{ik} + i\theta_b T_{ik}^b)(\delta_{jl} - i\theta_b T_{lj}^b)\, q_k\, \bar{q}_l \\
&\approx T_{ji}^a\, q_i\, \bar{q}_j + i\theta_b(T_{ji}^a T_{ik}^b \delta_{lj} - T_{ji}^a T_{lj}^b \delta_{ik})\, q_k\, \bar{q}_l \\
&= T_{ji}^a\, q_i\, \bar{q}_j + i\theta_b(T_{li}^a T_{ik}^b - T_{lj}^b T_{jk}^a)\, q_k\, \bar{q}_l \\
&= T_{ji}^a\, q_i\, \bar{q}_j + i\theta_b[T^a, T^b]_{lk}\, q_k\, \bar{q}_l = T_{ji}^a\, q_i\, \bar{q}_j + i\theta_b(if^{abc} T_{lk}^c)\, q_k\, \bar{q}_l \\
&= T_{ji}^a\, q_i\, \bar{q}_j + i\theta_b(-if^{bac} T_{lk}^c)\, q_k\, \bar{q}_l
\end{aligned} \tag{5.131}$$

Remembering that $\mathscr{T}_{ac}^b = -if_{bac}$ are the generators of the adjoint representation of dimension $N_C^2 - 1 = 8$, we have:

$$T_{ji}^a\, q_i\, \bar{q}_j \equiv v^a \rightarrow v^a + i\theta_b\, \mathscr{T}_{ac}^b\, v^c, \tag{5.132}$$

i.e. v^a transform like an octet of $SU(3)$.

Finally, for the baryon singlet we have:

$$\begin{aligned}
\varepsilon_{ijk}\, q_i\, q_j\, q_k &\rightarrow \varepsilon_{ijk}\, (\delta_{il} + i\theta_b T_{il}^b)(\delta_{jm} + i\theta_b T_{jm}^b)(\delta_{kn} + i\theta_b T_{kn}^b)\, q_l\, q_m\, q_n \\
&\approx \varepsilon_{ijk}\, q_i\, q_j\, q_k + \varepsilon_{ijk}\theta_b(T_{il}^b + T_{jm}^b + T_{kn}^b)\, q_l\, q_m\, q_n,
\end{aligned} \tag{5.133}$$

and notice that the expression in parentheses is symmetric under exchange of any of the ijk indexes, while the ε_{ijk} tensor is antisymmetric, and so the sum is zero. It is easy to show that this holds true also for a finite transformation U, since:

$$\varepsilon_{ijk}\, q_i\, q_j\, q_k \rightarrow \varepsilon_{ijk}\, U_{il}\, U_{jm}\, U_{kn}\, q_l\, q_m\, q_n = |\det U|\, \varepsilon_{lmn}\, q_l\, q_m\, q_n = \varepsilon_{ijk}\, q_i\, q_j\, q_k . \tag{5.134}$$

Solution

- The wave-function of the bound state $|h\rangle$ is given by

$$|h\rangle = \mathcal{N} \sum_{ij} \delta_{ij} |q_i \, \bar{q}_j\rangle, \qquad (5.135)$$

where δ_{ij} is the Kronecker tensor. The normalisation factor $\mathcal{N}$ can be readily obtained by requiring the normalisation condition:

$$1 = \langle h|h\rangle = \sum_{ijkl} |\mathcal{N}|^2 \delta_{ik}\delta_{jl}\delta_{ij}\delta_{kl} = |\mathcal{N}|^2 \mathrm{Tr}\left[\mathbb{1}_{N_C}\right] \Rightarrow \mathcal{N} = \frac{1}{\sqrt{N_C}} \qquad (5.136)$$

Therefore, the colour factor is:

$$f = \sum_{aijkl} \frac{\delta_{ij}}{\sqrt{N_C}} \frac{\delta_{kl}}{\sqrt{N_C}} T^a_{ki} T^a_{jl} = \frac{1}{N_C} \sum_{aijkl} T^a_{ji} T^a_{ij} = \frac{1}{N_C} \mathrm{Tr}\left[\sum_a T^a T^a\right] = \frac{1}{N_C} C_F N_C = C_F \qquad (5.137)$$

where we have used the identity $\mathscr{C}_F \equiv \sum_a T^a T^a = C_F \mathbb{1}_{N_C}$, and $\mathscr{C}_F$ is the Casimir of the fundamental representation. With the usual normalisation $T_R = 1/2$ of the Dinkin index:

$$f = C_F = \frac{N_C^2 - 1}{2N} = +\frac{4}{3}. \qquad (5.138)$$

The factor f is positive: the resulting potential is identical to the one generated by a single-photon exchange between a fermion-antifermion pair in QED, modulo the replacement $\alpha \to 4/3\alpha_s$. The potential is therefore *attractive*.
- The wave-function of the bound state is given by

$$|h; \, a\rangle = \mathcal{N} \sum_{ij} T^a_{ij} |q_i \, \bar{q}_j\rangle, \qquad (5.139)$$

where $T^a_{ij} = \lambda^a_{ij}/2$, with λ^a_{ij} the eight Gell-Mann matrices. The normalisation factor $\mathcal{N}$ can be obtained by requiring the normalisation condition:

$$1 = \langle h; \, a|h; \, a\rangle = |\mathcal{N}|^2 \sum_{ijkl} T^{a*}_{ij} T^a_{kl} \delta_{ij}\delta_{kl} = \sum_{ij} |\mathcal{N}|^2 T^a_{ji} T^a_{ij} = |\mathcal{N}|^2 \mathrm{Tr}\left[T^a T^a\right],$$

$$\Rightarrow \quad \mathcal{N} = \frac{1}{\sqrt{T_R}}, \qquad (5.140)$$

where we have used the normalisation condition $\text{Tr}\left[T^a T^b\right] = T_R \delta_{ab}$. Therefore:

$$f = \frac{1}{T_R} \sum_{bijkl} T_{ij}^a T_{kl}^{a*} T_{ki}^b T_{jl}^b = \frac{1}{T_R} \sum_{ijkl} T_{ij}^a T_{lk}^a \left[T_R (\delta_{kl}\delta_{ij} - \frac{1}{N_C}\delta_{ki}\delta_{jl}) \right] =$$

$$= (\text{Tr}\left[T^a\right])^2 - \frac{1}{N_C}\text{Tr}\left[T^a T^a\right] = -\frac{T_R}{N_C} = -\frac{1}{6}. \tag{5.141}$$

We have made use of the completeness relation:

$$\sum_b T_{ij}^b T_{kl}^b = T_R \left[\delta_{il}\delta_{jk} - \frac{1}{N_C}\delta_{ij}\delta_{kl} \right]. \tag{5.142}$$

The colour factor f is negative: the potential is therefore *repulsive*.

• The potential is generated by the exchange of a single gluon between two quarks, the third one behaving like a spectator. The two quarks are in the antisymmetric anti-triplet state: $3 \otimes 3 = 6 \oplus \bar{3}$. The wave-function of the bound state is given by

$$|h\rangle = \mathcal{N} \sum_{ijk} \varepsilon_{ijk} |q_i \, q_j \, q_k\rangle, \tag{5.143}$$

The normalisation factor $\mathcal{N}$ can be obtained by requiring the normalisation condition:

$$1 = \langle h|h\rangle = |\mathcal{N}|^2 \sum_{ijk,lmn} \varepsilon_{ijk}\,\varepsilon_{lmn}\,\delta_{il}\,\delta_{jm}\delta_{kn} = |\mathcal{N}|^2 \sum_{ijkl} \varepsilon_{ijk}\,\varepsilon_{ijk} = |\mathcal{N}|^2 N_C!,$$

$$\mathcal{N} = \frac{1}{\sqrt{N_C!}}. \tag{5.144}$$

Therefore:

$$f_C = \frac{1}{N_C!} \sum_{aijklmn} \varepsilon_{ijk}\,\varepsilon_{lmn}\, T_{li}^a \, T_{mj}^a \delta_{kn} = \frac{1}{N_C!} \sum_{aijlm} \left[\delta_{il}\,\delta_{jm} - \delta_{im}\,\delta_{jl} \right] T_{li}^a \, T_{mj}^a =$$

$$= -\sum_a \text{Tr}\left[T^a \, T^a\right] = -\frac{C_F \, N_C}{N_C!} = -\frac{2}{3}, \tag{5.145}$$

where we have used the relation $\sum_i \varepsilon_{ijk}\varepsilon_{imn} = \delta_{jm}\delta_{kn} - \delta_{jn}\delta_{km}$. The colour factor is negative. However, one should remember that this negative sign is relative to the fermion-fermion amplitude, which is repulsive for the QED case, and therefore the resulting potential is *attractive*.

Suggested Readings

For a more exahustive discussion on the potentials generated by a single-exchange of the force mediator, see Chap. 4.7 of Ref. [15]. More details on colour algebra can be found in Ref. [18]. The QCD potentials for bound states are also discussed in Ref. [12].

Problem 5.37 Calculate the $P_{qq}(z)$ and $P_{gq}(z)$ kernel functions using the explicit form of the related matrix elements.

Discussion

The kernel funcions, or *Altarelli-Parisi splitting functions*, can be derived by using the covariant formalism in dimensional regularisation, see e.g. Chap. 3.6 of Ref. [18]. Alternatively, one can follow the original approach by Altarelli and Parisi [31] which makes use of old-fashioned perturbation theory. The goal is to factorise the infinitesimal cross section $d\sigma$ for scattering of a lepton ℓ off the parton A which undergoes an almost *collinear* splitting $A \to B + C$, in the form:

$$d\sigma(\ell + A \to \ell' + C) \propto \int dz \left[\left(\frac{\alpha_s}{2\pi} \right) P_{BA}(z) \right] \times \left[d\sigma(\ell + B(z) \to \ell') \right],$$

$$(5.146)$$

i.e. as the convolution of a universal function $P_{BA}(z)$, that gives the probability that parton B takes a fraction z of the incoming parton A, with the cross section for the DIS scattering $\ell + B \to B'$. Given that the matrix element for collinear splitting is divergent, the cross section thus obtained will be proportional to an infinite factor $\int d\mathbf{p}_\perp^2 / \mathbf{p}_\perp^2$ which will be reabsorbed as a renormalisation of the parton pdf.

Solution

We consider the DIS process where a lepton with four-momentum ℓ scatters against a parton A with four-momentum p_A via a space-like photon γ^* carrying four-momentum q. Before interacting with γ^*, the parton splits via $A \to B + C$, where C is the parton for which we will consider the collinear limit:

$$p_B \to (1 - z)\, p_A, \qquad p_C \to z\, p_A. \qquad (5.147)$$

After interacting with γ^*, the parton B will have four-momentum $p_{B'} = p_B + q$. In old-fashioned perturbation theory, this process occurs at the second perturbative order, with an amplitude:

$$\mathcal{M}_{fi} = \int dv \frac{\mathcal{M}_{vi}^{NR} \cdot \mathcal{M}_{fv}^{NR}}{E_v - E_i}, \qquad (5.148)$$

where i, v, f are eigenstates of $\mathscr{H}_0$, the initial and final energies are equal, linear momentum is conserved at any vertex, and $\mathscr{M}_{i\to v}^{NR}$ is the non-relativistic quantum-mechanical transition amplitude. For the case of interest, we can write:

$$\mathscr{M}_{\ell+A\to\ell'+C} = \frac{\sum_B \mathscr{M}_{A\to B+C}^{R}\,\mathscr{M}_{\ell+B\to\ell'}^{R}}{2E_B(E_B+E_C-E_A)}, \tag{5.149}$$

where the factor $2E_B$ accounts for the relativistic wave-function normalisation $\sqrt{2E}$ used for the amplitudes at the numerator (using relativistic normalisation allows us to use Feynman rules for calculating specific amplitudes). We use a coordinate system such that:

$$p_A = (E_A, E_A, 0, 0), \qquad p_B = \left(zE_A + \frac{\mathbf{p}_\perp^2}{2zE_A},\, zE_A,\, 0,\, \mathbf{p}_\perp\right)$$

$$p_C = ((1-z)E_A + \frac{\mathbf{p}_\perp^2}{2(1-z)E_A},\, (1-z)E_A,\, 0,\, -\mathbf{p}_\perp) \tag{5.150}$$

The first equation fixes the reference frame. The second and third express momentum conservation along the longitudinal direction: z is then the momentum fraction taken by the parton C and $\mathbf{p}_\perp$ is the transverse momentum and is treated as the perturbative parameter on which all energies are expanded. The expression for the energies E_B and E_C is correct up to $\mathscr{O}(\mathbf{p}_\perp^2)$. Inserting the expressions above into Eq. (5.149), summing over the colour and spin polarisation of the final-state partons and averaging over those of the incoming parton:

$$\sum|\mathscr{M}_{\ell+A\to\ell'+C}|^2 = \frac{\sum_{A,B,C}|\mathscr{M}_{A\to B+C}|^2 \sum_B |\mathscr{M}_{\ell+B\to\ell'}|^2}{4\left(zE_A + \frac{\mathbf{p}_\perp^2}{2zE_A}\right)^2 \left(zE_A + \frac{\mathbf{p}_\perp^2}{2zE_A} + (1-z)E_A + \frac{\mathbf{p}_\perp^2}{2(1-z)E_A} - E_A\right)^2} =$$

$$= \frac{\sum_{ABC}|\mathscr{M}_{A\to B+C}|^2 \sum_B |\mathscr{M}_{\ell+B\to\ell'}|^2}{(\mathbf{p}_\perp^2)^2/(1-z)^2}. \tag{5.151}$$

The infinitesimal cross section $d\sigma$ is obtained by integrating over the phase space of the emitted parton:

$$d\sigma(\ell+A\to C+\ell') = \int \frac{d^3p_C}{(2\pi)^3 2E_C}\left(\frac{E_B}{E_A}\right) \times d\sigma(\ell+B(z\cdot p_A)\to\ell')\times$$

$$\times \frac{\sum_{A,B,C}|\mathscr{M}_{A\to B+C}|^2}{(\mathbf{p}_\perp^2)^2/(1-z)^2} \tag{5.152}$$

where the factor E_B/E_A accounts for the incoming flux ($d\sigma(\ell+B(z\cdot p_A)\to\ell')\propto 1/E_B$). The next step is to write the integration measure in terms of $\mathbf{p}_\perp$ and z:

$$\int \frac{d^3p_C}{(2\pi)^3 2E_C} = \int \frac{dp_L\,d^2\mathbf{p}_\perp}{(2\pi)^3 2(1-z)E_A} = \int \frac{1}{16\pi^2}\frac{dz}{(1-z)}\,d\,\mathbf{p}_\perp^2 \tag{5.153}$$

Hence, Eq. (5.152) becomes:

$$
d\sigma(\ell + A \to C + \ell') =
$$

$$
= \int d\,\mathbf{p}_\perp^2\, dz\, \frac{1}{16\pi^2}\frac{z}{(1-z)}\frac{(1-z)^2}{(\mathbf{p}_\perp^2)^2} \overline{\sum_{A,B,C}}|\mathcal{M}_{A \to B+C}|^2 \times d\sigma(\ell + B(z \cdot p_A) \to \ell')
$$

$$
= \int \left(\frac{d\,\mathbf{p}_\perp^2}{\mathbf{p}_\perp^2}\right)\int_0^1 dz \left[\left(\frac{\alpha_s}{2\pi}\right)\frac{z(1-z)}{2\,\mathbf{p}_\perp^2}\overline{\sum}|\mathcal{V}_{A \to B+C}|^2\right] \times d\sigma(\ell + B(z \cdot p_A) \to \ell'),
$$

$$
(5.154)
$$

where the coupling factor g_s has been factored out of the amplitude squared, and $\mathcal{V}_{A \to B+C}$ governs the pure collinear splitting. Comparing with Eq. (5.154), the splitting function can then be extracted:

$$
P_{BA} = \frac{z(1-z)}{2\,\mathbf{p}_\perp^2}\overline{\sum}|\mathcal{V}_{A \to B+C}|^2. \tag{5.155}
$$

Let's now consider the two specific cases $q \to g\,q$ and $g \to q\,\bar{q}$. For gluon emission out of a quark, $q \to g\,q$, the amplitude squared is given by:

$$
\overline{\sum}|\mathcal{V}_{q \to gq}|^2 = \frac{1}{2}C_F\sum_{\text{pol.}}\text{Tr}\{\gamma^\mu \not{p}_B \gamma^\nu \not{p}_A\}\varepsilon_\mu \varepsilon_\nu^* =
$$

$$
= 2C_F\,(p_B^\mu p_A^\nu + p_B^\nu p_A^\mu - g^{\mu\nu}p_A p_B)\left(\sum_{\text{pol.}}\varepsilon_\mu \varepsilon_\nu^*\right) =
$$

$$
= 2C_F\,(2p_A^i p_B^j + \delta^{ij}(p_A p_B))\delta_{ij}^\perp, \tag{5.156}
$$

where the transverse tensor $\delta_{ij}^\perp = \delta_{ij} - p_C^i p_C^j/\mathbf{p}_C^2$ has been introduced to project-out only the physical polarisation states, with $\delta_{ij}^\perp \delta^{ji} = 2$. We have:

$$
p_A^i p_B^j \delta_{ij}^\perp = \mathbf{p}_A \cdot \mathbf{p}_B - \frac{(\mathbf{p}_A \cdot \mathbf{p}_C)(\mathbf{p}_B \cdot \mathbf{p}_C)}{\mathbf{p}_C^2} =
$$

$$
= zE_A^2 - \frac{(1-z)E_A^2\left(z(1-z)E_A^2 - \mathbf{p}_\perp^2\right)}{(1-z)^2 E_A^2 + \mathbf{p}_\perp^2} = \frac{\mathbf{p}_\perp^2}{(1-z)^2} + \mathcal{O}(\mathbf{p}_\perp^2)
$$

$$
\delta^{ij}(p_A p_B)\delta_{ij}^\perp = 2\left(E_A\left(zE_A + \frac{\mathbf{p}_\perp^2}{2zE_A}\right) - zE_A^2\right) = \frac{\mathbf{p}_\perp^2}{z} \tag{5.157}
$$

Putting everything together:

$$
\overline{\sum}|\mathcal{V}_{q \to gq}|^2 = 2C_F\left(\frac{2\,\mathbf{p}_\perp^2}{(1-z)^2} + \frac{2\,\mathbf{p}_\perp^2}{2z}\right) = \frac{4\,\mathbf{p}_\perp^2}{2z}C_F\frac{(1+z)^2}{(1-z)^2}. \tag{5.158}
$$

By using Eq. (5.155), the splitting function can then be computed as:

$$P_{qq}(z) = C_F \left(\frac{1+z^2}{1-z} \right).$$
(5.159)

For gluon splitting, $g \to q\bar{q}$, we instead have:

$$\overline{\sum} |\mathcal{V}_{g\to q\bar{q}}|^2 = \frac{1}{2} T_R \sum_{\text{pol.}} \text{Tr}\{\gamma^\mu \not{p}_B \gamma^\nu \not{p}_C\}\varepsilon_\mu \varepsilon_\nu^* = 2T_R \, (2p_B^i p_C^j + \delta^{ij}(p_B p_C))\delta_{ij}^\perp,$$
(5.160)

with $\delta_{ij}^\perp = \delta_{ij} - p_A^i p_A^j / \mathbf{p}_A^2$. Carrying out the contraction:

$$p_B^i p_C^j \delta_{ij}^\perp = zE_A^2(1-z) - \mathbf{p}_\perp^2 - \frac{zE_A^2(1-z)E_A^2}{E_A^2} = -\mathbf{p}_\perp^2$$
(5.161)

$$\delta^{ij}(p_B p_C)\delta_{ij}^\perp = 2\left(\left(zE_A + \frac{\mathbf{p}_\perp^2}{2zE_A}\right)\left((1-z)E_A + \frac{\mathbf{p}_\perp^2}{2(1-z)E_A}\right) - z(1-z)E_A^2 + \mathbf{p}_\perp^2\right) =$$

$$= 2\mathbf{p}_\perp^2 \left[\frac{1-z}{2z} + \frac{z}{2(1-z)} + 1\right] = -\frac{2\mathbf{p}_\perp^2}{2z(1-z)}.$$
(5.162)

Putting everything together:

$$\overline{\sum} |\mathcal{V}_{g\to q\bar{q}}|^2 = 2T_R \, (2\mathbf{p}_\perp^2) \frac{(1-z)^2 + z^2}{2z(1-z)}.$$
(5.163)

By using Eq. (5.155), the splitting function can then be computed as:

$$P_{qg}(z) = T_R \left[(1-z)^2 + z^2\right].$$
(5.164)

To summarise, the regularised (LO) Altarelli-Parisi splitting functions are:

$$P_{qq}^{(0)}(z) = C_F \left(\frac{1+z^2}{1-z}\right)_+ = C_F \left[\frac{1+z^2}{(1-z)_+} + \frac{3}{2}\delta(1-z)\right] = P_{\bar{q}\bar{q}}^{(0)}(z)$$

$$P_{gq}^{(0)}(z) = P_{qq}^{(0)}(1-z) = P_{g\bar{q}}^{(0)}(z)$$

$$P_{qg}^{(0)}(z) = T_R \left[(1-z)^2 + z^2\right] = P_{\bar{q}g}^{(0)}(z)$$

$$P_{gg}^{(0)}(z) = 2C_A \left[\frac{z}{(1-z)_+} + \frac{1-z}{z} + z(1-z)\right] + \frac{11\,C_A - 4\,T_R\,n_f}{6}\delta(1-z).$$
(5.165)

We have used the "+" prescription defined as:

$$F(z)_+ = F(z) - \delta(1-z) \int_0^1 dy\, F(y), \qquad (5.166)$$

such that:

$$\int_x^1 dz\, g(z) \left[\frac{f(z)}{1-z} \right]_+ = \int_x^1 dz\, [g(z) - g(1)] \frac{f(z)}{1-z} - g(1) \int_0^x dz\, \frac{f(z)}{1-z} \qquad (5.167)$$

Suggested Readings

This exercise is vastly inspired by the original paper by Altarelli and Parisi [31].

Problem 5.38 Calculate the Mellin moments of the regularized Altarelli-Parisi splitting functions.

Discussion

Given a function $f(x)$ with $x \in [0, 1]$, its nth Mellin moment is defined as:

$$\tilde{f}(n) = \int_0^1 dx\, x^{n-1} f(x) \quad \Leftrightarrow \quad f(x) = \frac{1}{2\pi i} \int_{c-i\infty}^{c+i\infty} dn\, x^{-n} \tilde{f}(n). \qquad (5.168)$$

The Mellin moments allows to transform the integro-differential DGLAP evolution equation into a pure differential one:

$$\mu^2 \frac{\partial f}{\partial \mu^2}(x, \mu^2) = \frac{\alpha_s}{2\pi} \int_x^1 \frac{dz}{z} P^{(0)}(z) f\left(\frac{x}{z}\right)$$

$$= \frac{\alpha_s}{2\pi} \int_0^1 dz \int_0^1 dy\, P^{(0)}(z) f(y) \delta(x - z\,y). \qquad (5.169)$$

Taking the $n - 1$ moment on both sides of Eq. (5.169), we get:

$$\int_0^1 dx\, x^{n-1} \left(\mu^2 \frac{\partial f}{\partial \mu^2}(x, \mu^2) \right) = \frac{\alpha_s}{2\pi} \int_0^1 dx\, x^{n-1} \int_0^1 dz \int_0^1 dy\, P^{(0)}(z) f(y) \delta(x - z\,y)$$

$$\mu^2 \frac{\partial}{\partial \mu^2} \tilde{f}(n, \mu^2) = \frac{\alpha_s}{2\pi} \underbrace{\int_0^1 dz\, z^{n-1} P^{(0)}(z)}_{\gamma^{(0)}(n)} \underbrace{\int_0^1 dy\, y^{n-1} f(y)}_{\tilde{f}(n, \mu^2)}, \qquad (5.170)$$

Solution

For the case $g \rightarrow g(z) g(1-z)$, we have:

$$\gamma_{gg}^{(0)}(n) = \int_0^1 dx \, x^{n-1} P_{gg}(x) \tag{5.171}$$

The various terms that enter P_{gg} can be integrated to give:

$$\int_0^1 dx \, x^{n-1} \frac{x}{(1-x)_+} = \int_0^1 dx \frac{x^n - 1}{1-x} = \int_0^1 dx \frac{(x-1)(x^{n-1} + \cdots + 1)}{1-x} =$$

$$-\int_0^1 dx \,(x^{n-1} + \cdots + 1) = -\left[\frac{1}{n} + \frac{1}{n-1} + \cdots + 1\right] = -\sum_{m=1}^n \frac{1}{m} \tag{5.172}$$

$$\int_0^1 dx \, x^{n-1} \frac{1-x}{x} = \int_0^1 dx \,(x^{n-2} - x^{n-1}) = \frac{1}{n(n-1)} \tag{5.173}$$

$$\int_0^1 dx \, x^{n-1} x(1-x) = \int_0^1 dx \,(x^n - x^{n+1}) = \frac{1}{(n+1)(n+2)} \tag{5.174}$$

$$\int_0^1 dx \, \delta(1-x) = 1 \tag{5.175}$$

Putting everything together, we thus have:

$$\gamma_{gg}^{(0)}(n) = 2C_A \left[-\sum_{m=1}^n \frac{1}{m} + \frac{1}{n(n-1)} + \frac{1}{(n+1)(n+2)}\right] + \frac{11 C_A - 4 T_R n_f}{6}. \tag{5.176}$$

For the case $q \rightarrow q(z) g(1-z)$, we have:

$$\gamma_{qq}^{(0)}(n) = C_F \int_0^1 dx \, x^{n-1} \left(\frac{1+x^2}{1-x}\right)_+ = C_F \int_0^1 dx \,(x^{n+1} - 1) \frac{1+x^2}{1-x} =$$

$$= C_F \int_0^1 dx \frac{x-1}{1-x} (x^{n-2} + \cdots + 1)(1+x^2)$$

$$= C_F \left[-\int_0^1 dx \left[(x^{n-2} + \cdots + 1) + (x^n + \cdots + x^2)\right]\right] =$$

$$= -C_F \left[\frac{1}{n-1} + \cdots + 1 + \frac{1}{n+1} + \cdots + \frac{1}{3}\right] =$$

$$= -C_F \left[2\sum_{m=1}^{n-1} \frac{1}{m} + \frac{1}{n+1} + \frac{1}{n} - \frac{1}{2} - 1\right] = -C_F \left[2\sum_{m=1}^n \frac{1}{m} + \frac{1}{n+1} - \frac{1}{n} - \frac{3}{2}\right]$$

$$= C_F \left[\frac{1}{n(n+1)} + \frac{3}{2} - 2\sum_{m=1}^n \frac{1}{m}\right]. \tag{5.177}$$

For the case $g \to q(z)\bar{q}(1-z)$, we have:

$$\gamma_{qg}^{(0)}(n) = T_R \int_0^1 dx\, x^{n-1}\left[(1-x)^2 + x^2\right] = T_R \int_0^1 dx\, \left[2x^{n+1} - 2x^n + x^{n-1}\right] =$$

$$= T_R \left[\frac{2}{n+2} - \frac{2}{n+1} + \frac{1}{n}\right] = T_R \frac{n^2 + n + 2}{n(n+1)(n+2)}. \tag{5.178}$$

For the case $q \to g(z)q(1-z)$, we have:

$$\gamma_{gq}^{(0)}(n) = C_F \int_0^1 dx\, x^{n-1}\left(\frac{1 + (1-x)^2}{x}\right) = C_F \int_0^1 dx\, \left[2x^{n-2} - 2x^{n-1} + x^n\right] =$$

$$= C_F \left[\frac{2}{n-1} - \frac{2}{n} + \frac{1}{n-1}\right] = C_F \frac{n^2 + n + 2}{n(n+1)(n-1)} \tag{5.179}$$

We conclude by proving the necessity for the "+" prescription in the definition of the kernel functions. Let's study the implications of the DGLAP evolution functions to the non-singlet quark density $q - \bar{q} \equiv q^{\text{NS}}$[3]:

$$\begin{cases} \frac{\partial q_i}{\partial \ln Q^2} = \sum_j P_{q_i q_j} \otimes q_j + \sum_j P_{q_i \bar{q}_j} \otimes \bar{q}_j + P_{q_i g} \otimes g \\ \frac{\partial \bar{q}_i}{\partial \ln Q^2} = \sum_j P_{\bar{q}_i \bar{q}_j} \otimes \bar{q}_j + \sum_j P_{\bar{q}_i q_j} \otimes q_j + P_{\bar{q}_i g} \otimes g \end{cases} \tag{5.180}$$

Using flavour symmetry, the set of kernel functions at LO read as:

$$P_{q_i q_j}(z) = P_{\bar{q}_i \bar{q}_j}(z) \equiv P_{qq}^{\text{NS}}(z)\,\delta_{ij} + P_{qq}^{\text{S}}(z)$$
$$P_{q_i g}(z) = P_{\bar{q}_i g}(z) \equiv P_{qg}(z)$$
$$P_{q_i \bar{q}_j}(z) = P_{\bar{q}_i q_j}(z) \equiv P_{\bar{q}q}^{\text{NS}}(z)\,\delta_{ij} + P_{\bar{q}q}^{\text{S}}(z)$$

which allows to simplify Eq. (5.181) to:

$$\begin{cases} \frac{\partial q_i}{\partial \ln Q^2} = P_{qq}^{\text{NS}} \otimes q_i + P_{qq}^{\text{S}} \otimes \sum_j q_j + P_{\bar{q}q}^{\text{NS}} \otimes \bar{q}_i + P_{\bar{q}q}^{\text{S}} \otimes \sum_j \bar{q}_j + P_{qg} \otimes g \\ \frac{\partial \bar{q}_i}{\partial \ln Q^2} = P_{qq}^{\text{NS}} \otimes \bar{q}_i + P_{qq}^{\text{S}} \otimes \sum_j \bar{q}_j + P_{\bar{q}q}^{\text{NS}} \otimes q_i + P_{\bar{q}q}^{\text{S}} \otimes \sum_j q_j + P_{qg} \otimes g \end{cases} \tag{5.181}$$

Given that P_{qq}^{S} and $P_{\bar{q}q}$ start only at $\mathcal{O}(\alpha_s^2)$, taking the difference we have:

$$\frac{\partial q_i^{\text{NS}}}{\partial \ln Q^2} = P_{qq}^{\text{NS}} \otimes q_i^{\text{NS}}. \tag{5.182}$$

Since $\int_0^1 dx\, q_i^{\text{NS}}(x)$ must be conserved (it is the amount of valence quarks q_i in the hadron), it follows that $\int_0^1 dz\, P_{qq}^{\text{NS}} = 0$, since:

---

[3]for simplicity, we absorb the coefficient $\alpha_s/2\pi$ inside the splitting function.

$$\int dx\, x^n\, [f \otimes g](x) = \int dx\, x^n f(x) \cdot \int dx\, x^n g(x). \qquad (5.183)$$

Hence, the splitting function $P_{qq}(z)$ at order $\mathcal{O}(\alpha_s)$ derived by using perturbation theory is infact a distribution, and such that its integral is null. This amounts to replace:

$$C_F\left(\frac{1+z^2}{1-z}\right) \to C_F\left(\frac{1+z^2}{1-z}\right)_+ = C_F\left(\frac{1+z^2}{(1-z)_+} + \frac{3}{2}\delta(1-z)\right). \qquad (5.184)$$

The last equality can be proved in the sense of distributions:

$$\int_0^1 dz \left(\frac{1+z^2}{1-z}\right)_+ g(z) = \int dz\, [g(z) - g(1)]\frac{1+z^2}{1-z}$$

$$= \int dz\, \frac{g(z)(1+z^2) - 2g(1) + g(1) - z^2 g(1)}{1-z} = \int dz\, \frac{1+z^2}{(1-z)_+} g(z) + g(1)\int dz\,(1+z)$$

$$= \int dz\, \frac{1+z^2}{(1-z)_+} g(z) + \int dz\, \frac{3}{2}\delta(1-z)g(z) = \int dz \left[\frac{1+z^2}{(1-z)_+} + \frac{3}{2}\delta(1-z)\right]g(z),$$

$$(5.185)$$

which proves the equality in the sense of distributions. The requirement that the total hadron momentum is conserved further constrains the P_{qg} and P_{gg} splitting functions:

$$\int_0^1 dx\, x\left[g(x) + \sum_{f=q,\bar{q}} f(x)\right] = \text{const.} \quad \Rightarrow \quad \int_0^1 dx\, x\left[\frac{\partial g}{\partial \ln Q^2} + \sum_f \frac{\partial f}{\partial \ln Q^2}\right] = 0,$$

$$\int_0^1 dx\, x\left[P_{gg} + 2n_f P_{gq}\right] \otimes g + \int_0^1 dx\, x\left[P_{gq} + P_{qq}\right] \otimes \sum_f f = 0 \qquad (5.186)$$

Then, by the law of convolutions, we must have:

$$\int_0^1 dx\, x\left[P_{gg} + 2n_f P_{gq}\right] = 0, \qquad \int_0^1 dx\, x\left[P_{gq} + P_{qq}\right] = 0. \qquad (5.187)$$

A quick calculation reveals that this is indeed the case, thus justifying the coefficient of the δ function in the P_{gg} splitting function.

Suggested Readings

More details on the Mellin moments applied to the DGLAP equation can be found in Ref. [18].

5.3 Flavour Physics

From the structure of the SM Lagrangian of Eq. (5.39) complemented with the Yukawa sector (5.42), three neat consequences on the conservation of flavour arise. Neglecting the neutrino mass, which is an excellent approximation except when considering neutrino oscillations, we can state them in the form of theorems.

Theorem 5.1 *There is no flavour transition in the lepton sector.*

As seen in Problem 5.47, the minimal extension of the SM which allows for neutrino masses introduces three right-handed neutrinos N_R. Equation (5.42) should be then extended to include an additional term whose effect would be to misalign the lepton mass basis from the flavour basis by a the unitary matrix U, called *Pontecorvo-Maki-Nakagawa-Sakata* matrix (PMNS), in full analogy to the quark sector. Apart from neutrino oscillations, there are no other observable lepton-flavour violating effects caused by the presence of such right-handed particles: this is a consequence of the large mass of the right-handed neutrinos that suppresses any lepton-flavour violating amplitude to a negligible level.

Theorem 5.2 *In the quark sector, flavour transitions can only occur in the charged current interaction amplitude $d_j \to W^- u_i$ (and its complex conjugate), whose flavour dependence is encoded in the CKM matrix.*

The lack of flavour violation in the neutral-current interactions is a result of the neutral weak current being flavour-diagonal, whereas the charged-weak current mixes the isospin doublets thanks to the off-diagonal $T^\pm$ operators, see Eq. (5.62). This fact, together with the unitarity of the CKM matrix, is at the basis of the *Glashow-Iliopoulos-Maiani mechanism* (GIM), which explains the suppression of *flavour-changing neutral currents* (FCNC) in hadronic decays: besides being loop-suppressed, FCNC amplitudes would be identically zero if the quark masses were degenerate. For example, consider the diagram shown in Fig. 5.4 that contributes to the FCNC amplitude $t \to c\,\gamma$. For each down-type quark d_i running in the loop, one has a contribution proportional to $V_{ci} V_{ti}^*$:

$$\mathscr{A} = \sum_{i=u,s,b} f\left(\frac{m_i^2}{m_W^2}\right) V_{ci} V_{ti}^*. \tag{5.188}$$

Fig. 5.4 Diagram contributing to the FCNC $t \to c\gamma$ amplitude

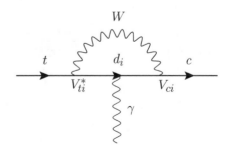

The unitarity of the CKM matrix implies $\sum_{i=u,s,b} V_{ci} V_{ti}^* = 0$, thus giving a vanishing amplitude in the limit of identical quark masses (see the discussion in Ref. [32] for further details).

Theorem 5.3 *Modulo a possible effect in the strong interactions (strong CP-problem), C P violation can only take place to the extent that the CKM matrix is complex.*

The strong CP-problem is related to the unnatural suppression of a tree-level CP-violating gluon-gluon interaction term of the form:

$$\mathscr{L}_{CP-\text{strong}} = \theta_{\text{QCD}} \sum_a \varepsilon_{\mu\nu\rho\sigma} G_{\mu\nu}^a G_{\rho\sigma}^a \qquad (5.189)$$

which, being renormalisable and of dimension $d = 4$, would naturally find its place in the minimal SM Lagrangian. Although this term is equivalent to a total derivative, hence it would give vanishing contributions to scattering amplitudes, it also gives to non-perturbative effects that contribute to CP-odd observables, like electric dipole moments. From current limits, one finds $\theta_{\text{QCD}} \lesssim 10^{-9}$.

Leaving aside the neutrino sector, the only source of CP-violation comes from the charged-weak interaction. Indeed, from the transformation properties under CP of the operators appearing in the SM Lagrangian, reported in Table 5.4, the only operator which is not invariant is exactly the one appearing in Eq. (5.43) to the extent that V cannot be made real by any re-phasing of the quark fields, as it is the case for three quark generations.

To this purpose, it is useful to remember Wolfenstein's parametrisation of the CKM matrix

Table 5.4 Transformation properties under CP of the operators appearing in the SM Lagrangian. Every Lorentz index with a tilde requires an overall $-$ sign if it is a space index or a $+$ sign if it is a time index

$\mathcal{O}$	$CP \, \mathcal{O} \, CP^\dagger$	Description
$G_\mu^a(\mathbf{x}, t)$	$-G_{\tilde{\mu}}^a(-\mathbf{x}, t)$	Gluon fields
$A_\mu(\mathbf{x}, t)$	$-A_{\tilde{\mu}}(-\mathbf{x}, t)$	Photon field
$Z_\mu(\mathbf{x}, t)$	$-Z_{\tilde{\mu}}(-\mathbf{x}, t)$	Z field
$W_\mu^\pm(\mathbf{x}, t)$	$-W_{\tilde{\mu}}^\mp(-\mathbf{x}, t)$	$W^\pm$ fields
$h(\mathbf{x}, t)$	$h(-\mathbf{x}, t)$	Higgs field
$\bar{\psi}\gamma_\mu\chi(\mathbf{x}, t)$	$-\bar{\chi}\gamma_{\tilde{\mu}}\psi(-\mathbf{x}, t)$	Vector current
$\bar{\psi}\gamma_\mu\gamma_5\chi(\mathbf{x}, t)$	$-\bar{\chi}\gamma_{\tilde{\mu}}\gamma_5\psi(-\mathbf{x}, t)$	Axial current
$\bar{\psi}\chi(\mathbf{x}, t)$	$\bar{\chi}\psi(-\mathbf{x}, t)$	Fermion mass
$F_{\mu\nu}F^{\mu\nu}$	$F_{\mu\nu}F^{\mu\nu}$	Field strength
$\varepsilon_{\mu\nu\rho\sigma}F_{\mu\nu}F_{\rho\sigma}$	$-\varepsilon_{\mu\nu\rho\sigma}F_{\mu\nu}F_{\rho\sigma}$	Dual field strength

$$
V = \begin{pmatrix} 1 - \lambda^2/2 & \lambda & \lambda^3 A(\rho - i\eta) \\ -\lambda & 1 - \lambda^2/2 & \lambda^2 A \\ \lambda^3 A(1 - \rho - i\eta) & -\lambda^2 A & 1 \end{pmatrix} + \mathcal{O}(\lambda^4) \qquad (5.190)
$$

which, besides making evident the presence of one irreducible phase,[4] provides the order-of-magnitude of the various entries in terms of the expansion parameter $\lambda = \sin\theta_C \approx 0.22$.

Experimentallty, CP-violation has been observed from effects due to 4-fermion interactions that change strangeness and beauty. Considering the case of $\Delta S \neq 0$ transitions, the effective lagrangian can be schematically decomposed as the sum of three pieces [14]:

$$
\mathscr{L}_{\text{eff.}} = \underbrace{A\,(\bar{s}_L\gamma_\mu d_L)(\bar{s}_L\gamma^\mu d_L)}_{\Delta S=2} + \sum_q \underbrace{B_q\,(\bar{s}_L\gamma_\mu d_L)(\bar{q}\,\gamma^\mu\, q)}_{\Delta S=1} + \sum_\ell \underbrace{C_\ell\,(\bar{s}_L\gamma_\mu d_L)(\bar{\ell}\,\gamma^\mu\,\ell)}_{\Delta S=1,\ \text{semilept.}}
$$

$$(5.191)$$

Violation of CP arises from an irreducible phase present among the coefficients A, B_q, and C_ℓ. An irreducible phase between A and any of the B_q is at the origin of the *indirect* violation of CP observed in the mass mixing of neutral kaons. A phase between the B_q operators, which are responsible for the flavour-changing decay of mesons, is instead at the origin of the *direct* violation of CP, observed in the K^0 and B^0 decay.

Problems

Bando n. 1N/R3/SUB/2005

Problem 5.39 Explain how the physics of K mesons has contributed to the comprehension of subnuclear physics in general.

Discussion

The K^0 and $\bar{K}^0$ are eigenstates of the strong hamiltonian. In the $SU(3)$ flavour representation, they are members of the pseudo-scalar meson octet with isospin $I = 1/2$ and strangeness $S = +1$ and $S = -1$, respectively. They are also one the antiparticle of the other. In the absence of weak interactions, they would be stable. The weak interaction introduces $\Delta S = 1/2$ and $\Delta S = 3/2$ transitions that make the neutral kaons decay predominantly to two or three pions, or via the semileptonic decays $K^0 \to \pi^- \ell^+ \nu_\ell$ and $\bar{K}^0 \to \pi^+ \ell^- \bar{\nu}_\ell$ (in the latter, the lepton charge is prescribed by the $\Delta S = \Delta Q$ rule). With the phase convention $CP\,|K^0\rangle = |\bar{K}^0\rangle$, the linear combinations

---

[4]The location of the phase in the CKM matrix is purely conventional, since only rephasing-invariant combinations of the CKM elements are truly observable.

$$|K_1^0\rangle = \frac{|K^0\rangle + |\bar{K}^0\rangle}{\sqrt{2}}, \qquad |K_2^0\rangle = \frac{|K^0\rangle - |\bar{K}^0\rangle}{\sqrt{2}} \qquad (5.192)$$

are eigenstates of CP with eigenvalue ± 1, respectively. If the strong and weak Hamiltonian is postulated to be CP-invariant, then the mass eigenstates must also be eigenstates of CP. Hence, under the assumption that CP is conserved, the effective Hamiltonian matrix $\tilde{\mathscr{H}}$ in the $K^0/\bar{K}^0$ basis reads as:

$$\tilde{\mathscr{H}} = P\left(\mathscr{H}_0 + \mathscr{R}\right)P = \begin{pmatrix} M - i\frac{\Gamma}{2} & M_{12} - i\frac{\Gamma_{12}}{2} \\ M_{12} - i\frac{\Gamma_{12}}{2} & M - i\frac{\Gamma}{2} \end{pmatrix} = U^{-1}\begin{pmatrix} m_1 - i\frac{\Gamma_1}{2} & 0 \\ 0 & m_2 - i\frac{\Gamma_2}{2} \end{pmatrix}U,$$

$$\text{with} \quad U = \frac{1}{\sqrt{2}}\begin{pmatrix} 1 & 1 \\ -1 & 1 \end{pmatrix}, \qquad M = \frac{m_1 + m_2}{2}, \qquad \Gamma = \frac{\Gamma_1 + \Gamma_2}{2}, \qquad (5.193)$$

where P is the projector to the strong eigenstates basis, and m_i and Γ_i are the mass and width of the two mass eigenstates $|K_{1,2}^0\rangle$, respectively. When neutral kaons are produced by the strong interaction, like $\pi^- p \rightarrow \Lambda^0 K^0$, they start their life as eigenstates of the strong Hamiltonian, and then propagate according to the full Hamiltonian.

$$|\Psi(\tau)\rangle = \frac{1}{\sqrt{2}}\left[e^{-i\left(m_1 - i\frac{\Gamma_1}{2}\right)\tau}|K_1^0\rangle + e^{-i\left(m_2 - i\frac{\Gamma_2}{2}\right)\tau}|K_2^0\rangle\right], \qquad (5.194)$$

where τ is the proper time. Owing to the different mass, the physical state, that propagates by virtue of Eq. (5.193), oscillates between the $K^0/\bar{K}^0$ states, giving rise to a time-dependent transition amplitude with frequency in the rest frame given by $\Delta m/\hbar$. Furthermore, since the K_2^0 has $CP = -1$, it can only decay to three-pions or semileptonically, with a total decay rate that is a factor of about 1.7×10^{-3} smaller than for K_1. This hierarchy in lifetimes allows to distinguish between the two mass eigenstates by separating the short- and long-lived components of $K^0/\bar{K}^0$-initiated beams.

As shown by Eq. (5.193), the mass difference Δm is related to the $\Delta S = 2$ amplitude $\langle \bar{K}^0|\mathscr{H}|K^0\rangle = M_{12}$ by the relation $\Delta m = 2|M_{12}|$. This amplitude is described, at leading order, by a diagram like the one shown in Fig. 5.5. The CKM matrix elements are such that this amplitude is mostly given by the diagram corresponding to the exchange of charm quarks:

Fig. 5.5 Leading-order diagram contributing to the $\Delta S = 2$ transition $|\bar{K}^0\rangle \leftrightarrow |K^0\rangle$

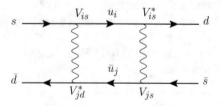

$$\mathscr{L}_{\Delta S=2} \approx (V_{cs} \, V_{cd}^*)^2 \, F_{cc} \, (\bar{d}_L \gamma^\mu s_L)(\bar{d}_L \gamma_\mu s_L), \quad \text{with} \quad F_{cc} \approx \frac{G_F^2 \, m_c^2}{\pi^2}. \tag{5.195}$$

However, gluonic corrections to this amplitude are not under perturbative control, so that this operator can only provide an estimate of the mass difference [14].

Solution

The mixing of the two neutral kaons can be studied in a more general framework, where the weak interaction determines the new mass eigenstates as a linear combination of K^0 and $\bar{K}^0$, namely:

$$|K_{S,,L}^0\rangle = \frac{1}{\sqrt{|p|^2 + |q|^2}} \left(p|K^0\rangle \pm q|\bar{K}^0\rangle \right). \tag{5.196}$$

These new states, now defined K-short and K-long, are the eigenvectors of the full Hamiltonian. They are CP-eigenstates to the extent that $|p/q| = 1$, which is for example the case in Eq. (5.192). Indeed, the effective Hamiltonian matrix of Eq. (5.193) must have identical diagonal elements $\tilde{\mathscr{H}}_{11} = \tilde{\mathscr{H}}_{22} = \tilde{\mathscr{H}}_d$ by CPT-invariance. This fully determines the eigenvectors in terms of the off-diagonal elements as:

$$|K_{S,L}^0\rangle = \frac{\sqrt{\tilde{\mathscr{H}}_{12}}|K^0\rangle \pm \sqrt{\tilde{\mathscr{H}}_{21}}|\bar{K}^0\rangle}{\sqrt{|\tilde{\mathscr{H}}_{12}| + |\tilde{\mathscr{H}}_{21}|}} \equiv \frac{(1+\varepsilon)|K^0\rangle + (1-\varepsilon)|\bar{K}^0\rangle}{\sqrt{2(1+|\varepsilon|^2)}}, \tag{5.197}$$

where the complex parameter ε defined by

$$\frac{1-\varepsilon}{1+\varepsilon} \equiv \sqrt{\frac{\tilde{\mathscr{H}}_{21}}{\tilde{\mathscr{H}}_{12}}} \tag{5.198}$$

has been introduced. Assuming the more generic transformation property $|K^0\rangle \rightarrow e^{i\delta}|\bar{K}^0\rangle$ and $|\bar{K}^0\rangle \rightarrow e^{-i\delta}|K^0\rangle$, invariance under CP of the Hamiltonian implies

$$\tilde{\mathscr{H}}_{12} = \langle \bar{K}^0|\tilde{\mathscr{H}}|K^0\rangle = \left(\bar{K}^0|(CP)^\dagger \, \mathscr{H} \, CP|K^0\right) = e^{2i\delta}\langle K^0|\mathscr{H}|\bar{K}^0\rangle$$
$$= e^{2i\delta}\,\tilde{\mathscr{H}}_{21}. \tag{5.199}$$

Hence, the off-diagonal elements of the Hamiltonian matrix of Eq. (5.193) must be relatively real, and the states in Eq. (5.197) come out as eigenvectors of CP.

The off-diagonal elements receive contributions from both virtual transitions via $\Delta S = 2$ operators like the one depicted in Fig. 5.5 and by $\Delta S = 1$ transitions via $|\bar{K}^0\rangle \rightarrow (2\pi)_I \rightarrow |K^0\rangle$, see Eq. (5.191): an irreducible phase between the two, which can be ultimately traced to the irreducible phase in the CKM matrix, can thus make $\tilde{\mathscr{H}}_{12}$ and $\tilde{\mathscr{H}}_{21}$ not relatively real and turn K_S^0 and K_L^0 into an admixture of

K_1 and K_2. With the phase-convention $\delta = 0$, the parameter ε governs the amount of K_2 (K_1) present in K_S (K_L). As discussed in Sect. 5.3, the fact that $|\varepsilon| \neq 0$ is an indirect manifestation that CP is broken, because the mass eigenstates are *not* pure CP-eigenstates. Instead, direct violation of CP would manifest itself through interfereing amplitudes sensitive to different combinations of the CKM elements. This is for example the case of $K_L^0 \rightarrow \pi^+ \pi^- e^+ e^-$ decays, where the relative orientation between the decay planes $\mathbf{e}_\pi$ and $\mathbf{e}_e$ of the charged pions and of the electron-positron pair receives contributions from both CP-even and CP-odd amplitudes, thus offering the opportunity to measure CP-violation. This is experimentally the case, with a measured asymmetry [9]

$$A = \frac{N_{\mathbf{e}_\pi \cdot \mathbf{e}_e > 0} - N_{\mathbf{e}_\pi \cdot \mathbf{e}_e < 0}}{N_{\mathbf{e}_\pi \cdot \mathbf{e}_e > 0} + N_{\mathbf{e}_\pi \cdot \mathbf{e}_e < 0}} = (13.7 \pm 1.5)\%. \tag{5.200}$$

The traditional CP-violation parameters for the neutral kaons are defined as

$$\eta_{+-} \equiv \frac{\langle \pi^+ \pi^- | \mathscr{H}_{\text{ewk}} | K_L^0 \rangle}{\langle \pi^+ \pi^- | \mathscr{H}_{\text{ewk}} | K_S^0 \rangle} = \varepsilon + \varepsilon', \qquad \eta_{00} \equiv \frac{\langle \pi^0 \pi^0 | \mathscr{H}_{\text{ewk}} | K_L^0 \rangle}{\langle \pi^0 \pi^0 | \mathscr{H}_{\text{ewk}} | K_S^0 \rangle} = \varepsilon - 2\,\varepsilon' \tag{5.201}$$

The parameter ε' controls the direct violation of CP. It can be related to the EWK phase of the amplitude for K^0 transitions to two-pion states in $I = 0$ and $I = 2$ configurations, which can be proved to be identical if CP is conserved by using unitarity and CPT invariance, see Ref. [11] for more details. Experimentally, η_{+-} can be measured by studying the time-dependent decay rate of neutral kaons into $\pi^+ \pi^-$. For example, from an initial K^0 beam, one has:

$$\Gamma_{\pi^+ \pi^-}(\tau) \sim e^{-\Gamma_S \tau} + 2|\eta_{+-}| e^{-\frac{\Gamma_S + \Gamma_L}{2}\tau} \cos(\phi_{+-} - \Delta m\,\tau) + |\eta_{+-}|^2 e^{-\Gamma_L \tau}. \tag{5.202}$$

Experimentally, one finds $|\eta_{+-}| = (2.232 \pm 0.011) \times 10^{-3}$ and a consistent value for $|\eta_{00}|$ [9]. The real part of ε can be measured from the time-integrated right-to-wrong sign ratio

$$A_L = \frac{\Gamma(K_L^0 \rightarrow \pi^- \mu^+ \nu_\mu) - \Gamma(K_L^0 \rightarrow \pi^+ \mu^- \bar{\nu}_\mu)}{\Gamma(K_L^0 \rightarrow \pi^- \mu^+ \nu_\mu) + \Gamma(K_L^0 \rightarrow \pi^+ \mu^- \bar{\nu}_\mu)} = 2\,\text{Re}\,\varepsilon, \tag{5.203}$$

with an experimental result $A_L = (3.32 \pm 0.06) \times 10^{-3}$ [9]. The ε' parameter can be instead constrained by measuring a deviation from unity of the double ratio of neutral-to-charged pion decay for K_L^0 and K_S^0, see Eq. (5.201). The results from the NA48 experiment at CERN gave a value of

$$\left| \frac{\eta_{00}}{\eta_{+-}} \right| \approx 1 - 6\,\text{Re}\,\frac{\varepsilon'}{\varepsilon} = (0.9950 \pm 0.0007), \tag{5.204}$$

thus giving a neat evidence of direct CP-violation through a non-zero value of ε'.

Suggested Readings

The physics of the $K^0/\bar{K}^0$ system has been the object of an intensive theoretical and experimental research work over several decades. There are lots of textbooks where this topic can be studied in great details, see e.g. Ref. [8]. Chapter 7 of Ref. [11] is certainly a good starter, since it provides a coincise overview of the underlying theory and summarises the main experimental results. The reader is also addressed to study in more details the experimental setup used by the NA48 experiment to simultaneously measure the two-pion decays of both K_S^0 and K_L^0.

Bando n. 1N/R3/SUB/2005

Problem 5.40 What is the K_S^0 regeneration? How can it be explained?

Solution

When two beams of K^0 and $\bar{K}^0$ of the same momentum $|\mathbf{p}|$ propagate through a thickness L of material, their phase at the exit has undergone a different shift because of the different elastic-scattering amplitude. This can be understood by the *Optical theorem*, which relates the forward scattering amplitude to the total cross section. The latter is necessarily different for K^0 and $\bar{K}^0$: for example, the reaction $\bar{K}^0 p \rightarrow \pi^+ \Lambda$ is an allowed $\Delta S = 0$ reaction, whereas the same is suppressed for K^0, since $\Delta S = +1$. For a wave propagating in matter with density N, the extra phase taken up at the proper time τ because of elastic scattering with the nucleons is

$$|\mathbf{p}|(n-1)\,\ell = \frac{2\pi\,N\,\beta\gamma\,\tau}{|\mathbf{p}|} f(0), \qquad (5.205)$$

where $f(0)$ takes different values for K^0 and $\bar{K}^0$. In terms of the Hamiltonian, this shift is equivalent to adding an extra momentum. Considering an eigenstate of the free Hamiltonian of energy $E_{\mathbf{p}}$ and momentum $\mathbf{p}$, the interaction with the medium induces a shift in the momentum $|\mathbf{p}| \rightarrow (1+\delta n)\,|\mathbf{p}|$. In terms of the rest mass, this is equivalent to a shift $m \rightarrow m\,(1-\delta n)$. Hence, the mass matrix of Eq. (5.193) gets modified to:

$$\tilde{\mathscr{H}} = \begin{pmatrix} M - i\frac{\Gamma}{2} - \frac{2\pi\,N\,\beta\gamma}{|\mathbf{p}|} f_0 & M_{12} - i\frac{\Gamma_{12}}{2} \\ M_{12} - i\frac{\Gamma_{12}}{2} & M - i\frac{\Gamma}{2} - \frac{2\pi\,N\,\beta\gamma}{|\mathbf{p}|} \bar{f}_0 \end{pmatrix} \equiv \begin{pmatrix} a - \varepsilon_1 & b \\ b & a - \varepsilon_2 \end{pmatrix}$$
$$(5.206)$$

The eigenvalues and eigenvectors of this matrix are given by:

$$\lambda_{1,2} = a \pm b - \left(\frac{\varepsilon_1 + \varepsilon_2}{2}\right) + \mathcal{O}(\varepsilon_i), \qquad |K_{1,2}^{0'}\rangle \propto \left(\begin{array}{c} \frac{\lambda_{1,2} - a + \varepsilon_2}{b} \\ 1 \end{array}\right) \equiv \left(\begin{array}{c} \pm 1 + 2r \\ 1 \end{array}\right)$$

$$\text{with} \quad r = -\frac{\varepsilon_2 - \varepsilon_1}{4b} = \frac{\pi N \beta \gamma}{|\mathbf{p}|} \frac{f_0 - \bar{f}_0}{m_1 - m_2 - \frac{i}{2}(\Gamma_1 - \Gamma_2)}. \tag{5.207}$$

To first order in r, the new eigenvectors can be decomposed in terms of the CP-eigenstates as:

$$|K_1^{0'}\rangle = |K_1^0\rangle + r|K_2^0\rangle, \qquad |K_2^{0'}\rangle = -r|K_1^0\rangle + |K_2^0\rangle. \tag{5.208}$$

Hence, even starting from a pure K_2^0 beam, an amount of K_1^0 of order r can be regenerated.

Suggested Readings

The regeneration of the short-lived neutral kaon from a beam of pure K_2^0 passing through an iron plate was first observed at the Bevatron in 1960 [33]. Interestingly, there is another sources of collinear K_1^0: diffractive production via $K_2^0\, p \to K_1^0\, p$. The purely regenerated and diffractive components could be separated from an analysis of the angle between the incoming K_2^0 beam and the reconstructed $\pi^+ \pi^-$ momentum, which is more strongly peaked at zero for coherent regeneration. See Chap. 7 of Ref. [11] for more details.

Bando n. 1N/R3/SUB/2005

Problem 5.41 Consider the decays $D^0 \to \bar{K}^0 \pi^0$ and $D^0 \to K^0 \pi^0$. Draw the Feynman diagrams of the two decays and estimate the ratio between the two decay amplitudes.

Solution

The flavour-changing decay $D^0 \to \bar{K}^0 \pi^0$ proceeds at parton-level through the branchings $c \to s\, W^+$ followed by $W^+ \to u\, \bar{d}$: the amplitude is therefore proportional to $V_{cs}^* V_{ud}$. See Fig. 5.6 for a graphical representation of the decay amplitude. The decay $D^0 \to K^0 \pi^0$ requires instead the transitions $c \to d\, W^+$ followed by $W^+ \to u\, \bar{s}$: the amplitude is therefore proportional to $V_{cd}^* V_{us}$. See Fig. 5.6 for a graphical representation of the decay amplitude. The ratio between the two decay widths can be then estimated to be:

$$\frac{\Gamma(D^0 \to K^0 \pi^0)}{\Gamma(D^0 \to \bar{K}^0 \pi^0)} = \frac{|V_{cs}^* V_{ud}|^2}{|V_{cd}^* V_{us}|^2} = \tan^4 \theta_C \approx 2.6 \times 10^{-3}. \tag{5.209}$$

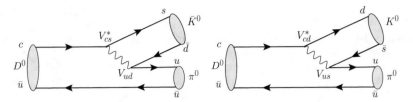

Fig. 5.6 Partonic diagrams describing the decays $D^0 \to \bar{K}^0 \pi^0$ and $D^0 \to K^0 \pi^0$

The decay $D^0 \to K^0 \pi^0$ is thus doubly Cabibbo-suppressed.

Suggested Readings

The doubly Cabibbo-suppressed decay $D^0 \to K^+ \pi^-$, which has its favoured-counterpart $D^0 \to K^- \pi^+$ in analogy with the decay considered in this exercise, has been measured at the BaBar experiment at PEP-II [34]. The measured value of $(0.303 \pm 0.019)\%$ is in agreement with the expectation from CKM.

Bando n. 1N/R3/SUB/2005

Problem 5.42 Why are the $D^0/\bar{D}^0$ oscillations more difficult to be observed than for $B^0/\bar{B}^0$?

Solution

The time-dependent oscillation in the decays of K^0, D^0, and B^0 (and their charge-conjugate states) is a consequence of mixing between the strong-eigenstates by the electroweak interaction, as discussed in Problem 5.39. The mass difference between the eigenstates Δm is determined by a box diagram that changes flavour by two units, analogous to the one depicted in Fig. 5.5. For K^0, B^0_s, and B^0_d, the amplitude is proportional to

$$(V_{cs} V_{cd}^*)^2 \sim \lambda^2, \qquad (V_{tb} V_{ts}^*)^2 \sim \lambda^4, \qquad (V_{tb} V_{td}^*)^2 \sim \lambda^6, \qquad (5.210)$$

respectively, see Eq. (5.190). For the case of D^0 mixing, the amplitude is instead proportional to $(V_{cb} V_{ub}^*)^2 \sim \lambda^{10}$, hence suppressed by a factor of $\lambda^4 \approx 2 \times 10^{-3}$ compared to e.g. the B^0_d case. Therefore, the mass difference between the two D^0-eigenstates is expected to be small, hence oscillations slower and more difficult to be observed.

Suggested Readings

The first evidence for $D^0/\bar{D}^0$ oscillations in the decay channel $D^0 \to K^{\mp} \pi^{\pm}$ has been established by the BaBar experiment at PEP-II [34].

Bando n. 18211/2016

Problem 5.43 What is the unitarity triangle? Indicate at least one process that allows to measure one of its angles and sides.

Discussion

Unitarity of the CKM matrix implies a number of relations between its elements. Considering for example the matrix by columns, one has:

$$\sum_{i=u,c,t} V_{ij} V_{ik}^* = \delta_{jk}, \quad j,k \in \{d,s,b\}. \tag{5.211}$$

The case $j = k$ gives three independent equations, while for $j \neq k$ one has three equations in the complex space, equivalent to six independent equations in real space.

Solution

Out of the three equations in Eq. (5.211), only one is of phenomenological interest because the three terms are of comparable magnitude (of order λ^3) and thus it can be experimentally verified to good accuracy:

$$V_{ud} V_{ub}^* + V_{cd} V_{cb}^* + V_{td} V_{tb}^* = 0 \quad \Leftrightarrow \quad 1 + \frac{V_{ud} V_{ub}^*}{V_{cd} V_{cb}^*} + \frac{V_{td} V_{tb}^*}{V_{cd} V_{cb}^*} = 0. \tag{5.212}$$

According to Eq. (5.212), the three terms at the left-hand side can be represented by as many vectors in the complex plane that define the contour of a closed triangle, also called the *unitarity triangle*. From Wolfenstein parametrisation of the CKM matrix in Eq. (5.190), it follows that $-V_{ud} V_{ub}^*/V_{cd} V_{cb}^* = \rho + i\eta$: the low vertices of the unitarity triangle are therefore located in $(0,0)$ and $(1,0)$, and the upper one in $\rho + i\eta$, see Fig. 5.7.

Starting from the upper vertex and moving clockwise, the three angles of the triangle are conventionally denoted by α, β, and γ, respectively, so that:

Fig. 5.7 The unitarity triangle in the complex (ρ, η) plane (from Ref. [9])

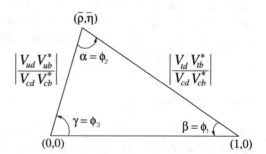

$$\alpha = \arg\left(-\frac{V_{td} V_{tb}^*}{V_{ud} V_{ub}^*}\right), \quad \beta = \arg\left(-\frac{V_{cd} V_{cb}^*}{V_{td} V_{tb}^*}\right), \quad \gamma = \arg\left(-\frac{V_{ud} V_{ub}^*}{V_{cd} V_{cb}^*}\right). \quad (5.213)$$

Since these angles are related to an irreducible phase between the CKM elements, they must be related to CP-violating observables.

The angle β can be measured from the time-dependent oscillations in the decay $B^0(t) \to J/\Psi \, K_S^0$, where $B^0(t)$ is an admixture of B^0 and $\bar{B}^0$ in analogy with the K^0 system. Indeed, the time-dependent amplitude squared for this decay is proportional to:

$$|\langle J/\Psi \, K_S^0 | \mathscr{H} | B^0(t) \rangle|^2 \propto e^{-\Gamma_B |t|} \left[1 - \sin 2(\beta + \phi_{\text{wk}}) \sin(\Delta m_B \, t)\right]. \quad (5.214)$$

The mechanism responsible for Eq. (5.214) can be traced back to the interference between two amplitudes: the first proceeds through the tree-level branching of the B^0 component: $\bar{b} \to \bar{c} \, W^+$ followed by $W^+ \to c \, \bar{s}$, which is proportional to $V_{cd} V_{cb}^*$; the second amplitude involves a transition $B^0 \to \bar{B}^0$ by means of a $\Delta B = 2$ effective operator proportional to $(V_{td} V_{tb}^*)^2$. The relative phase between the two is thus related to the angle β and gives rise to time-dependent oscillations of frequency Δm_B. The current world average is $\sin 2\beta = 0.691 \pm 0.017$ [9].

The angle α can be measured by combining measurements of the time-dependent oscillations in the decays $B^0(t) \to \pi\,\pi, \, \rho\,\pi, \, \rho\,\rho$. Indeed, such decays can proceed both through a tree-level transition $\bar{b} \to \bar{u} \, W^+$, proportional to $V_{ud} V_{ub}^*$, and through the transition $B^0 \to \bar{B}^0$, proportional to $(V_{td} V_{tb}^*)^2$, followed by $b \to u \, W^-$, which is proportional to $V_{ub} V_{du}^*$: the total phase is thus $\beta + \gamma = \pi - \alpha$. The current world average is $\alpha = 87.6^{+3.5°}_{-3.3}$ [9].

Finally, the angle γ can be measured from the interference between the CKM-favoured $B^+ \to \bar{D}^0 K^+$ and the CKM-disfavoured $B^+ \to D^0 K^+$ amplitudes in the resonant decays $B^+ \to K_S^0 \pi^+ \pi^- K^+$. The current world average is $\gamma = 73.2^{+6.3°}_{-7.0}$ [9].

The two upper sides of the unitarity triangle have length proportional to $|V_{ud} V_{ub}^*|$ and $|V_{td} V_{tb}^*|$. The former can be constrained by the measurement of V_{ub} from the suppressed semileptonic decays $B \to X_u \, \ell \, \nu$. The latter is instead related to the $B^0/\bar{B}^0$ mass mixing through a calculable FCNC process, and can thus be measured from the oscillation frequency in B_d^0 and B_s^0 decays.

Suggested Readings

For an overview of CKM-related measurements, the reader is addressed to Chap. 15 of Ref. [11].

Bando n. 18211/2016

Problem 5.44 Consider the two leptonic decays of the B mesons:

$$B^+ \to \mu^+ \nu_\mu \qquad B^0 \to \mu^+ \mu^- \qquad\qquad (5.215)$$

Do you expect the two processes to have a similar branching ratio? If not, which one do you expect to be the largest?

Solution

The charged-current decay width for $B^+ \to \mu^+ \nu_\mu$ is analogous to the $\pi^+ \to \mu^+ \nu_\mu$ case of Eq. (5.50), modulo the replacement $|V_{ud}| \to |V_{ub}|$ and $f_{\pi^+} \to f_{B^+}$, with $f_{B^+} \approx 190$ MeV [9]. Since the CKM element for $b \to u W^-$ is small ($V_{ub} \approx 4.5 \times 10^{-3}$), and since the decay is chirality-suppressed, the branching ratio is tiny compared to the semileptonic decay $B \to D \ell \nu_\ell$. By using Eq. (5.50) with the appropriate replacements, one gets:

$$\Gamma(B^+ \to \mu^+ \nu_\mu) \approx 3.5 \times 10^{-6} \text{ s}^{-1}, \qquad\qquad (5.216)$$

Given that the B^+ lifetime is $\tau_{B^+} = 1.6$ ps, we get an expected branching ratio of about 5×10^{-7}, consistent with the current best upper limit of 1.0×10^{-6} s at 90% CL [9]. Presently, out of the three leptonic decays of the B^+ mesons, only the decay $B^+ \to \tau^+ \nu_\tau$ has been observed at the $\Upsilon(4S)$ factories.

The decay $B^0 \to \mu^+ \mu^-$ is a FCNC process with $|\Delta B| = 1$, hence it is loop-suppressed, although more CKM-favourable compared to the leptonic decay, since the amplitude is proportional to $V_{tb} V_{ts}^*$, i.e. it involves a transition from the third to the second generation. The expected SM branching ratio are however smaller than for the leptonic decay by at least two orders of magnitude, namely

$$\begin{aligned} \text{BR}(B_s^0 \to \mu^+ \mu^-) &= (3.7 \pm 0.2) \times 10^{-9} \\ \text{BR}(B_d^0 \to \mu^+ \mu^-) &= (1.1 \pm 0.1) \times 10^{-10}. \end{aligned} \qquad (5.217)$$

These two decays have been observed at the LHC with a significance in excess of 6σ and 3σ, respectively. The measurements are in agreement with the SM expectation.

Suggested Readings

See the PDG review [9] dedicated to leptonic decays of charged pseudo-scalar mesons. The observation of the rare $B_{d,s}^0 \to \mu^+ \mu^-$ decays is documented in Ref. [35].

Bando n. 18211/2016

Problem 5.45 Discuss whether the partonic process $b \to s \gamma$ can be represented by a tree-level diagram in the SM.

Solution

The partonic transition $b \to s \gamma$ is a FCNC process, hence in the SM it cannot occur at tree-level. It is generated at the one-loop level via a Penguin-diagram similar to the one shown in Fig. 5.4. Experimentally, this amplitude can be studied from e.g. the decay $B^+ \to K^{*+} \gamma$.

Suggested Readings

The reader is addressed to Chap. 11 of Ref. [11] for some more examples on this subject.

Bando n. 18211/2016

Problem 5.46 In order to study CP violation in the B meson system, asymmetric $e^+ e^-$ colliders with centre-of-mass. energy equal to the $\Upsilon(4S)$ mass are generally used. Motivate this particular choice.

Solution

The $\Upsilon(4S)$ is an excited bottomium state. It has a mass of 10.579 GeV and decays mostly to $B^+ B^-$ and $B^0 \bar{B}^0$ with nearly equal branching ratio. Since it has $J^{PC} = 1^{--}$, it can be produced via s-channel $e^+ e^-$ scattering. The Q-value for both decays is small, giving:

$$Q = m_\Upsilon - 2 m_B \approx 20 \text{ MeV} \quad \Rightarrow \quad \beta^* \gamma^* \approx \sqrt{\frac{T_B}{2 m_B}} = \sqrt{\frac{Q}{m_B}} \approx 0.06. \quad (5.218)$$

Given that the B mesons have $c\tau$ of about 450 μm, the mean distance of flight in the laboratory for $\Upsilon(4S)$ produced at rest is $\beta^* \gamma^* c\tau \approx 30$ μm, which is at the limit for the experimental detection of the B decay vertex. The latter is necessary to study CP violation by studying the time-dependent decay probability of the physical B^0 mesons as a function of its time of flight.

In order to overcome this limitation, asymmetric $e^+ e^-$ colliders (B-factories) have been built at SLAC and KEK. By colliding beams of different energies E_1 and E_2, but such that $\sqrt{2E_1 E_2} = m_\Upsilon$, the average $\beta\gamma$ factor of B mesons in the laboratory frame is enhanced. For example, by using two beams of energies $E_1 = 9$ GeV and $E_2 = 3$ GeV, the average γ factor of the B mesons in the laboratory frame is:

$$\langle \gamma \rangle = \gamma^* \sqrt{\frac{|E_1 - E_2|^2}{m_\Upsilon^2} + 1} = 1.15 \quad \Rightarrow \quad \langle \beta\gamma \rangle \approx 0.57, \quad (5.219)$$

corresponding to an average decay distance of about 250 μm.

Suggested Readings

The reader is addressed to Chap. 15 of Ref. [11] for more details on this subject.

Bando n. 1N/R3/SUB/2005

Problem 5.47 The decays $\mu \to e\gamma$ and $\tau \to \mu\gamma$ violate the conservation of the lepton number.

- Draw at least one diagram that can explain such decays.
- Describe which are the irreducible backgrounds.

Discussion

In the SM with Dirac-type neutrinos, the lepton number L and the individual lepton numbers L_i, with $i = e, \mu, \tau$, are "accidental" global symmetries of the SM Lagrangian. Both can be broken in minimal extensions of the SM. Indeed, if the PMNS matrix is non-diagonal, as implied by the evidence of their oscillation, individual lepton numbers can be violated. However, the rate of decays that violate the individual lepton numbers are negligibly small due to the heaviness of the right-handed neutrino mass. A diagram like the one shown in Fig. 5.8, where right-handed neutrinos mediate the transition between lepton families, contribute by an effective operator:

$$\mathscr{L}_{\mu \to e\gamma} \approx A\,(\bar{\mu}\sigma_{\mu\nu}e)F^{\mu\nu}, \quad \text{with} \quad A < e\frac{\alpha}{\pi}\frac{m_\mu}{m_W^2}\frac{m_\nu}{m_W}. \qquad (5.220)$$

The smallness of the observed neutrino mass, $m_\nu \lesssim 1$ eV, which is due to the largeness of the right-handed neutrino mass M^R, makes this amplitude totally negligible. Indeed, the branching ratio corresponding to this amplitude is:

$$\text{BR}(\mu \to e\gamma) = \frac{3\alpha}{32\pi}\left|\sum_i U_{\mu i}^*\left(\frac{m_{\nu_i}^2}{m_W^2}\right)U_{ei}\right|^2 \lesssim 10^{-50}, \qquad (5.221)$$

Fig. 5.8 A diagram contributing to the $\mu \to e\gamma$ transition. The crosses denote appropriate mass insertions

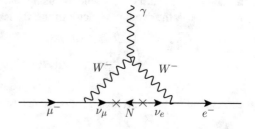

where U_{ij} are the appropriate elements of the PMNS matrix responsible for flavour mixing of neutrinos. In extensions of the SM, like the Minimal Supersymmetric Standard Model (MSSM), the amplitude receives additional, and potentially much larger, contributions from diagrams where supersymmetric particles are exchanged in the triangle: the amplitude is no longer suppressed by the large mass scale M^R, rather by the naturally smaller scale of supersymmetry breaking [36], overall giving branching ratios of the order of $10^{-15} \div 10^{-14}$.

Solution

A diagram contributing to the $\mu \rightarrow e\gamma$ transition is shown in Fig. 5.8. An analogous diagram holds for $\tau \rightarrow \mu\gamma$. Here, the transition is mediated by a right-handed neutrino.

Muon decays via $\mu^\pm \rightarrow e^\pm \gamma$ can be searched by stopping muon beams in matter and studying the kinematics of the electron and photon produced in their decay. Besides being back-to-back, in a two-body decay the electron and photon energies are fixed, while in the ordinary three-body decay (*Michel decay*), the same electron energy is only taken at the endpoint of the spectrum. In order to tag a signal event, the electron and photon are required to be in coincidence as to reduce the main background arising from the accidental pileup of a photon produced in a different muon decay $\mu \rightarrow e\nu_\mu \nu_e + \gamma$ (*radiative decays*) or from the interaction of the electron with the material upstream of the calorimeter. The second source of background is represented by a pure radiative decay, where the photon and electron are instead produced in coincidence. The key experimental challenges are therefore the time resolution for measuring the coincidence and the energy/momentum resolution on both electrons and photons to reduce the contamination from radiative decays. The MEG experiment at PSI has put the most stringent upper limit on this decay, corresponding to BR($\mu^+ \rightarrow e^+ \gamma$) $< 4.2 \times 10^{-13}$ at 90% CL [37].

Tau decays via $\tau^\pm \rightarrow \mu^\pm \gamma$ are best searched by producing τ leptons in high-energy collisions. The most stringent limits are obtained in e^+e^- colliders, in particular by the BaBar and Belle experiments [38, 39], where τ leptons are produced in pairs. The main background arises from standard τ decays $\tau \rightarrow \mu \nu_\mu \nu_\tau$ where a photon accidentally overlaps with the muon. The tightest upper bound has been measured by the BaBar experiment at PEP-II, giving BR($\tau \rightarrow \mu\gamma$) $< 4.4 \times 10^{-8}$ at 90% CL [38].

Suggested Readings

Reference [14] provides an elegant and coincise theoretical overview on the subject. The most recent results from the MEG experiment at PSI have been published in Ref. [37]. See Ref. [38, 39] for the details on searches for the lepton flavour violating decay $\tau \rightarrow \mu\gamma$.

5.4 Higgs Boson

By adding a kinetic Lagrangian for Φ to the right-hand side of Eq. (5.39) , with the usual $\partial_\mu \leftrightarrow D_\mu$ replacement, the Higgs field propagates and interacts with the gauge fields. More generally, the scalar sector can be supplemented with a gauge-invariant potential $V(\Phi, \Phi^*)$. In order for the theory to be renormalizable, no operators with dimension $d > 4$ should be present at tree-level, which singles-out a limited number of possible terms in $V(\Phi, \Phi^*)$. The most general scalar Lagrangian is then given by:

$$\mathscr{L}_S = (D^\mu \Phi^\dagger)(D_\mu \Phi) - \mu^2 \Phi^\dagger \Phi - \lambda (\Phi^\dagger \Phi)^2, \qquad (5.222)$$

where λ needs to be positive for the potential to be bounded from below and μ^2 is a mass term for the Φ field.

The ground state of the theory (vacuum) is defined as the state where the energy density is at a minimum. The minimum of the scalar potential in Eq. (5.222) depends on the sign of μ^2. If $\mu^2 > 0$, then the ground state $|0\rangle$ satisfies: $\langle 0|\Phi|0\rangle = \mathbf{0}$. On the contrary, for negative values of μ^2, the vacuum state on which the expectation value of $V(\Phi, \Phi^*)$ is at a minimum is away from zero. By defining the *vacuum expectation value* of the Higgs doublet as

$$v \equiv \left(\frac{-\mu^2}{\lambda} \right)^{1/2}, \qquad (5.223)$$

all states on which the expectation value of Φ is related to $(0, v/\sqrt{2})$ by a rotation of $SU(2)_L$, sit on a minimum of the potential The ground state is clearly no longer invariant under an arbitrary transformation of the $SU(2)_L \times U(1)_Y$ sub-group: this symmetry of the Lagrangian has been spontaneously broken. The neutral component of the doublet is chosen to develop $v \neq 0$, so that the electric charge can be conserved. By virtue of the *Brout-Englert-Higgs* (BEH) mechanism [40, 41], all generators of $SU(2)_L \times U(1)_Y$ *but* the combination $T_3 + Y/2$ are broken, and three Goldstone bosons appear in the spectrum. It is only in a gauge theory that these scalar bosons get reabsorbed by the gauge bosons as their longitudinal degrees of freedom. Thus, the BEH mechanism can give mass to three of the gauge bosons ($W^\pm$, Z^0), while the fourth (A) remains massless. It still remains to generate masses for the fermions. This happens by virtue of the Yukawa interactions in Eq. (5.43): by replacing Φ with $(0, 1/\sqrt{2}(v + H))$, a Dirac mass term $m_i \, \bar{f} f$, with

$$m_i \equiv \frac{\lambda_i v}{\sqrt{2}}, \qquad (5.224)$$

is generated for a fermion of type i. Expanding the Higgs doublet around the minimum, Eq. (5.222) becomes

$$\mathcal{L}_H = \frac{1}{2}(\partial_\mu H)(\partial^\mu H) - \lambda v^2 H^2 - \lambda v H^3 - \frac{\lambda}{4} H^4, \tag{5.225}$$

which implies that the physical Higgs boson has a mass $m_H = \sqrt{2\lambda}\, v$, and a cubic and quartic self–interaction with a vertex

$$g_{3H} = 3\frac{m_H^2}{v} \quad \text{and} \quad g_{4H} = 3\frac{m_H^2}{v^2}. \tag{5.226}$$

The μ and λ parameters of the bare Lagrangian can be traded for the Higgs boson mass and the vacuum expectation value (v). The latter can be put in relation with the $W^\pm$ boson mass and with an other experimental observable, the Fermi constant G_F, giving:

$$v = \frac{1}{(\sqrt{2}\, G_F)^{1/2}} \approx 246 \text{ GeV}. \tag{5.227}$$

The only unknown parameter of the Higgs sector is therefore the mass of the physical Higgs boson. The Yukawa couplings in Eq. (5.42) determine the strength of the interaction between the Higgs boson and the fermions:

$$g_{Hff} = \frac{m_f}{v} \equiv (\sqrt{2}\, G_F)^{1/2}\, m_f, \tag{5.228}$$

while the couplings of the Higgs boson to gauge vectors can be extracted from the covariant derivative:

$$g_{HVV} = -2\frac{m_V^2}{v} \equiv -2(\sqrt{2}\, G_F)^{1/2}\, m_V^2, \quad g_{HHVV} = -2\frac{m_V^2}{v^2} \equiv 2\sqrt{2}\, G_F\, m_V^2. \tag{5.229}$$

Equations (5.228) and (5.229) show that the Higgs boson couples to the SM particles with strength proportional to the particle mass for fermions, and to the mass squared for gauge bosons. The Higgs boson in the SM is a definite $CP = 1$ eigenstate, and is assigned the quantum numbers $J^{CP} = 0^{++}$.

Problems

Bando n. 1N/R3/SUB/2005

Problem 5.48 Order by decreasing cross section value the following reactions.

1. At a hadron collider:

- inclusive Z^0 production;
- top quark pair-production;
- inclusive Higgs boson production;
- inclusive b quark production;
- elastic pp scattering;

Fig. 5.9 The theoretical
proton-(anti)proton cross
sections as a function of $\sqrt{s}$
for total scattering, inclusive
b quark production, $W^{\pm}$ and
Z^0 boson production, t
quark production, and
inclusive Higgs boson
production (from Ref. [42])

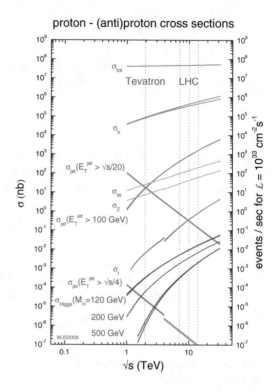

2. At the LHC, the following Higgs production mechanisms:

- $q\bar{q},\ gg \rightarrow H t\bar{t}$
- $q\bar{q} \rightarrow H Z^0$
- $gg \rightarrow H$
- $qq \rightarrow H qq$
- $q\bar{q},\ gg \rightarrow H b\bar{b}$

Solution

The theoretical proton-(anti)proton cross section for the processes listed in the exercise is shown in Fig. 5.9 for different values of $\sqrt{s}$. The theoretical cross section for Higgs boson production, broken up by different production mechanism, is shown in Fig. 5.10.

Discussion

As shown by Fig. 5.9, the total proton-proton cross section varies slowly with $\sqrt{s}$. This is in agreement with the *Froissart limit*, which asserts that the total cross section must be bounded from above by some constant multiplied by $\ln^2 s$ [44]. The Higgs boson production cross sections increase with $\sqrt{s}$, although with different slopes. In

Fig. 5.10 Standard Model Higgs boson production cross sections in proton-proton collisions as a function of the centre-of-mass energy (from Ref. [43])

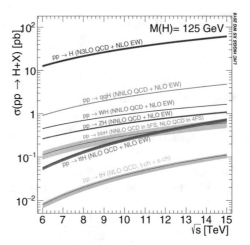

particular, the steeper increase of the vector-boson fusion (VBF) cross section is a reminiscence of the longitudinal boson scattering amplitude growing with energy: the partonic cross section increases like $\hat{\sigma} \propto \ln \hat{s}$, see e.g. Ref. [20] for a review on this subject. For example, for the case $q\,q' \to H\,q\,q'$ in weak-boson fusion, an approximate expression for the total cross section is given by [45, 46]:

$$\hat{\sigma} = \frac{1}{16\,m_W^2} \left(\frac{\alpha}{\sin^2 \theta_W} \right)^3 \left[\left(1 + \frac{m_H^2}{\hat{s}} \right) \ln \frac{\hat{s}}{m_H^2} - 2 + 2\frac{m_H^2}{\hat{s}} \right] \tag{5.230}$$

For $t\bar{t}H$, the fast increase with energy is mostly due to the larger phase-space available for producing three massive particles.

Bando n. 1N/R3/SUB/2005

Problem 5.49 What is the ratio BR_b/BR_τ for a Higgs boson mass of 125 GeV?

Discussion

The Born-level decay width of the Higgs boson to a fermion-antifermion pair is given by the general formula:

$$\Gamma_{\text{Born}}(H \to f\,\bar{f}) = \frac{G_F\,N_C}{4\sqrt{2}\pi}\,m_H\,m_f^2 \left(1 - \frac{4\,m_f^2}{m_H^2} \right)^{\frac{3}{2}}, \tag{5.231}$$

where N_C is the number of colours, see e.g. Ref. [20] for a compendium of decay formulas. For quarks, the higher-order corrections are important and can be re-asborbed, to a large extent, inside the running quark mass m_q. The latter is defined by a RGE:

$$\frac{dm(\mu^2)}{d\ln\mu^2} = -\gamma_m(\alpha_s)\,m \quad \Rightarrow \quad m(\mu^2) = m(\mu_0^2)\,\exp\left[-\int_{\alpha_s(\mu_0^2)}^{\alpha_s(\mu^2)} d\alpha_s\,\frac{\gamma_m(\alpha_s)}{\beta(\alpha_s)}\right],$$

$$(5.232)$$

where γ_m is called *anomalous dimension*, which, in a mass-independent renormalisation scheme, can be expanded as a series in α_s:

$$\gamma_m(\alpha_s) = \alpha_s\,\gamma_0 + \alpha_s^2\,\gamma_1 + \cdots, \quad \text{with} \quad \gamma_0 = \frac{3\,C_F}{4\pi} = \frac{1}{\pi}, \tag{5.233}$$

see e.g. Chap. 3.4 of Ref. [18]. Since $\gamma_0 > 0$, the running mass is a decreasing function of μ^2. With the introduction of the running mass, the decay width becomes:

$$\Gamma(H \to q\bar{q}) = \frac{3\,G_F}{4\sqrt{2}\pi}m_H\,m_q^2(m_H)\left[1 + \Delta_{qq} + \Delta_H^2\right], \tag{5.234}$$

where Δ_{qq} and Δ_H are additional radiative corrections of $\mathcal{O}(\alpha_s)$.

Solution

For bottom quarks, the running mass $m_b(m_H^2)$ at the Higgs boson mass scale is a factor of about 0.85 smaller compared to the pole mass $m_b(m_b^2)$, see e.g. Fig. 2.3 of Ref. [19], whereas the running of the τ lepton mass between the pole mass and m_H is negligible. Given Eq. (5.231), and using a value of 4.18 GeV for the b quark pole mass, we should therefore expected a ratio of:

$$\frac{\Gamma(H \to \tau^+\tau^-)}{\Gamma(H \to b\bar{b})} = \frac{1}{N_C}\frac{m_\tau^2}{m_b^2(m_H)} \approx \frac{1}{3}\left(\frac{1.7}{4.18\cdot 0.85}\right)^2 \approx 0.08. \tag{5.235}$$

A complete calculation [43] of the partial widths, which includes higher-order corrections to both decays, yields ratio of 0.108, in fair agreement with the above estimate.

Suggested Readings

For a comprehensive review of the Higgs boson physics, the reader is addressed to Ref. [20].

Bando n. 1N/R3/SUB/2005

Problem 5.50 What is the BR and what the main background for $H \to \gamma\gamma$?

Discussion

The Higgs boson is a gauge singlet. Since the spontaneous breaking of the $SU(2)_L \times U(1)_Y$ symmetry leaves unbroken both $U(1)_{em}$ and $SU(3)_C$, the Higgs boson is neutral under the electromagnetic and strong interactions. As such, it only couples to the photon and gluons via loop-induced operators of dimension five:

$$\mathcal{L}_{H\gamma\gamma} = c_\gamma \frac{H}{v} F_{\mu\nu} F^{\mu\nu}, \qquad \mathcal{L}_{Hgg} = c_g \frac{H}{v} G^a_{\mu\nu} G^{a\,\mu\nu}, \qquad (5.236)$$

where v is the vacuum-expectation-value of the Higgs field, see Eq. (5.227). The coupling to photons is mostly induced by loops of t quarks and of W boson. The coupling to gluons is instead largely dominated by top quark loops. The Lorentz structure of the top-mediated loop is identical between the Hgg and $H\gamma\gamma$ vertex, differing only on the coupling strength ($g_s \to e$) and on the colour algebra. In the limit $m_t \to \infty$, the effective operator generated by top quarks gives:

$$c_\gamma = \frac{e^2}{18\pi^2}, \qquad c_g = \frac{g_s^2}{48\pi^2}, \qquad (5.237)$$

see e.g. Ref. [47]. For the photon case, the leading-order amplitude receives contributions of comparable size from two interfering diagrams. The resulting decay width is given by:

$$\Gamma(H \to \gamma\gamma) = \frac{G_F \alpha^2 m_H^3}{128\sqrt{2}\,\pi^3} \left| N_C \left(\frac{2}{3}\right)^2 A_t \left(\frac{m_H^2}{4m_t^2}\right) + A_W \left(\frac{m_H^2}{4m_W^2}\right) \right|^2,$$

$$\text{with} \quad A_i(\tau) = \begin{cases} +2\left[\tau + (\tau - 1)\arcsin^2\sqrt{\tau}\right]\tau^{-2} & t \\ -\left[2\tau^2 + 3\tau + 3(2\tau - 1)\arcsin^2\sqrt{\tau}\right]\tau^{-2} & W \end{cases} \qquad (5.238)$$

Numerically, A_W is a factor of about 4.6 larger. They interfere destructively.

Solution

The $H \to \gamma\gamma$ decay is loop-mediated and it involves weak couplings. It should be therefore suppressed by rougly a factor of $(\alpha/\alpha_s)^2 \approx 4 \times 10^{-3}$ compared to the analogous $H \to gg$ decay. However, the $H \to \gamma\gamma$ amplitude receives large contributions from W loops, as discussed above, increasing the decay width by a factor of about 12. Overall, the branching ratio into two photons turns out to be $\text{BR}(H \to \gamma\gamma) \approx 2.27 \times 10^{-3}$ at $m_H = 125$ GeV [43], which should be compared to a branching ratio of $\text{BR}(H \to gg) \approx 8.19 \times 10^{-2}$.

The dominant experimental backgrounds in proton-proton collisions arise from prompt and non-prompt di-photon production. Prompt photons can be either produced from hard parton interaction or from the fragmentation of final state partons.

Non-prompt and high-p_T isolated photons mostly arise from $\pi^0 \to \gamma\gamma$ decays.

Suggested Readings

The basic formulas for the $H \to \gamma\gamma$ decay amplitude can be found in Ref. [20]. The experimental measurement of the $H \to \gamma\gamma$ cross section at the LHC is documented in Refs. [48, 49].

Problem 5.51 At a hadron collider, the Higgs boson cross section can only be measured in exclusive production and decay channels in the combination $\sigma_i \cdot \mathrm{BR}_j$. These measurements can be related to their SM expectations via coupling strength modifiers κ_i, such that $\sigma_i = \kappa_i^2 \sigma_i^{\mathrm{SM}}$ and $\Gamma_j = \kappa_j^2 \Gamma_j^{\mathrm{SM}}$, and by an unknown branching ratio into exotic particles, $\mathrm{BR}_{\mathrm{BSM}}$. Show that the κ_i modifiers extracted from a comprehensive set of cross section measurements are bounded from above, but not from above.

Solution

In terms of the strength modifiers, the generic cross section for exclusive production and decay channels is given by:

$$\sigma(i \to H \to j) \equiv (\sigma_i \cdot \mathrm{BR}_j) = \frac{\sigma_i \, \Gamma_j}{\Gamma_H} = \left[\sigma_i^{\mathrm{SM}} \frac{\Gamma_j^{\mathrm{SM}}}{\Gamma_H^{\mathrm{SM}}} \right] \frac{\kappa_i^2 \, \kappa_j^2}{\sum_j \kappa_j^2 \, \mathrm{BR}_j^{\mathrm{SM}}} (1 - \mathrm{BR}_{\mathrm{BSM}})$$

$$(5.239)$$

From Eq. (5.239), it is easy to verify that a simultaneous measurements of multiple $\sigma_i \cdot \mathrm{BR}_j$ cannot constrain the κ_i into a closed confidence level. Indeed, a simultaneous re-scaling

$$\kappa_i \to \lambda \kappa_i \quad \text{and} \quad \mathrm{BR}_{\mathrm{BSM}} \to \frac{\mathrm{BR}_{\mathrm{BSM}} + (\lambda^2 - 1)}{\lambda^2}, \qquad (5.240)$$

with $\lambda > 0$, leaves all of the $\sigma_i \cdot \mathrm{BR}_j$ unchanged. Hence, a likelihood function of the $\sigma_i \cdot \mathrm{BR}_j$ measurements will admit a flat direction in the $(\kappa_i, \mathrm{BR}_{\mathrm{BSM}})$ space. Conversely, any of $\sigma_i \cdot \mathrm{BR}_j$ measurements yields a lower bound on κ_i and κ_j. Indeed:

$$(\sigma_i \cdot \mathrm{BR}_j) = (\sigma_i \cdot \mathrm{BR}_j)^{\mathrm{SM}} \frac{\kappa_i^2 \, \kappa_j^2}{\sum_j \kappa_j^2 \, \mathrm{BR}_j^{\mathrm{SM}}} (1 - \mathrm{BR}_{\mathrm{BSM}})$$

$$\leq (\sigma_i \cdot \mathrm{BR}_j)^{\mathrm{SM}} \frac{\kappa_i^2 \, \kappa_j^2}{\kappa_j^2 \, \mathrm{BR}_j^{\mathrm{SM}}} = \sigma_i^{\mathrm{SM}} \kappa_i^2 \quad \Rightarrow \quad \kappa_i \geq \sqrt{\frac{(\sigma_i \cdot \mathrm{BR}_j)}{\sigma_i^{\mathrm{SM}}}}. \quad (5.241)$$

Discussion

If the total width Γ_H can be constrained from above (for example, from the invariant mass distribution of the Higgs decay products, or from off-shell interference with the background), both the strength modifiers and the exotic branching ratio are clearly also limited from above. Alternatively, if one admits that the exotic decays consist of invisible particles only, which can be detected through transverse momentum imbalance, then an additional measurement is possible, namely $\sigma_i \cdot \mathrm{BR}_{BSM} = \sigma_i^{SM} \cdot (\kappa_i^2 \, \mathrm{BR}_{BSM})$, which removes the degeneracy of Eq. (5.240). However, is the exotic decays can give rise to undetectable events, then the degeneracy remains.

If one wishes to set a bounded confidence interval on the strength modifiers in the lack of a direct measurement of the total width, additional model-dependent assumptions are needed. The simplest one is clearly to assume no exotic decays, i.e. $\mathrm{BR}_{BSM} = 0$. In this case, the degeneracy of Eq. (5.240) is lifted. Alternatively, one can still allow for $\mathrm{BR}_{BSM} \geq 0$, but assume that at least one of the κ_i is limited from above, so that the flat direction is again removed (for example, a physics-motivated case is provided by the assumption $\kappa_{Z,W} \leq 1$).

Suggested Readings

For more details on the Higgs coupling extraction from LHC data, the reader is addressed to Ref. [50].

Problem 5.52 Show that, differently from what happens at a hadron collider, at an $e^+ e^-$ collider operating above the $Z H$ threshold, the Higgs boson width can be measured in a model-independent fashion from the combination of two cross section measurements.

Solution

At a lepton collider with $\sqrt{s} > m_H + m_Z$, the *inclusive* Higgsstrahlung cross section σ_{ZH} can be measured by studying the recoil mass distribution of the Z^0:

$$ p_{e^+} + p_{e^-} = p_Z + p_H \quad \Rightarrow \quad s + m_Z^2 - 2 p_Z (p_{e^+} + p_{e^-}) \equiv m_{\text{recoil}}^2 = m_H^2. \tag{5.242}$$

Notice that this is not the case for a hadron collider, where the kinematics of the initial state is not known *a priori*. If the cross section $\sigma_{ZH} \cdot \mathrm{BR}_Z$ can be also measured, the total width can be measured as:

$$ (\sigma_{ZH} \cdot \mathrm{BR}_Z) = \sigma_{ZH} \frac{\Gamma_Z}{\Gamma_H} = \sigma_{ZH} \frac{\Gamma_Z^{SM} \cdot \sigma_{ZH}}{\sigma_{ZH}^{SM} \Gamma_H} \quad \Rightarrow \quad \Gamma_H = \frac{(\sigma_{ZH})^2}{(\sigma_{ZH} \cdot \mathrm{BR}_Z)} \frac{\Gamma_Z^{SM}}{\sigma_{ZH}^{SM}}. \tag{5.243}$$

Here, we have used the fact that $\sigma_{ZH} \sim \Gamma_{ZH}$ by first principles, so that we can use a particular model (for example, the SM), to get the coefficient of proportionality. Hence, the total width can be measured with the same precision as $(\sigma_{ZH})^2/(\sigma_{ZH} \cdot \mathrm{BR}_Z)$.

Discussion

Notice that the key point here is to measure a channel where the same couplings are probed at the *production* and at the *decay* level, so that one can relate Γ_i to σ_i without additional unknown couplings. The same argument can be therefore applied to e.g. the VBF production channel $W^* W^* \to H$, followed by $H \to W W^*$. In this latter case, the inclusive cross section σ_{WWH} proves more difficult to be measured regardless of the Higgs decay channel because of the final-state neutrinos. However, one can still profit from multiple measurements in the Higgsstrahlung channel and derive σ_{WWH} from a suitable combination of them.

Suggested Readings

There is a growing literature about the potentialities of future lepton colliders on the Higgs boson physics, see e.g. Ref. [51] for a future circular collider (FCC). A summary of expected performances for the various options (CLIC, ILC, CEPC, TLEP) can be found in Refs. [52, 53].

Bando n. 18211/2016

Problem 5.53 Which prospects would a muon collider offer and what are the main technological challenges for its realisation?

Solution

A muon collider operating at a centre-of-mass energy in excess of 100 GeV would offer the possibility of direct s-channel Higgs production, a reaction with a totally negligible cross section at a conventional e^+e^- colliders, but becomes significant for μ-fusion. Indeed, for unpolarised muon beams the cross section at the Higgs peak is given by:

$$\sigma(s = m_H^2) = \frac{16\pi}{m_H^2} \frac{1}{(2s_1 + 1)(2s_2 + 1)} \mathrm{BR}_\mu \approx$$

$$\approx \frac{4\pi}{(125)^2} \cdot 2 \times 10^{-4} \cdot 0.389 \,\text{mbarn} \approx 70 \,\text{pb}. \qquad (5.244)$$

A scan over the muon beam energy would allow to measure directly the Higgs boson width Γ_H. In a BSM perspective, a fine-step energy scan could allow to detect the presence of additional neutral resonances with mass splitting of order of the SM width. Given that synchrotron radiation is not a problem for muon beams of order 100 GeV, see Eq. (3.95), an appealing aspect of muon collider is the contained accelerator size: for a beam energy $E = 60$ GeV, a radius $R = 100$ m would be sufficient for a magnetic field $B \sim 2$ T, see Eq. (3.45).

The main challenges posed by muon colliders are the luminosity and the energy spread of the beam. The luminosity target requires to overcome the problem of generating and collecting high intensity muon beams ($> 10^{12}$ muons/bunch) with reduced emittance. The finite muon lifetime sets constraints on the accelerating stage and provides a beam lifetime of order 1 msec. In order for a muon collider to serve as Higgs factory, the beam energy needs to be calibrated to better than 10^{-4}, and the beam spread needs also to be of the same level to maintain the largest possible luminosity.

Suggested Readings

For an overview on the subject, the reader can start from the PDG review [9] and references therein.

Bando n. 18211/2016

Problem 5.54 What aspects of a detector are crucial for a precision measurement of the Higgs boson mass?

Solution

The ATLAS and CMS combined measurement of the Higgs boson mass is mostly determined by the measurements in the $H \to \gamma\gamma$ and $H \to 4\ell$ channels, each contributing by a similar amount to the final accuracy.

In the $H \to \gamma\gamma$ channel, the mass resolution depends on the energy and angular resolution of the two photons, see Problem 5.55. The angular resolution arises from the finite size of the calorimetric cells and from uncertainty on the vertex position in the presence of multiple pile-up events. In this respect, a fine granularity of the calorimeter, together with an efficient vertex identification capability, are of key importance. The energy resolution at $E_\gamma \sim 60$ GeV receives large contributions also from the noise and uniform terms: the electromagnetic calorimeters thus require good energy resolution, large gains, and precise calibration.

Among the $H \to 4\ell$ channels, the $H \to 4\mu$ one provides the smallest mass resolution. The muon momentum scale can be calibrated to high accuracy by using standard candles. The main limitation comes from the limited statistics given that $BR(H \to 4\mu) = 3.2 \times 10^{-5}$ [43], thus requiring the highest possible acceptance.

The result from the Run 1 of the LHC is $m_H = 125.09 \pm 0.24$ GeV [54]. The relative uncertainty of 0.2% is still statistically limited.

Suggested Readings

See Ref. [54] for a discussion on the measurement of the Higgs boson mass at the LHC.

Bando n. 13705/2010

Problem 5.55 One of the possible channels for detecting a light Higgs boson H at the LHC is the rare decay $H \rightarrow \gamma \gamma$. Assume a mass $m_H = 120$ GeV and a longitudinal momentum along the beam axis $|\mathbf{p}| = 200$ GeV. Compute the range of photon energies in the laboratory frame and the minimum opening angle between the two photons. In the case of equally energetic photons, compute the di-photon mass resolution by using an electromagnetic calorimeter with energy resolution of $3\%/\sqrt{E}$ and angular resolution of 5 mrad.

Solution

The photon energy in the centre-of-mass frame is $E^* = m_H/2 = 60$ GeV. The γ and β factors of the boost to the laboratory frame are given respectively by:

$$\gamma = \sqrt{1 + \left(\frac{|\mathbf{p}|}{m_H}\right)^2} = 1.94, \qquad \beta = \sqrt{1 - \frac{1}{\gamma^2}} = 0.857. \tag{5.245}$$

The range of photon energy in the laboratory is given by Eq. (1.109):

$$E_{1,2} \in E^* \left[\gamma - \sqrt{\gamma^2 - 1}, \gamma + \sqrt{\gamma^2 - 1}\right] \approx [16.6, 217] \text{ GeV}. \tag{5.246}$$

The minimum opening angle corresponds to an emission at a polar angle $\theta^* = \pi/2$ in the centre-of-mass frame, which gives the same energy and polar angle in the laboratory to both photons, see Problem 1.12. By using Eq. (1.59), we obtain a minimum opening angle:

$$\Delta\theta = \mathrm{acos}\left(2\beta^2 - 1\right) = 1.08 \text{ rad} = 61.9° \tag{5.247}$$

The condition of identical photon energies in the laboratory frame implies that the opening angle between the two photons is given by Eq. (5.247). Indeed:

$$E_1 = E_2 \quad \Leftrightarrow \quad E^*(1 + \beta \cos\theta^*) = E^*(1 - \beta \cos\theta^*) \quad \Rightarrow \quad \theta^* = \pi/2 \tag{5.248}$$

From Eq. (5.248), the photon energy corresponding to an emission at an angle $\theta^* = \pi/2$ is $E_1 = E_2 = E^*$. The invariant mass of the photon pair is given by:

$$m_{\gamma\gamma} = \sqrt{2 E_1 E_2 (1 - \cos\Delta\theta)} = 2\sqrt{E_1 E_2} \sin\frac{\Delta\theta}{2} \tag{5.249}$$

By using the linear error propagation of Eq. (4.73), we estimate the relative uncertainty $\delta m_{\gamma\gamma}/m_{\gamma\gamma}$ on the photon-pair mass as:

$$
\frac{\delta m_{\gamma\gamma}}{m_{\gamma\gamma}} = \left(\frac{1}{2}\frac{\delta E_1}{E_1}\right) \oplus \left(\frac{1}{2}\frac{\delta E_2}{E_2}\right) \oplus \left(\frac{\delta \sin(\frac{\Delta\theta}{2})}{\sin(\frac{\Delta\theta}{2})}\right) = \left(\frac{1}{\sqrt{2}}\frac{\sigma_{E^*}}{E^*}\right) \oplus \left(\frac{\delta(\Delta\theta)}{2\tan(\frac{\Delta\theta}{2})}\right)
$$

$$
= \left(\frac{1}{\sqrt{2}}\frac{\sigma_{E^*}}{E^*}\right) \oplus \left(\frac{\sigma_{\theta_1} \oplus \sigma_{\theta_2}}{2\tan(\frac{\Delta\theta}{2})}\right) = \left[\frac{1}{2}\left(\frac{0.03}{\sqrt{60}}\right)^2 + \frac{2\cdot(5\times 10^{-3})^2}{4\cdot\tan^2 0.54}\right]^{\frac{1}{2}} = 0.65\%
$$

$$(5.250)$$

Notice that the angular measurement gives the dominant contribution to the mass uncertainty (0.59%), while the energy measurement contributes by 0.27%.

Discussion

The angular resolution in the di-photon mass arises from the finite size of the calorimetric cells and from uncertainty on the vertex position in the presence of multiple pile-up events. Besides, at the LHC experiments the energy resolution at $E \sim 60\,\mathrm{GeV}$ receives non-negligible contributions from the noise and uniform terms, see Problem 2.33, so one should consider the $3\%/\sqrt{E}$ uncertainty as a lower bound. The combined LHC Run 1 measurement of the Higgs boson mass is mostly determined by the measurements in the $H \to \gamma\gamma$ and $H \to 4\ell$ channels, each contributing by a similar amount to the final accuracy.

Suggested Readings

See Ref. [54] for a discussion on the measurement of the Higgs boson mass at the LHC.

References

1. C.N. Yang, Phys. Rev. **77**, 242 (1960). https://doi.org/10.1103/PhysRev.77.242
2. A. Di Giacomo, G. Paffuti, P. Rossi, *Problemi di Fisica teorica* (Edizioni ETS, 1992)
3. C. McGrew et al., Phys. Rev. D **59**, 052004 (1999). https://doi.org/10.1103/PhysRevD.59.052004
4. Super-Kamiokande Collaboration (2017). arXiv:1705.07221 [hep-ex]
5. W.R. Leo, *Techniques for Nuclear and Particle Physics Experiments*, 2nd edn. (Springer, Berlin, 1993)
6. The ALEPH Collaboration, Phys. Lett. B **374**, 319–330 (1996). https://doi.org/10.1016/0370-2693(96)00300-0
7. W.M. Gibson, B.R. Pollard, *Symmetry principles, Particle Physics* (Cambridge University Press, Cambridge, 1980)

8. D.H. Perkins, *Introduction to Hig Energy Physics*, 4th edn. (Cambridge University Press, Cambridge, 2000)
9. C. Patrignani et al., Particle data group. Chin. Phys. C **40**, 100001 (2016)
10. P.L. Connolly, Phys. Rev. Lett. **10**, 371 (1963)
11. R. Cahn, G. Goldhaber, *The Experimental Foundations of Particle Physics*, 2nd edn. (Cambridge University Press, Cambridge, 1989)
12. D.J. Griffiths, *Introduction to Elementary Particles* (Wiley, New York, 2008)
13. H.L. Andreson et al., Phys. Rev. **85**, 936 (1952). https://doi.org/10.1103/PhysRev.85.936
14. R. Barbieri, Lectures on the Electroweak interactions (Edizioni della Normale, 2007)
15. M. Peskin, D.V. Schroeder, *An introduction to Quantum Field Theory* (Advanced Book Program, Westview Press, Boulder, 1995)
16. LHCb Collaboration, Phys. Rev. Lett. **115**, 072001 (2015). https://doi.org/10.1103/PhysRevLett.115.072001
17. D0 Collaboration, Phys. Rev. Lett. **117**, 022003 (2016). https://doi.org/10.1103/PhysRevLett.117.022003
18. G. Dissertori, I.G. Knowles, M. Schmelling, *Quantum Chromodynamics: High Energy Experiments and Theory* (Oxford University Press, Oxford, 2002)
19. R.K. Ellis, W.J. Stirling, B.R. Webber, *QCD and Collider Physics* (Cambridge Monographs, Cambridge, 1996)
20. A. Djouadi, Phys. Rept. **457**, 1–216 (2008)
21. D.A. Bryman et al., Phys. Rep. **88**, 151–205 (1982). https://doi.org/10.1016/0370-1573(82)90162-4
22. A.M. Bernstein, B.R. Holstein, Rev. Mod. Phys. **85**, 81–84 (2013). https://doi.org/10.1103/RevModPhys.85.49
23. H.W. Atherton, Phys. Lett. B **158**, 81–84 (1985). https://doi.org/10.1016/0370-2693(85)90744-0
24. T. Kinoshita, Phys. Rev. Lett. **2**, 477 (1959). https://doi.org/10.1103/PhysRevLett.2.477
25. A. Pich, Int. J. Mod. Phys. A **15**, 157–173 (2000), eConf C990809:157-173, 2000
26. G. Altarelli et al., Int. J. Mod. Phys. A **13**, 1031 (1998). https://doi.org/10.1142/S0217751X98000469
27. A. Blondel, Nuovo Cim. A **109**, 771 (1996). https://doi.org/10.1007/BF02731711
28. G. Altarelli et al., Phys. Rept. **403–404**, 189–201 (2004). https://doi.org/10.1016/j.physrep.2004.08.013
29. UA 1 Collaboration, Phys. Lett. **122B**, 103 (1983). https://doi.org/10.1016/0370-2693(83)91177-2
30. D. Buttazzo et al., JHEP **12**, 082 (2013). https://doi.org/10.1007/JHEP12(2013)089
31. G. Altarelli, G. Parisi, Nucl. Phys. B **126**, 298 (1977). https://doi.org/10.1016/0550-3213(77)90384-4
32. J.A. Aguilar-Saavedra, Acta Phys. Polon B **35**, 2695–2710 (2004)
33. F. Muller et al., Phys. Rev. Lett. **4**, 418 (1960) https://doi.org/10.1103/PhysRevLett.4.418
34. BaBar Collaboration, Phys. Rev. Lett. **98**, 211802 (2007). https://doi.org/10.1103/PhysRevLett.98.211802
35. CMS Collaboration and LHCb Collaboration, Nature **522**, 68–72 (2015). https://doi.org/10.1038/nature14474
36. R. Barbieri at al., Nucl. Phys. B **445** 219–251 (1995). https://doi.org/10.1016/0550-3213(95)00208-A
37. M.E.G. Collaboration, Eur. Phys. J. C **76**, 434 (2016). https://doi.org/10.1140/epjc/s10052-016-4271-x
38. BaBar Collaboration, Phys. Rev. Lett. **104**, 021802 (2010). https://doi.org/10.1103/PhysRevLett.104.021802
39. Belle Collaboration, Phys. Lett. B **666**, 16–22 (2008). https://doi.org/10.1016/j.physletb.2008.06.056
40. P. Higgs, Phys. Rev. Lett. **12**, 132–133 (1964). https://doi.org/10.1103/PhysRevLett.13.508

41. F. Englert, R. Brout, Phys. Rev. Lett. **13**, 321–323 (1964). https://doi.org/10.1103/PhysRevLett.13.321
42. https://mstwpdf.hepforge.org/plots/plots.html
43. D. de Florian et al., CERN Yellow Reports: Monographs Volume 2/2017 (CERN-2017-002-M). https://doi.org/10.23731/CYRM-2017-002
44. M. Froissart, Phys. Rev. **123**(3), 1053–1057 (1961). https://doi.org/10.1103/PhysRev.123.1053
45. R.N. Cahn, S. Dawson, Phys. Lett. **136B**, 196 (1984). https://doi.org/10.1007/BF02909135
46. G. Altarelli et al., Nucl. Phys. B **287**, 205–224 (1987). https://doi.org/10.1016/0550-3213(87)90103-9
47. A. Azatov, A. Paul, JHEP **2014**, 14 (2014). https://doi.org/10.1007/JHEP01(2014)014
48. ATLAS Collaboration, Phys. Rev. D. **90**, 112015 (2014). https://doi.org/10.1103/PhysRevD.90.112015
49. C.M.S. Collaboration, Eur. Phys. J. C **74**, 3076 (2014). https://doi.org/10.1140/epjc/s10052-014-3076-z
50. JHEP **08**, 045 (2016). https://doi.org/10.1007/JHEP08(2016)045
51. The TLEP working group (2013). arXiv:1308.6176
52. M. Peskin, SLACPUB15178 (2013). arXiv:1207.2516
53. A. Ajaib et al., Snowmass report (2014). arXiv:1310.8361
54. ATLAS and CMS Collaborations, Phys. Rev. Lett. **114**, 191803 (2015). https://doi.org/10.1103/PhysRevLett.114.191803

Index

© Springer International Publishing AG 2018
L. Bianchini, *Selected Exercises in Particle and Nuclear Physics*,
UNITEXT for Physics, https://doi.org/10.1007/978-3-319-70494-4

Printed in the United States
By Bookmasters